Physical Principles of Biological Membranes

Physical Principles of Biological Membranes

Proceedings of the Coral Gables Conference on
Physical Principles of Biological Membranes
Center for Theoretical Studies, University of Miami
December 1968

Edited by

F. SNELL
State University of New York at Buffalo

J. WOLKEN
Carnegie-Mellon University

G. IVERSON
Center for Theoretical Studies

J. LAM
National Research Council, Ottawa

GORDON AND BREACH SCIENCE PUBLISHERS
New York   London   Paris

Copyright © 1970 by GORDON AND BREACH, SCIENCE PUBLISHERS, INC.
150 Fifth Avenue, New York, N.Y. 10011

Library of Congress catalog card number: 75-94077

Editorial office for the United Kingdom:

Gordon and Breach, Science Publishers Ltd.
12 Bloomsbury Way
London W.C.1

Editorial office for France:

Gordon & Breach
7–9 rue Emile Dubois
Paris 14[e]

All rights reserved. No part of this book may be reproduced or utilized in any form or by any means, electronic or mechanical, including photocopying, recording, or by any information storage and retrieval system, without permission in writing from the publishers. Printed in Great Britain by Robert MacLehose and Co Ltd The University Press, Glasgow

Introduction

THIS VOLUME represents in large part the proceedings of a conference entitled the 'Coral Gables Conference on the Physical Principles of Biological Membranes' held at the Center for Theoretical Studies of the University of Miami December 18–20, 1968. The conference was attended by over 40 distinguished biologists, biophysicists, biochemists, neurophysiologists, theoretical biologists, and physicists, all having prime interest in the structure and function of biological membranes. There is little question that the elusive detail structure of living membranes has attracted wide attention among researchers derived from a variety of fields and disciplines. The elusiveness remains a fact. In spite of a multiplicity of approaches and the application of a variety of elegant sophisticated techniques, the convergence of views is still beyond the acuity of perception.

There is little question that the elusive nature of the functional role of living membranes, nerve, mitochondrial, and others continues to defy the probing inquirer—experimentalists and theorists alike. Progress has certainly been made, but real understanding still remains blurred by opacities and ignorance.

With the advantage of a retrospective view at the time of this writing, we can state that the results and the 'conclusions' drawn from these results presented at this conference perhaps better be considered only as a single frame in a whole cinema of development.

Partial support of the conference by the National Aeronautics and Space Administration is greatly acknowledged.

F. SNELL

Conference Committee

SIDNEY FOX *University of Miami*

ELLIOTT MONTROLL *University of Rochester*

LARS ONSAGER *Yale University*

MORRIS ROCKSTEIN *University of Miami*

FRED SNELL *State University of New York at Buffalo*

JEROME J. WOLKEN *Carnegie-Mellon Institute*

ARNOLD PERLMUTTER (conference secretary) *Center for Theoretical Studies, University of Miami*

BEHRAM KURSUNOGLU (conference chairman) *Center for Theoretical Studies, University of Miami*

Contents

	Introduction	v
1	Dielectric properties of living membranes (K. S. COLE) . . .	1
2	Optical and electrophysiological evidence for conformational changes in membrane macromolecules during nerve excitation (I. TASAKI, W. BARRY and L. CARNAY)	17
3	Theory of nerve excitation as a cooperative cation exchange in a two-dimensional lattice (G. ADAM)	35
4	A Non-linear term in the ion transport across membranes (R. SCHLÖGL)	69
5	Ion distribution equilibria in bulk phases and the ion transport properties of bilayer membranes produced by neutral macrocyclic antibiotics (G. SZABO, G. EISENMAN and S. M. CIANI) . . .	79
6	Possible mechanisms of ion transit (L. ONSAGER)	137
7	Physical principles in monolayer and membrane permeation (M. BLANK and J. S. BRITTEN)	143
8	Current rectification and action potentials across thin lipid membranes (D. R. KALKWARF, D. L. FRASCO and W. H. BRATTAIN) . .	165
9	Dispersion forces and stability of lipid bilayers (S. OHKI) . .	175
10	Ion transport across lipid bilayer membranes (D. LÄUGER) . .	227
11	Role of water structure in various membrane systems (W. DROST-HANSEN)	243
12	Structural and functional properties of bacterial cell membranes (M. R. J. SALTON)	259

13	*Structure of the mitochondrial cristael membrane* (D. E. GREEN and G. VANDERKOOI)	287
14	*Conformational basis of energy transductions in biological membranes* (D. E. GREEN and R. A. HARRIS)	315
15	*Contrasting protein architectures of plasma and mitochondrial membranes* (D. F. H. WALLACH, A. S. GORDON, J. M. GRAHAM and B. R. FERNBACH)	345
16	*Cell and photoreceptor membranes* (J. J. WOLKEN) . . .	365
17	*The indication of a light induced electrical field by pigments incorporated in chloroplast membranes* (W. JUNGE, H. M. EMRICH and H. T. WITT)	383
18	*Concept of the reactive site in biological transport* H. N. CHRISTENSEN	397
19	*Membrane-like properties in microsystems assembled from synthetic protein-like polymer* (S. W. FOX, R. J. MCCAULEY, P. O'B. MONTGOMERY, T. FUKUSHIMA, K. HARADA and C. R. WINDSOR) .	417
	List of participants	433

PAPER 1

Dielectric properties of living membranes

KENNETH S. COLE
National Institutes of Health, Bethesda, Maryland 20014

AT THE Spring meeting of the American Physical Society in 1923, Hugo Fricke announced his measurement of 0.81 $\mu F/cm^2$ for the electrical capacity of the red blood cell membrane. On the assumption of a lipid structure with a dielectric constant of three he arrived at a thickness of 33 Å and so he proposed that the membrane was as pauci as possible—monomolecular. At the least it then became highly probable that this membrane was very thin.

I had the singular good luck to spend the following summer in Fricke's laboratory and, for these 45 years since, I have been mostly concerned with living membranes. About the only thing we have been able to do is to make a separation of the dielectric and the conductive properties whose independence seems highly probable, and to accumulate some experimental information on these electrical properties for quite a few membranes. But I think I am entitled to say that, since Fricke, we have had nothing except a disgusting array of unanswered questions about the dielectric behavior of each and every membrane that we have looked at.

Until 1937 a few of us tried very hard to find some correlation between the membrane capacity and cell function. It was then found that the membrane conductance was thoroughly involved in the propagation of impulses—first in *Nitella* and then in the squid giant axon (Cole and Curtis, 1938, 1939). As electricians and physiologists have gone more deeply into this aspect they have rather come to look on the capacity as a one microfarad nuisance. Now that the cell membrane is becoming one of the more popular and promising frontiers of molecular biology, theoretical and experimental scientists of every ilk are increasingly involved in both the structure and function of membranes. Even if structure and function may be quite independent, they

Reproduction of this paper for purpose of United States Government is permitted.

are at least complementary properties: successful theoretical descriptions of them cannot conflict with each other and indeed some facts about one may be crucial factors for an understanding of the other. I shall review some of the results which may be ascribed to the dielectric behavior of living membranes. We shall find that they become increasingly sparse and uncertain as they go through the three stages of experiment, analysis, and interpretation. Also, the references have not been well assembled; the available tables (Cole, 1942; Spector, 1946; Cole and Curtis, 1950; Schwan and Cole, 1960) are not complete and may be difficult to understand; and I do not attempt to document each and every statement I make.†

MEASUREMENTS

Many of the measurements are old, and were made between 1 kc and 10 Mc (Fricke, 1925; Cole and Curtis, 1937) but the precision needed and obtained was admired at Bell Telephone Laboratories. The range for steady state

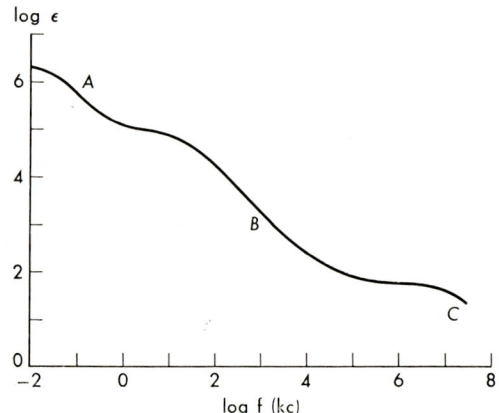

FIGURE 1 Dispersion regions of a red blood cell suspension expressed as apparent dielectric constant, ε, vs. frequency, f. The central, B, dispersion is explained by the dielectric capacity of the cell membranes. The A and C regions are attributed to a surface conductance and to intracellular components.

measurements has since been extended to a complete coverage from less than a cycle per second up to at least 10 kMc (Schwan, 1963).

In general all cells and tissues may be expected to show three major dispersion regions. These are most impressively presented as the equivalent

† In some cases more complete discussions and references are available (Cole, 1968, referred to as MII).

dielectric constant component of the measured impedance as shown in Figure 1 for suspensions of red blood cells (Schwan, 1963). The central frequencies of each region are very dependent upon the individual cell dimensions and to a lesser extent upon the cell volume concentration.

DISPERSION REGIONS

It was the central, most accessible, part of the spectrum which Fricke measured and analyzed as a one $\mu F/cm^2$ membrane capacity and interpreted as a dielectric polarization. An extension of Maxwell's result for the resistance of a suspension of spheres (Cole, 1928) may be expressed by the simple equivalent circuit shown in Figure 2 (Fricke and Morse, 1925). There

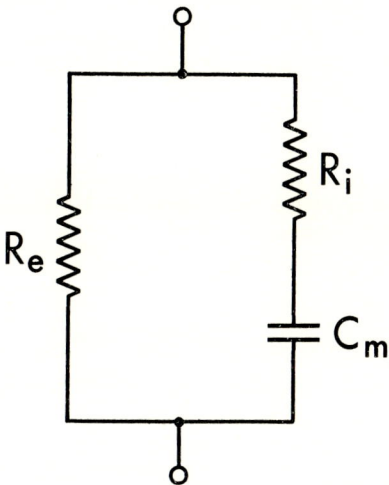

FIGURE 2   An equivalent circuit for theory and experiment on simple cell suspension impedances.

have been complaints that such circuits explain nothing and are only used by electricians who know no other language. To this can be added the observation that where such a device represents results of experiment or of analysis, it may be so convenient and powerful a means for the comparison of fact with theory that it should not be dismissed or even avoided. The circuit which the Maxwell analysis produces must give the semi-circular relation of Figure 3 (Cole, 1937) between the dissipative and conservative components of the suspension impedance as the applied frequency is varied.

This impedance representation applies quite well for the moderate frequency data on a suspension of starfish eggs (Cole and Cole, 1936) cf. Figure 4. But, as is seen here, there is usually another dispersion at high frequencies for which the effect of the membrane capacity becomes negligible. I tried to blame some of this on nuclear membranes before they were seen to be so full

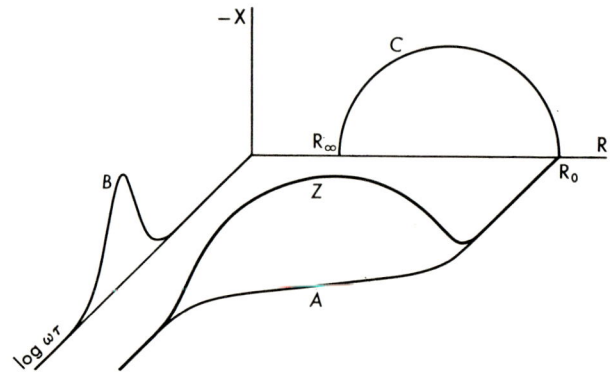

FIGURE 3 The resistive, R, and reactive, X, components of the impedance, Z, for a simple cell suspension as functions of its time constant, τ, and the frequency $\omega = 2\pi f$.

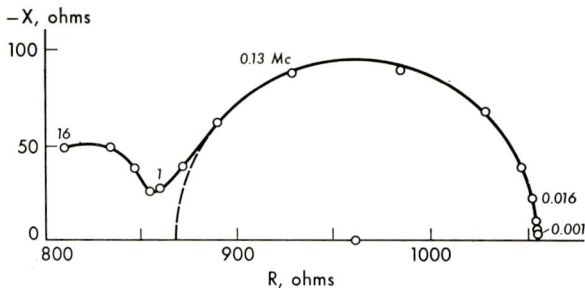

FIGURE 4 Complex impedance locus for starfish egg suspension showing the dielectric capacity dispersion at lower frequencies.

of holes as to sink any such explanation (Loewenstein, 1964). Probably it is better to blame the behavior of water and the relaxation times for orientation of internal proteins (Schwan, 1957), but this has not been widely investigated.

As more measurements were made at lower frequencies—a kilocycle and less—amazingly high capacities of ten to perhaps a thousand microfarads per square centimeter had to be added to the equivalent circuits for red blood cells, non conducting particles, the squid axon, and muscle fibers. The starfish eggs also had the same kind of behavior. It does not appear in Figure 4, but the conductance plot of Figure 5 clearly shows a break at low frequencies where the capacity increases without a comparable change in conductance.

An interesting explanation based on the time for the establishment of an electrokinetic surface conductance (Schwartz, 1962) is most promising for the red cells and for glass and plastic spheres. An empirical explanation as the consequence of a delay of membrane conductance change is, temporarily at least, adequate to describe (Cole, 1947, 1949) not only the addition of a large

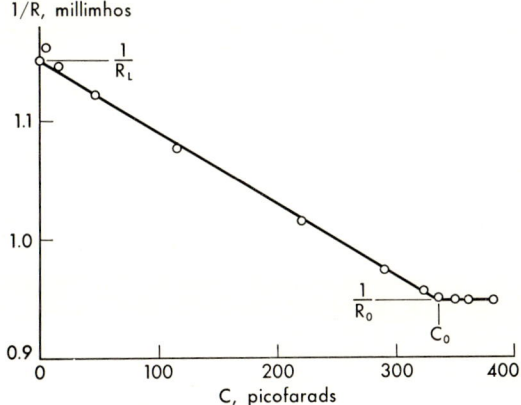

FIGURE 5 An inverse plot, parallel conductance, $1/R$, vs. capacity, C, of starfish egg data, showing dielectric capacity dispersion at higher frequencies.

capacity to the equivalent circuit of the squid axon, but also the almost incredible replacement of it by a massive inductive element (Cole and Baker, 1941b) under many conditions. Yet another explanation for the low frequency, long time behavior in muscle has been proposed (Fatt, 1964), namely that it represents the transverse tubule system which transmits orders to the contractile elements from the outer membrane. Spectacular support has come from the coincidence that this electrical component, and the tubules, disappear with glycerol treatments (Eisenberg and Eisenberg, 1968).

DIELECTRIC CAPACITY

Certainly the first question I should attempt to answer is whether or not the nominal $1 \mu F/cm^2$ membrane which predominates and is measured in the kilocycle and low megacycle or millisecond and microsecond ranges, can or must be explained as a dielectric phenomenon. The other principal suggestions are an electrode polarization (Segal, 1967) and a time dependent conductance (Tasaki, 1968). The relative order of magnitude and stability of the membrane capacity makes both alternatives highly improbable, and the appearance of a capacitive response without a change of conductance rules out the second. Conversely, such a membrane capacity seems likely to be the one and the only measure of the dielectric properties of the membrane. So it may reasonably be called the dielectric capacity.

Our next question is the ubiquity and the values of the membrane capacity. It has been measured for cells and organelles over wide ranges in size, from a centimeter plant *Valonia* to the submicron pleuro-pneumonia-like-organism, PPLO (Schwan and Morowitz, 1962), and in function, from mitochondria (Pauly, *et al*. 1960) to egg cells to invertebrate nerve and skeletal muscle. The nominal capacity of one microfarad per square centimeter is a crude generalization. In any one type of cell, some measurements may vary by nearly a factor of two or all of a series may be within a few percent of each other. I guess that of perhaps a hundred different membranes, the median value may be $1.5\,\mu F/cm^2$ with less than half outside the range of 1.0 to 2.0 $\mu F/cm^2$. The most embarrassing exceptions are the high capacities—approaching $10\,\mu F/cm^2$—for crustacean muscle (MII, p. 97) and nodes of medullated axons (Hodgkin, 1964). Although the actual membrane areas are uncertain, this is not an entirely satisfactory explanation.

Further, the presentation of a membrane as an ideal classical dielectric would be almost an abstraction without the measurements on several marine eggs. This simple capacity well represents the data on *Arbacia* and *Asterias* eggs and on phospholipid bilayers. At the other extreme we find many tissues for which the data are well represented by depressed circles, as in Figure 6

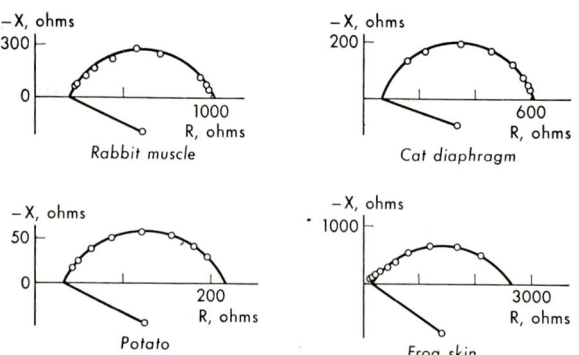

FIGURE 6 Complex impedance loci of early data on some tissues.

(Cole, 1932). To the accuracy with which the arcs are truly circular, the Maxwell capacity in the equivalent circuit must be replaced by one having a dielectric loss of the type long known for many solid insulators (Cole and Cole, 1941). Such behavior might be produced by a distribution of size, shape or capacities among the population. Orientation of red blood cells may produce the small effect which Fricke found and worried about (1953). But Fatt showed that the size distribution of fibers in sartorius muscle was far too narrow to account for the data, and this may well be true for many other tissues. An alternative explanation is that the lossy behavior was a characteristic of the membrane of each and every cell. This was indeed found to be true for the single cells of *Valonia* (Blinks, 1936), *Nitella* and in three quite different, independent and careful experiments, on the squid axon (Curtis and Cole, 1938; Hodgkin, *et al.* 1952; Taylor, 1965). Yet Falk and Fatt (1964) found no evidence for any deviation from simple ideal dielectric behavior in single muscle fibers! Thus we are tempted to blame the complicated current flow through such a highly concentrated maze of cells for this behavior. Starts have been made by analyses of arrays of close packed angular cells (Ranck, 1963; Nicholson, 1965; MII, p. 93) but they are so far more hopeful than conclusive. When we return to single cells we cannot guarantee that each and every square micron has the same characteristics. But even if they are essentially identical, our problem is not solved.

There are two problems of this dielectric capacity which seem to have been resolved (Rothschild, 1957). It was found for some marine eggs that the calculated capacities increased several fold upon fertilization. This was to be explained as an increase of the area of membrane as cortical granules entered it (MII, p. 35). In unfertilized eggs the capacity decreased regularly as they were swollen. Here the area was unchanged as the crinkles and wrinkles were flattened out.

There have not been many or extensive measurements of temperature coefficients of membrane capacities, probably because they have seemed rather dull and not very important. The reported values are: an increase at about a percent per degree for squid axon as seen in Figure 7 (Taylor, 1965) and $\frac{1}{3}$ per cent for red cells (Schwan, 1948), while the effect of temperature was given as small for frog node (Tasaki, 1955) and muscle (Fatt, 1964), and on earlier squid axon measurements (Curtis and Cole, 1938). Thus it seems safe to assume that within the range of 0° to 40°C there is no phase change such as those which produce dramatic effects in many dielectrics.

Further, I know of no clear effects of changes in the chemical composition of either external or internal media on the dielectric capacity. The external concentrations of potassium and calcium ions have most dramatic effects on the low frequency conductance and dispersion of several nerve membranes

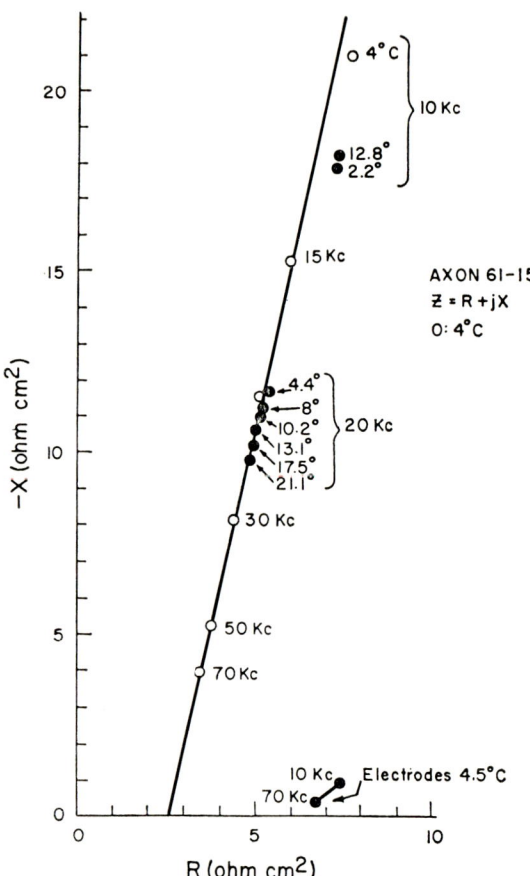

FIGURE 7 Complex impedance locus of squid axon membrane as functions of frequency and temperature.

(MII, p. 84). There have not been careful or systematic measurements on the dielectric capacity. It can only be said that the effects, if any, are relatively small and it is not now possible to set limits. Early measurements (MII, p. 40) on the kelp *Laminaria* in pure sodium chloride did not clearly indicate any major change of a dielectric capacity. Data on frog muscle in chloroform could be interpreted as about a fifty per cent increase of capacity as complete ion permeability and death were approached (Guttman, 1939). The first squid axon measurements gave no indication of a membrane capacity change as the axon deteriorated far beyond loss of excitability and to a conductance increase of a thousand times normal.

Dielectric properties of living membranes

The constancy of the squid axon capacity during deterioration (Curtis and Cole, 1938) was a good hint, but it was not really anticipated that the separation of capacity and conductance would be as complete as it was during a propagating impulse. The experiments were not easy and their accuracy and analysis was not certain, but the average capacity decrease of two per cent seemed completely negligible by comparison with the forty fold increase of conductance (Cole and Curtis, 1939). The steady state capacity change came out to be even less during changes of the membrane potential from about -60 mV to probably $+50$ mV, but there was a definite indication (Cole and Baker, 1941a) of a capacity decrease for hyperpolarization to potentials considerably more negative than -60 mV.

There have been several membrane models in which considerable conformational changes have been postulated during activity. However I could only guess that experiment required that these should not result in a capacity decrease of more than about 5 per cent. Now, about 30 years later, Moore has preliminary on-line records of the membrane capacity which show no change during propagating impulses to within an experimental accuracy of 2 per cent. Further, Adelman and Palti also have the preliminary results of a similar limit to the capacity changes during voltage clamp experiments.

Probably one of the most impressive features of most cell membranes is that, even at rest, the nominal potential difference of 60 mV and thickness of 100 Å give an average field of 6×10^4 V/cm and we should not be surprised o find that local fields may be well in excess of 10^4 V/cm. This is a widely used value for dielectric breakdown and indeed there are a few reports (MII, p. 529) on squid and lobster, of currents increasing irreversibly over tens of milliseconds for internal negative membrane potentials of more than 150 mV. A reversible current increment has been found in *Nitella* in the same potential range (Coster, 1965).

It would be quite misleading to allow an impression that all of the measurements and their analyses are not to be criticized or improved. The amount of information gained from the squid axon and the use of it is truly astounding. Yet it is quite certain that the freshly excised axon is significantly different from what it was in a normal animal (Moore and Cole, 1960). Even without the indignities of voltage clamping equipment and internal perfusion, the axon can only maintain a steady state for a short time—if at all. These tools force it to perform far beyond anything it is called upon to do in propagating a normal impulse. Just as use of the voltage clamp has expanded so amazingly, so have the techniques for its application. At every step there have been hazards that were difficult to recognize and avoid (Cole and Moore, 1960; Taylor, *et al.* 1960; Moore and Cole, 1963), but beyond this the analyses of the data may be uncertan. On the outside of the squid axon there are several micra of connective tissue and a dense layer of Schwann cells. Then both outside and inside there are probably layers of protein and polar groups in which we reach the effective boundaries of the dielectric capacity. It has become reasonably clear that there is always a resistance of a few ohm cm^2 in series with that capacity (MII, pp. 249, 277). Thus during current flow our potential measurements may be in error by tens of millivolts.

More and more biological membranes have appeared to have a capacity of about one, or at most a few, microfarads per square centimeter and almost without exception. This capacity has remained relatively unaffected by the high field strengths and wide variations of current density which occur in normal function, but also for considerable extensions beyond these usual ranges. It seems to change only a little, if at all, under several other highly abnormal conditions—such as an ionic environment, anesthesia, and approaching cell death. Indeed the correspondence with the unit membrane (Robertson, 1960), the three layer or railroad tracks seen by electron microscopy, is very strong evidence that the structure responsible for this membrane capacity survives fixation, staining, monomer imbedding, polymerization, sectioning and electron bombardment.

It seems clear that the solution of the problems of membrane structure and function has one necessary condition of outstanding importance—the universality and stability of the dielectric capacity.

INTERPRETATION

For a long time the principal question was how thick or thin a membrane might be, and the corresponding electrical question was 'What is the dielectric constant?' Fricke never modified his original value of three, Danielli (1935) gave a broad discussion of the uncertainties, and Mullins (1961) proposed a highly specific layered structure. Even with the X-ray diffraction, optical, and electron microscope information there is no clear and simple answer, and also the question has lost most of its importance as we move to the molecular level. But I do not know if it is yet feasible to calculate a membrane capacity for a bimolecular model—taking into account, without *ad hoc* assumptions, the lipid-sterol ratios and arrangements, the fraction of unsaturated bonds, the joggling of hydrocarbon chains, the freedom of polar groups, and the effect of surface proteins.

There seems ample evidence that the membrane is not crystalline. The broad X-ray diffraction band at 4.8 Å (Finean, 1966), the nuclear magnetic resonance of myelin (Lecar, *et al.* 1967) and with tagged spinners on living cells (Hubbell and McConnell, 1968), optical studies with lamillar models (Chapman, 1966) strongly suggest a liquid crystal. Howarth, *et al.* (1968)

have invoked an entropy change in the membrane to explain the heat measurements during a nerve impulse. This was calculated from the temperature coefficient of the capacity as a decrease of order produced by the electric field. A Maxwellian interpretation by R. H. Cole is that this indicates a relatively well ordered structure. But what effect should this have on the dielectric capacity and what is the relation to the dielectric loss? At least it seems probable that the central hydrocarbon region is loose enough to permit ion passage.

We have more problems as we turn to those membranes which may have the more complicated behavior shown in Figure 7 for the impedance of the squid axon between internal and external electrodes. If we accept the infinite frequency extrapolation as a series electrolyte resistance, the capacity is then the same as that empirically described long ago for the frequency dependent component of many solid dielectrics. Such behavior may be expressed as a distribution of Debye relaxations, but this is not yet a molecular explanation. A stochastic relaxation at short range only suggests that an adequate analysis of long range cooperation may give an answer (Cole, 1965). Alternatively Taylor (1965) suggested that there may be a dispersion centered at higher frequencies such as ice might produce. Whether or not living membranes will submit to adequate tests is not yet to be predicted.

The most powerful model yet available is the artificial bilayer (Mueller and Rudin, 1963) but so much is being done with it in so many places that I cannot give more than a very superficial and mostly uninformed discussion of it. In its most primitive form the bilayer is made of phospholipids and might be expected to display much the same dielectric behavior as found in living cell membranes. However it is only 50 Å thick, is quite impermeable to ions, and has a simple dielectric capacity of $0.5 \mu F/cm^2$ (Thompson, 1964). This capacity is lower than for any—perhaps all—cell membranes that have been well measured. So the suggestion has been made (Hanai, *et al.* 1965) that a one per cent volume of ion conductive channels of water or ice would make up the difference. Unfortunately it has not been possible, nor does it seem feasible, to eliminate the living cell conductance as a test of the hypothesis; so we must rely on the artificial membranes. From these it seems probable that specific proteins may reach through the bilayers to give spectacular imitations of the conductance and negative conductance found in some living membranes. But I do not know that such an activated bilayer has an increase of capacity, or if it shows a dielectric loss behavior, or what the before and after temperature coefficients may be. From optical changes in proteins at living cell surfaces it has been suggested (Wallach and Zahler, 1966) that lipophilic polypeptide chains may reach through the membrane. Onsager (1967) has proposed proton assisted ion conduction along such a chain. Among other problems are the results (Hubbell and McConnell, 1968) with tagged electron magnetic resonance tumblers which indicate a tightening of the hydrocarbon component in high calcium ion environments and anesthesia. Should any of these make detectible changes in the dielectric capacity?

The bilayer model with conductive channels is a very attractive and correspondingly popular model at present. However I know of no experiment which requires such an interpretation of the data. Until—if ever—there is conclusive evidence for channels we must not forget the possibility that a biological membrane is a continuum, somewhat accessible everywhere to each and every ion. The oldest and best known model is that of Planck, as based on Nernst, for a lipid liquid junction. To this may be added fixed charges or mobile carriers (Eisenman, *et al.* 1967). And indeed as this has been a powerful approach to semi-conductor theory, so it may also be possible to apply some of the semi-conductor techniques and results to membranes. As a single illustration, Mauro (1962) predicted the observed (Schwartz and Case, 1964) one microfarad capacity at the junction between anion and cation exchangers, and this has been extended to the reversible hyperpolarization current in *Nitella* as a Zener punch through (Coster, 1965).

CONCLUSION

Each model needs to be carefully examined for its ability to explain the dielectric capacity—first as to the magnitude of $1-2\,\mu F/cm^2$ and second, its constancy in the face of spectacular changes in biological and physical conditions. However, it seems likely that if a decision is to be made between channels and a continuum, the much more spectacular conductive behavior will carry the most weight.

References

BLINKS, L. R. 1936, The polarization capacity and resistance of *Valonia*. I. Alternating current measurements. *J. Gen. Physiol.* **19**, 673–691.

CHAPMAN, D. 1966, Liquid crystals and cell membranes. *Ann. N.Y. Acad. Sci.*, **137**, 745–754.

COLE, K. S. 1928, Electric impedance of suspensions of spheres. *J. Gen. Physiol.*, **12**, 29–36.

COLE, K. S. 1932, Electric phase angle of cell membranes. *J. Gen. Physiol.* **15**, 641–649.

COLE, K. S. 1937, Electric impedance of marine egg membranes. *Trans. Faraday Soc.*, **33**, 966–972.

COLE, K. S. 1942, Impedance of single cells. *Tabulae Biologicae* (*Cellula*), **19**, 24–27.

COLE, K. S. 1947, Four Lectures on Biophysics, Institute of Biophysics, University of Brazil.

COLE, K. S. 1949, Some physical aspects of bioelectric phenomena. *Proc. Natl. Acad. Sci.* (*USA*), **35**, 558–566.
COLE, K. S. 1968, (MII) Membranes, ions and impulses. Berkeley, University of California Press.
COLE, K. S., and BAKER, R. F. 1941a, Transverse impedance of the squid giant axon during current flow. *J. Gen. Physiol.*, **24**, 535–549.
COLE, K. S., and BAKER, R. F. 1941b, Longitudinal impedance of the squid giant axon. *J. Gen. Physiol.*, **24**, 771–788.
COLE, K. S., and COLE, R. H. 1936, Electric impedance of *Asterias* eggs. *J. Gen. Physiol.*, **19**, 609–623.
COLE, K. S., and COLE, R. H. 1941, Dispersion and absorption in dielectrics. I. Alternating current characteristics. *J. Chem. Physics*, **9**, 341–351.
COLE, K. S., and CURTIS, H. J. 1937, Wheatstone bridge and electrolytic resistor for impedance measurements over a wide frequency range. *Rev. Sci. Instr.*, **8**, 333–339.
COLE, K. S., and CURTIS, H. J. 1938, Electric impedance of *Nitella* during activity. *J. Gen. Physiol.*, **22**, 37–64.
COLE, K. S., and CURTIS, H. J. 1939, Electric impedance of the squid giant axon during activity. *J. Gen. Physiol.*, **22**, 649–670.
COLE, K. S., and CURTIS, H. J. 1950, Bioelectricity: Electric Physiology. In (Glasser, O., editor) *Medical Physics*, 2 Chicago, Yearbook Publishers, pp. 82–90.
COLE, K. S., and MOORE, J. W. 1960, Ionic current measurements in the squid giant axon. *J. Gen. Physiol.*, **44**, 123–167.
COLE, R. H. 1965, Relaxation processes in dielectrics. *J. Cell. Comp. Physiol.*, **66** (Suppl. 2), 13–20.
COSTER, H. G. L. 1965, A quantitative analysis of the current-voltage relationships of fixed charge membranes and the associated property of 'punch-through'. *Biophys. J.*, **5**, 669–686.
CURTIS, H. J., and COLE, K. S. 1938 Transverse electric impedance of the squid giant axon. *J. Gen. Physiol.*, **21**, 757–765.
DANIELLI, J. F. 1935, The thickness of the wall of the red blood corpuscle. *J. Gen. Physiol.*, **19**, 19–22.
EISENBERG, B., and EISENBERG, R. S. 1968, Transverse tubular system in glycerol-treated skeletal muscle. *Science*, **160**, 1243–1244.
EISENMAN, G., SANDBLOM, J. P., and WALKER, J. L. JR. 1967, Membrane structure and ion permeation. *Science*, **155**, 965–974.
FALK, G., and FATT, P. 1964, Linear electrical properties of striated muscle fibres observed with intracellular electrodes. *Proc. Roy. Soc.* (London), **B 160**, 69–123.
FATT, P. 1964, An analysis of the transverse electrical impedance of striated muscle. *Proc. Roy. Soc.* (London), **B 159**, 606–651.
FINEAN, J. B. 1962, The nature and stability of the plasma membrane. *Circulation*, **26**, 1151–1162.
FRICKE, H. 1923, The electric capacity of cell suspensions. *Phys. Rev.*, **21**, 708.
FRICKE, H. 1925. The electric capacity of suspensions with special reference to blood. *J. Gen. Physiol.*, **9**, 137–152.
FRICKE, H. 1953, The Maxwell-Wagner dispersion in a suspension of ellipsoids. *J. Phys. Chem.*, **57**, 934–937.

FRICKE, H., and MORSE, S. 1925, The electric resistance and capacity of blood for frequencies between 800 and $4\frac{1}{2}$ million cycles. *J. Gen. Physiol.*, **9**, 153–167.

GUTTMAN, R. 1939, The electrical impedance of muscle during the action of narcotics and other agents. *J. Gen. Physiol.*, **22**, 567–591.

HANAI, T., HAYDON, D. A., and TAYLOR, J. 1965, Some further experiments on bimolecular lipid membranes. *J. Gen. Physiol.*, **48** (5, 2), 59–63.

HOWARTH, J. V., KEYNES, R. D., and RITCHIE, J. M. 1968, The origin of the initial heat associated with a single impulse in mammalian non-myelinated nerve fibres. *J. Physiol.*, **194**, 745–793.

HODGKIN, A. L. 1964, The conduction of the nervous impulse. Springfield, Thomas.

HODGKIN, A. L., HUXLEY, A. F., and KATZ, B. 1952, Measurement of current-voltage relations in the membrane of the giant axon of *Loligo*. *J. Physiol.*, **116**, 424–448.

HUBBELL, W. L., and McCONNELL, H. M. 1968, Spin-label studies of the excitable membranes of nerve and muscle. *Proc. Natl. Acad. Sci.* (U.S.A.), **61**, 12–16.

LECAR, H., EHRENSTEIN, G., and STILLMAN, I. 1967, Nuclear magnetic resonance absorption in myelin. *Biophys. Soc. Abstr.*, **7**, 65.

LOEWENSTEIN, W. R. 1964, Permeability of the nuclear membrane as determined by electrical methods. Protoplasmatologia, Handbuch der Protoplasmaforschung, Vienna, Springer, pp. 26–34.

MAURO, A. 1962, Space charge regions in fixed charge membranes and the associated property of capacitance. *Biophys. J.*, **2**, 179.

MUELLER, P., and RUDIN, D. O. 1963, Induced excitability in reconstituted cell membrane structure. *J. Theoret. Biol.*, **4**, 268–280.

MULLINS, L. J. 1961, The macromolecular properties of excitable membranes. *Ann. N. Y. Acad. Sci.*, **94**, 390–404.

MOORE, J. W., and COLE, K. S. 1963, Voltage Clamp Techniques. In (Nastuk, W. L. (ed.)) *Physical Techniques in Biological Research*, **6**, New York, Academic Press, pp. 263–321.

MOORE, J. W., and COLE, K. S. 1960, Resting and action potentials of the squid giant axon *in vivo*. *J. Gen. Physiol.*, **43**, 961–970.

NICHOLSON, P. W. 1965, Specific impedance of cerebral white matter. *Exptl. Neurol.*, **13**, 386–400.

ONSAGER, L. 1967, Ion passages in lipid bilayers. *Science*, **156**, 541.

PAULY, H., PACKER, L., and SCHWAN, H. 1960, Electrical properties of mitochondrial membranes. *J. Biophys. Biochem. Cytol.*, **7**, 589–601.

RANCK, J. B., JR. 1963, Analysis of specific impedance of rabbit cerebral cortex. *Exptl. Neurol.*, **7**, 153–174.

ROBERTSON, J. D. 1960, The molecular structure and contact relationship of cell membranes. *Prog. Biophysics*, **10**, 343.

ROTHSCHILD, LORD. 1957, The membrane capacitance of the sea urchin egg. *J. Biophys. Biochem. Cytol.*, **3**, 103–110.

SCHWAN, H. P. 1948, Die Temperaturabhängigkeit der Dielektrizitätskonstante von Blut bei Neiderfrequenz. *Zs. f. Naturf.*, **3b**, 361–367.

SCHWAN, H. P. 1957, Electrical properties of tissue and cell suspensions. In (Tobias, C. A. and Lawrence, J. H., (eds.)) *Advances in Biological and Medical Physics*, Vol. 5, New York, Academic Press, pp. 147–209.

SCHWAN, H. P. 1963, Determination of biological impedances. In (W. L. Nastuk (ed.)) *Physical Techniques in Biological Research* Vol. 6, New York, Academic Press, pp. 323–407.

SCHWAN, H. P., and COLE, K. S. 1960, Bioelectricity: Alternating current admittance of cells and tissues. In (O. Glasser (ed.)) *Medical Physics*, **3**, 52–56, Chicago, Yearbook Publishers.

SCHWAN, H. P., and MOROWITZ, H. J. 1962. Electrical properties of the membranes of pleuropneumonia-like organism A5969. *Biophys. J.*, **2**, 395–407.

SCHWARTZ, G. 1962, A theory of the low frequency dispersion of colloidal particles in electrolyte solutions. *J. Phys. Chem.*, **66**, 2636–2642.

SCHWARTZ, M. and CASE, C. T. 1964, Electric impedance and rectification of fused anion-cation membranes in solution. *Biophys. J.*, **4**, 137–149.

SEGAL, J. R. 1967, Electrical capacitance of ion-exchange membranes. *J. Theo. Biol.*, **14**, 11–34.

SPECTOR, W. S. 1956, Handbook of Biological Data. Philadelphia, Saunders.

TASAKI, I. 1955, New measurements of the capacity and the resistance of the myelin sheath and the nodal membrane of the isolated frog nerve. *Am. J. Physiol.*, **181**, 639–650.

TASAKI, I. 1968, Nerve excitation. Springfield, Thomas.

TAYLOR, R. E. 1965, Impedance of the squid axon membrane. *J. Cell. Comp. Physiol.*, **66** (Suppl. 2, II), 21–25.

TAYLOR, R. E., MOORE, J. W., and COLE, K. S. 1960, Analysis of certain errors in squid axon voltage clamp measurements. *Biophys. J.*, **1**, 161–202.

THOMPSON, T. E. 1964, The properties of bimolecular phospholipid membranes. In 'Cellular membranes in development' pp. 83–96, New York, Academic Press.

WALLACH, D. F. H., and ZAHLER, P. H. 1966, Protein conformations in cellular membranes. *Proc. Natl. Acad. Sci.* (U.S.A.), **56**, 1552–1559.

DISCUSSION

KALKWARF   My colleague, Walter Grafton has measured phospholipid bilayers at Mueller and Rudin's Laboratory in Philadelphia. He finds that the capacities are somewhat higher than you have given, about $0.7\,\mu\text{F}/\text{cm}^2$, and that there is no difference between a membrane in its low conductance form and in its high conductance form after it has interacted with proteins.

COLE   Thank you for this information which is new to me.

EISENMAN   Thompson has found something like ten molecules of the solvent for each phospholipid molecule in the bilayers. Since the solvent is usually decane or a similar aliphatic hydrocarbon with a dielectric constant of about 2, it may be the principal cause of the differences in capacity between the artificial bilayers and the living membranes. Does anyone know what happens to the bilayer capacity, if an unsaturated solvent such as decene with a higher dielectric constant is used?

KALKWARF   D-α-tocopherol is the principal solvent I have used myself and the capacitance is around $0.8\,\mu\text{F}\,\text{cm}^{-2}$. Maybe that's a move in the right direction.

LAUGER   We have made some experiments with decene having one double bond as a solvent, and we observed about a 10 per cent increase of the bilayer capacitance.

COLE   It should be very interesting to measure artificial bilayers which are essentially solvent-free.

EISENMAN   One would expect an increase of the bilayer dielectric constant with substantial additions of the highly polar cyclic-polyethers or monactin.

PAPER 2

Optical and electrophysiological evidence for conformational changes in membrane macromolecules during nerve excitation

I. TASAKI, W. BARRY, and L. CARNAY

Laboratory of Neurobiology National Institute of Mental Health

---

I INTRODUCTION

THERE HAS been a strong tendency in the past to describe the nerve impulse in terms of its electrical concomitants (1). This tendency may be attributed to the relative ease with which electrical measurements may be made, and to the physico-chemical complexity of the intact axon in its natural state which makes an analysis of the molecular biology of nerve excitation very difficult. However, it has become increasingly clear that a purely electrical description of the events associated with nerve excitation, valuable as it has been, is rather limited in its vocabulary. We have attempted to define some of the non-electrical concomitants of the nerve impulse, in the hope that such an approach might expand our present knowledge, gained largely through electrical measurements, of the basic molecular events underlying nerve excitation.

The technique of intracellular perfusion of squid giant axons, developed about six years ago, has proved a powerful tool in investigating the excitation process. By this technique, it was possible to remove the protoplasm almost completely without affecting the ability of the axon to produce action potentials. The excitation process in the axon membrane could be analyzed with simplified, well-defined solutions on both sides of the membrane. The conditions necessary and sufficient for the maintenance of excitability were clarified (2, 3). One of the important conclusions derived from these studies was that the process of excitation is accompanied by a drastic change in the macromolecular conformation of the membrane, triggered by cooperative cation exchange at fixed negative sites in the membrane.

Recently, another powerful technique has been developed (4, 5) which promises to increase further our understanding of the mechanism of production of the nerve impulse. By the use of specifically designed optical and electronic equipment, it has been possible to detect changes in the optical properties of various invertebrate nerves during the nerve impulse (4, 5). Specifically, changes in turbidity, birefringence, and fluorescence (in stained preparations) have been detected. These optical studies now offer additional evidence that there is a significant change in the conformation of the nerve membrane macromolecules during excitation. These and other observations strongly suggest that a cooperative change in membrane conformation is the primary event involved in nerve excitation (2, 6, 7).

In this article, we wish to review first our present knowledge concerning the optical changes in nerve excitation. Next, we present some of the electrophysiological data in support of the view that there is a cooperative ion-exchange process in the nerve membrane. Finally, in order to relate the electrophysiological data with the optical phenomena, the effect of electric currents on the optical properties of the nerve is briefly discussed.

II OPTICAL CHANGES IN NERVE DURING EXCITATION

A. *Changes in Light Scattering During Nerve Activity*

Neurophysiologists have known for some time that repetitive stimulation of an invertebrate nerve produces a slow persistent change in the intensity of the light scattered by the nerve (8, 9). This optical signal represents a gradual increase in the turbidity of the nerve produced by a large number of nerve impulses. Recently, Cohen, Keynes and Hille (4) discovered that the light scattered by the nerve increases almost simultaneously with the onset of the action potential. To record optical signals of this type, occurring with individual action potentials, a sensitive light-detecting device with a high time resolution is required.

In our studies of light scattering in crab nerves (5), quasi-monochromatic light in the spectrum range between 360 and 600 mμ was used. A 3 to 6 mm long portion of the nerve (immersed in sea water) was illuminated by the incident light. The photomultiplier tube for detecting changes in light intensity was placed at either 0 or 90 degrees.

With all the wave lengths examined, there was an increase in the intensity of the scattered light detected at 90 degrees. The ratio of the increase in light intensity at the peak of the action potential to the intensity of light scattered by the nerve at rest was 10^{-4} to 10^{-5} in crab nerves. To improve the signal-to-noise ratio in our recording, a digital computer was used for averaging multiple signals. The light detected at 0 degrees was found to decrease during nervous activity, as expected from the results obtained at 90 degrees.

The duration of the optical signal observed under these conditions was usually longer than that of the action potential; the rising phase of the optical signal of this type was sharp; but the falling phase was long and often very complex.

B. *Changes in Birefringence during Nerve Activity*

Small changes in birefringence of various intracellular structures can be detected by inserting the cell between a polarizer and an analyzer in a crossed position (see e.g. Inoue and Sato, 10). When the optical axis of a birefringent material is placed at 45 degrees relative to the plane of polarization of the incident light, a change in the intensity of the transmitted light is a very sensitive measure of the variation of the refractive indices of the material (see p. 696 in Born and Wolf, 11). This technique has been used successfully to demonstrate changes in birefringence associated with the nerve impulse in invertebrate nerve fibers (4, 5).

Our instrument for detecting birefringence changes during nerve activity consists of the following parts: (1) a light source, (2) a lens for obtaining parallel beams of light, (3) an interference filter for 550 mμ, (4) a collimating lens, (5) a polarizer (polaroid sheet), (6) an analyzer in a crossed position, and (7) a photomultiplier tube. A nerve was placed between the analyzer and the polarizer in the focal plane of the collimating lens. With the long axis of the nerve placed at 45 degrees to the plane of polarization of the incident light, a relatively large decrease in the photo-current was observed when the nerve was stimulated electrically. In crab nerve trunks, the ratio of the change at the peak of the action potential to the intensity of the light in the resting state was of the order of 10^{-4}. The corresponding value for the squid giant axon was roughly 10^{-5}.

The optical signal observed under these conditions may be interpreted as indicating a decrease in the birefringence of the resting nerve during nerve activity. This interpretation is supported by the fact that the sign of the optical signal can be reversed by insertion of a compensator between the nerve and the analyzer (see references 4 and 5).

C. *Changes in Fluorescence of Dye-Stained Crab Nerves*

Under natural (unstained) conditions, nerves from spider crabs do not show any detectable fluorescence in the visible spectrum. When nerves vitally stained with proper dyes are irradiated with quasi-monochromatic light, emission of fluorescent light by the nerve can be detected. In certain ranges of dye concentrations, vitally stained nerves are found to maintain their ability to produce action potentials for many hours. When these stained nerves are stimulated electrically, a small change is frequently found in the intensity of the fluorescent light at the time when the nerve impulse arrives at the irradiated portion of the nerve.

To date we have tested approximately 20 dyes, and have found that the following 10 dyes can be used to detect fluorescence changes during nerve activity: acridine orange (456), 8-anilino-naphthlene sulfonic acid (365), pyronin B (550), rhodamine B (550), rhodamine G (500), auramine (450), erythrosine B (500), rose bengal (550), phosphine (500), and L.S.D. (365). The numerals in the brackets above indicate the wave length of the irradiating light expressed in mμ. With the exception of pyronin B and rhodamine B, the intensity of the fluorescent light from dye-stained nerves was found to increase during excitation. Nerves stained with either pyronin B or rhodamine B gave a decrease in fluorescence during an action potential.

Examples of the results obtained with acridine orange, pyronin B and rhodamine B are presented in Figure 1. In these experiments, vital staining was performed by immersing a portion of a nerve in a dye solution (containing

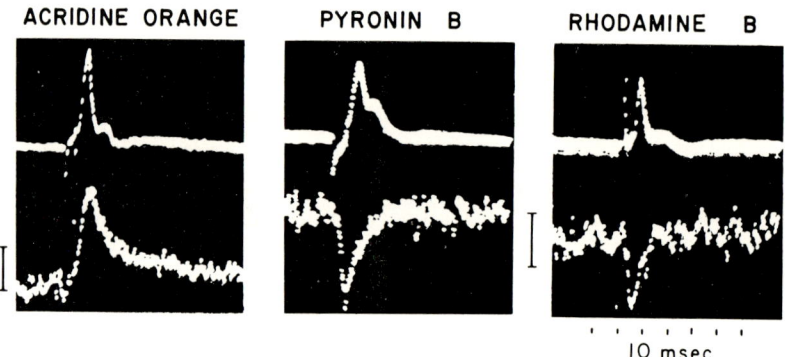

FIGURE 1 Action potentials (top) and fluorescence changes (bottom) associated with propagated nerve impulses in crab nerves stained with acridine orange, pyronin B and rhodamine B. A CAT computer was used to record both action potentials and optical signals. The amplitude of the action potential was approximately 2 mV. An upward deflection of the lower trace indicates an increase in the intensity of fluorescence. The vertical bars represent a change of 5×10^{-5} times the fluorescence intensity before stimulation. The time scale below applies to all three records. Note the difference in the sign of optical signal obtained with different dyes. Temperature 22 °C.

0.1 mg of dye in 1 ml of sea water) for a period of 10 to 15 minutes. Subsequently, the nerve was washed thoroughly with sea water free of dye, and the stained portion of the nerve was irradiated with a beam of light (from a 100 watt quartz-iodine lamp) passed through an interference filter. Changes in the intensity of the fluorescent light were detected with a photomultiplier tube placed at 90 degrees to the incident light. An optical filter was inserted between the nerve and the photomultiplier, allowing light waves 40 mμ or longer in wave length than the incident light to reach the photomultiplier. The nerve was stimulated with a pair of platinum electrodes near one end of the nerve (away from the illuminated area) and the electric responses to stimuli were recorded with another pair of electrodes near the site of optical recording. The change in the photo-current observed following nerve stimulation was of the order of 10^{-4} to 10^{-5} times the intensity of the background photo-current.

The change in fluorescence was found to start upon the arrival of the nerve impulse at the site of optical recording. The duration of the optical signal was comparable to that of the intracellularly recorded action potential. (The biphasic action potentials in Figure 1 are shorter in duration than intracellularly recorded action potentials.)

The observed variations in the intensity of the fluorescent light during nervous activity may be considered as indicative of changes in physicochemical properties of the micro-environment of the dye molecules bound to the nerve membrane. At present, however, the factors affecting the absorption of the incident light and the quantum yield of light emission by the dye molecules are only poorly understood, so that a direct statement of the nature of micro-environment change occurring in the nerve membrane cannot be made. It is well known, however, from studies of relatively pure proteins stained with fluorescent dyes, that a change in the molecular conformation of the protein often drastically changes the intensity of fluorescence of the bound dye molecules (12).

III LOCALIZATION OF STRUCTURAL CHANGES ACCOMPANYING NERVE EXCITATION

The experimental findings described in the preceding sections are consistent with the view that the nerve membrane undergoes a structural change when stimulated. The magnitudes of the observed optical signals are very small. However, if only a small fraction of the cellular elements in the nerve is involved in nerve excitation, the observed magnitude of the optical signals does not necessarily indicate that the extent of the structural change is very small.

Electrophysiological and electron-microscopic evidence shows that the axonal membrane, which is about 10 mμ in thickness, is the site where the primary process of nerve excitation takes place (see Cole, 13). Each axon (nerve fiber) is surrounded by a layer of Schwann cells, and is wrapped further by a layer of connective tissue. The interior of the axon is filled with protoplasm which is totally dispensable to the process of nerve excitation. In a nerve trunk consisting of axons 40 μ in diameter, it is estimated that the axonal membrane occupies roughly 0.1 per cent of the total volume of the nerve. In the squid giant axon, this percentage is expected to be even smaller. It is safe to assume that the structural change in the axon at the onset of an action potential is restricted to the membrane layer of about 10 mμ in thickness and its immediate vicinity. Then, the extent of the structural change in the membrane during nerve excitation, as revealed by changes in optical properties of the whole nerve, has to be considered as very great.

IV ELECTROPHYSIOLOGICAL EVIDENCE FOR COOPERATIVE STRUCTURAL CHANGE IN NERVE MEMBRANE

Although the optical studies mentioned in preceding sections have yielded strong evidence for structural changes in the nerve membrane, the optical data alone provide limited information concerning the physico-chemical nature of the process underlying nerve activity. Our previous electrophysiological studies offer a reasonable basis on which we can interpret some of our optical data. In this section, we briefly summarize the major electrophysiological findings obtained by the use of our technique of intracellular perfusion (2).

A. *Cation-Exchanger Properties of the Resting Axon Membrane*

Under continuous intracellular perfusion with an isotonic solution containing potassium fluoride (or phosphate) and sucrose (pH 7.3), a squid giant axon maintains its ability to develop action potentials for many hours in an isotonic medium (pH 8) containing $CaCl_2$ and NaCl. Replacement of chloride ion in the medium with bromide or methylsulfate does not affect the normal function of the axon. This fact is consistent with the view that anions are excluded from the membrane by virtue of its negative fixed charge.

In the resting state with no electric current through the membrane, the trans-membrane fluxes of Na, K, Ca, Cl, and F ions in such an axon should satisfy the following conditions:

$$J_{Na} + J_K + 2J_{Ca} - J_{Cl} - J_F = 0 \tag{1}$$

If the membrane has cation-exchanger properties, fluxes of anions should be small; consequently, Eq. (1) becomes

$$J_{Na} + J_K + 2J_{Ca} \approx 0 \tag{2}$$

Radio-tracer studies carried out under these experimental conditions revealed that the relationship given by Eq. (2) is approximately correct. Furthermore, it was found that the flux of the divalent cation, J_{Ca}, is far smaller than the fluxes of univalent cations; therefore,

$$J_{Na} + J_K \approx 0 \tag{3}$$

When an axon is immersed in an isotonic solution containing $MgCl_2$ (or $CaCl_2$) and sucrose, the membrane potential changes with the salt concentration in the external medium. The sign and magnitude of this potential change agree with those seen in an ideal cation exchanger membrane, where the potential change associated with variation of the external salt concentration is due to a change in the Donnan potential (2). This finding indicates that the anions in the external medium are nearly perfectly excluded from the membrane. The negative fixed charges responsible for this anion exclusion are considered to derive from the phosphate and carboxyl groups in the membrane macromolecules which are most likely to be on the alkaline side of their isoelectric point.

B. *Axon Membrane in the Active (Excited) State*

The membrane potential at the peak of nerve activity is insensitive to the chemical nature or the valence of the anions in the external medium. This fact is consistent with the view that the axon membrane retains its cation-exchanger properties in its active state. Radiotracer studies indicate that the anion fluxes are far smaller than the fluxes of univalent cations and, consequently, that Eq. (3) is approximately valid at the peak of nervous activity.

When NaCl (or NaBr) is the major salt component in the external medium, the action potential amplitude varies with the external Na-ion concentration. (Note however that the presence of divalent cation is required in the medium to maintain excitability.) The excitability can be maintained with Na-ions in the medium completely replaced with hydrazinium, guanidinium or other favorable univalent cations; the amplitude of the action potential varies with the concentration of these polyatomic cations under these conditions. All these findings strongly support the view that the axon membrane has a high density of fixed negative charge in the active state.

C. *Importance of Divalent Cations*

It is well known that complete removal of the divalent cations from the external fluid medium bathing an axon leads to a loss of excitability. Ordinarily, the concentrations of the major external univalent cation species (e.g. Na^+) are greater than those of the divalent cations (Ca^{++} and Mg^{++}), and the action potential amplitude is strongly affected by the univalent cation.

However, complete elimination of the univalent cation in the external medium does not always lead to a complete loss of excitability.

Recently, the ability of the squid giant axons to produce all-or-none action potentials was examined under continuous intracellular perfusion with an isotonic solution containing the salt of one of the following cations: Li, Na, Cs, Rb, K, guanidinium, tetraethylammonium, etc. It was found (see references 2 and 14) that action potentials can be elicited from these axons in an external medium containing $CaCl_2$ ($CaBr_2$ or $SrCl_2$) as the sole electrolyte. (With K- or Rb-phosphate internally, a steady polarizing current was required to demonstrate all-or-none action potentials in axons immersed in media free of univalent cations.)

D. Abrupt Depolarization as a Sign of Cooperative Ion-Exchange Process

In squid giant axons with Cs- (or Na-) phosphate internally and $CaCl_2$ externally, addition of the salt of a univalent cation (K^+, Rb^+, Cs^+, Na^+, etc.) to the external medium produces an abrupt rise in the intracellular potential (15). This abrupt potential rise (i.e., abrupt depolarization) is accompanied by a simultaneous increase in the membrane conductance (see

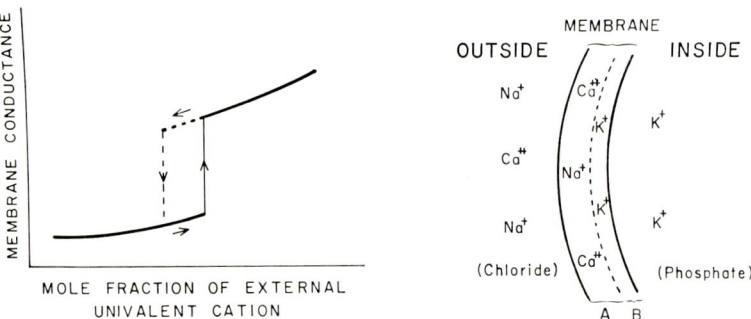

FIGURE 2 Left: Simplified diagram showing abrupt changes in the membrane conductance produced by continuous changes in mole fraction of the external univalent cation. This diagram is based on the results from experiments with squid giant axons internally perfused with CsF or Na-phosphate. Note the existence of a hysteresis loop in this diagram.

Right: Schematic diagram showing the ion distribution in- and outside a squid giant axon internally perfused with K-phosphate and immersed in a solution containing NaCl and $CaCl_2$. The cation distribution in the outer (A) and inner (B) layer of the axon membrane is also indicated. The negative fixed charges of the membrane are omitted. The ion distribution in unperfused nerve fibers can be represented (approximately) by this diagram.

Figure 2). The membrane conductance is determined by the mobility-concentration products of the ions within the membrane. Therefore, an abrupt rise in the membrane conductance is an indication of a sudden rise in the concentration and/or mobility of the cations in the membrane. A plot of the membrane conductance as a function of the mole fraction of the external univalent cation shows a steep, S-shaped curve. This fact suggests that the ion-exchange process involved is cooperative.

Following abrupt depolarization, a further increase in the mole fraction of the univalent cation produced only a continuous change in the membrane potential and conductance. When the mole fraction is decreased gradually, repolarization occurs at a level lower than that required to induce abrupt depolarization (see the broken line in Figure 2). This hysteresis can be regarded as a sign of the appearance of a metastable state in the vicinity of the point where the abrupt transition takes place.

E. *Electric Stimulation*

In the theory of nerve excitation proposed by Hodgkin and Huxley, it is assumed that a stimulating current brings about an increase in the 'permeability' of the membrane to Na-ions (see reference 1). Since the Na-ion concentration in the external medium is normally higher than that in the protoplasm, a high Na-permeability induced by the stimulus is considered to lead to a rise in the intracellular potential and eventually to an action potential.

Quite recently, however, we have encountered a number of facts which would seem to require a significant modification of this theory. Squid giant axons internally perfused with a Na-phosphate solution are found to remain excitable in a medium containing $CaCl_2$ (or $SrBr_2$) as the sole electrolyte. If there is a rise in the Na-permeability following electric stimulation, the intra cellular potential is expected to fall under these conditions with a reversed Na-concentration gradient across the membrane. Actually, however, the sign of the action potential is the same as that of an axon immersed in a Na-rich medium.

The condition necessary for the maintenance of excitability in squid axons is the presence of the salt of a divalent cation in the exterior and the salt of a univalent cation in the interior. A variety of univalent cations (e.g., Na, Li, choline, tetramithylammonium, etc.) can be used internally for this purpose. In other words, there is no absolute ion-specificity in the axon membrane. As has been mentioned earlier, the salt of Ca, Sr or Ba ion has been used to maintain excitability under these 'bi-ionic' conditions. It is evident that the analysis of these 'bi-ionic' action potentials is relatively simple as compared with the analysis of 'normal' action potentials, which involves more than 2 cations. (The traditional theory of nerve excitation is based on an analysis of 'multi-ionic' action potentials.)

From an analysis of excitation under 'bi-ionic' conditions, we have proposed an alternative theory of the ionic events underlying nerve excitation. A stimulating current is directed outwards through the axon membrane. This current tends to transport internal univalent cations into the axon membrane. This electrophoretic effect of the stimulating current increases the mole fraction of the univalent cation in the membrane. This increase in the mole fraction is capable of inducing abrupt depolarization (see the left-hand diagram in Figure 2). Thus, it is inferred that the onset of a bi-ionic action potential is nothing but abrupt depolarization caused by the electrophoresis of the internal univalent cation into the membrane by the stimulating current.

The right-hand diagram in Figure 2 shows the ion distribution in the simplest experimental condition under which a very brief (normal) action potential can be produced. The outer layer of the membrane (which has a high density of negative fixed charges) is considered to contain cations derived from the external medium (Na^+ and Ca^{++}) and the inner layer contains cations derived from the internal medium (K^+). A stimulating (i.e., outward-directed) current carries the internal cations (K^+) into the outer layer, displacing the divalent cations in this layer. Therefore, abrupt depolarization (i.e., initiation of an action potential) is expected to take place when the strength of the stimulating current reaches a critical level (i.e., threshold). A further discussion on the process of initiation and termination of action potentials may be found elsewhere (2, 3, 14).

V PLAUSIBLE INTERPRETATION OF THE RESULTS OF FLUORESCENCE STUDIES

Basing on the theory of nerve excitation described in the preceding sections, it is possible to offer a reasonable interpretation of some of the optical data obtained with fluorescent dyes. When the axon membrane undergoes a conformational change induced by cooperative ion-exchange following electric stimulation, the micro-environment of individual dye molecules bound to the membrane would be drastically altered. Such an alteration of the micro-environment could lead to a change in the fluorescence of the dye molecules.

When the axon membrane undergoes a transition from the resting state of the active (excited) state, there is a sudden rise in the membrane conductance. This rise in conductance induces an increase in the rate of interdiffusion of cations across the membrane. Consequently, the mole fraction of the internal univalent cation (K^+ under ordinary experimental conditions) in the outer membrane layer rises. Simultaneously, the mole fraction of the external cations (Ca^{++} and Na^+) in the inner membrane layer goes up when a nerve impulse travels along the axon. The changes in fluorescence induced by nerve stimulation could then be related to changes in the membrane molecular conformation induced by these changes in the population of the counter-ions in the membrane.

The population of the counter-ions in the membrane can be altered also by application of long 'polarizing' current pulses to the nerve. An outward-directed current increases the fraction of the negative sites in the outer membrane layer neutralized by K ions; and an inward-directed current increases the number of Ca and Na ions in the inner membrane layer (see the right hand diagram in Figure 2). Therefore, a study of the effect of the polarizing current upon the intensity of the fluorescence could lead us to a plausible interpretation of the fluorescence data in nerves during the nerve impulse.

In the following experiments on the effect of polarizing currents on fluorescence, nerve trunks from crabs were used. A short portion of the nerve (about 3 mm long) was stained with one of the dyes used in our previous studies and was subsequently irradiated with quasi-monochromatic light. The fluorescent light was detected at 90 degrees (as in the experiment illustrated in Figure 1). When a near threshold electric current pulse of approximately 50 msec in duration was delivered to the site of optical recording, the intensity of the fluorescent light was found to either increase or decrease, depending on the dye used and also on the direction of the applied current.

It was found that the sign of the fluorescence change observed with a cathodal (outward-directed) current pulse was, as a rule, the same as that of the optical signal associated with a propagated nerve impulse. In nerve fibers stained with acridine orange, for example, arrival of a nerve impulse at the site of optical recording increases the fluorescence intensity (see Figure 1); a cathodal current also increases the fluorescence in this case (see Figure 3). In nerves stained with pyronin B, on the other hand, a decrease in fluorescence occurs during an action potential evoked by electrical stimulation. A cathodal pulse also produces a negative optical signal in this case (compare Figures 1 and 3). The effect of an anodal (i.e., inward-directed) current pulse on fluorescence is opposite to and smaller than that of a cathodal pulse under these experimental conditions.

The result of these observations may be interpreted as indicating that the dye molecules in the nerve stained with acridine orange or pyronin B are located predominantly in the outer membrane layer (see the part of the membrane marked A in Figure 2). In this layer, the state of the membrane macromolecules during an action potential is similar to that under cathodal polarization. Consequently, the sign of the optical signal associated with a nerve impulse is the same as that induced by a cathodal current. It is of interest to note that cathodal and anodal currents of near-threshold intensity produced changes in the turbidity and birefringence of the nerves, and the direction of the change produced by a cathodal current was the same as that which occurred during an action potential.

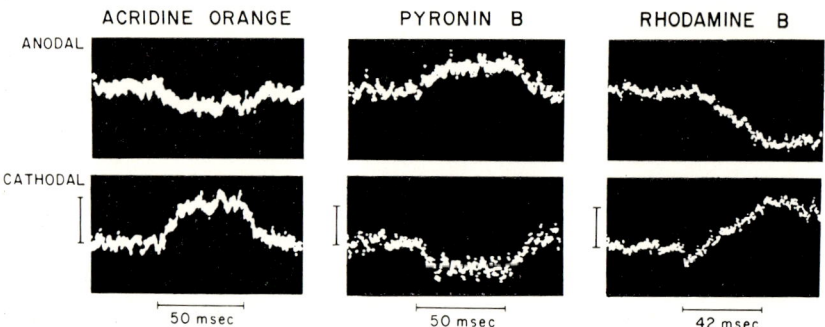

FIGURE 3 The effect of anodal (top) and cathodal (bottom) current on fluorescence of crab nerves. In nerves stained with acridine orange and pyronin B, the sign of the optical signals observed with cathodal current was the same as that associated with a propagated nerve impulse: the optical signals observed with anodal current were opposite in sign and of smaller intensity. The optical signals coincident with cathodal current in nerves stained with rhodamine B were biphasic. The vertical bars represent a change of 5×10^{-5} times the background fluorescence. Temperature 22°C.

The dependence of the fluorescence intensity upon various physico-chemical factors of the environment is complex. Even in purified macromolecules (stained with dyes), the fluorescence intensity is not a simple monotonic function of the pH of the medium or of the concentrations of salts added (see e.g., Edelhoch and Steiner, 12; Albers, 15). Thus far, we have not found any general rule which enables us to predict the signs of the optical signals of new untested dyes. An additional complicating factor is that local ionic changes also affect the fluorescence of the dye bound to connective tissues and Schwann cells adjacent to the axon membrane.

VI COMPLICATIONS ENCOUNTERED WITH RHODAMINE B

The effects of polarizing currents upon the fluorescence intensity of a nerve stained with rhodamine B were very complex. A near-threshold cathodal current pulse produced a diphasic optical signal in this case (see Figure 3).

The brief first phase was small and negative (representing a decrease in fluorescence); the second phase was long and positive. With an anodal current pulse, the first phase was practically absent and the second phase was negative. The first phase in the optical signal observed with a cathodal current appeared to become obscure when the current intensity was increased.

One might suggest that the dye molecules are distributed in both layers of the membrane in this case. A strong anodal current raises the concentrations of Ca and Na ions in the inner layer. A similar rise in the concentration of these cations may be brought about by a propogated nerve impulse and initially by a near-threshold cathodal pulse; this might explain the sign of the optical signal associated with a propagated nerve impulse. The late, positive signal arising during a cathodal pulse would then represent a change in the outer layer, due to an increase in the K ion concentration in that area. However, this analysis is speculative at the present time.

An interesting complication was encountered when the effect of urea was examined in nerves stained with rhodamine B (see Figure 4). Addition of 1.5 M urea to the surrounding sea water was found to completely eliminate the first phase of the optical signal induced by a cathodal current pulse. The

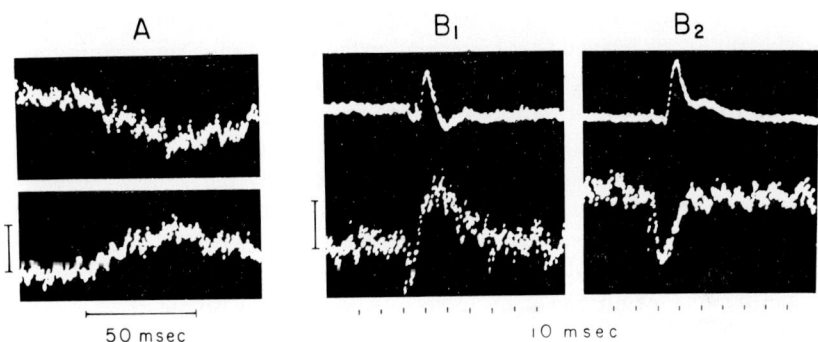

FIGURE 4   Records showing the effect of 1.5 M urea added to the surrounding sea water upon the fluorescence and the action potential in nerves stained with rhodamine B. Record A shows fluorescence changes produced by anodal (top) and cathodal (bottom) current pulses of 50 msec in duration applied to a stained crab nerve. Note the difference between the left lower photograph in this figure and the right lower photograph in Figure 3. Record B_1 shows the electric (top) and optical (bottom) signals associated with conduction in a stained crab nerve in urea sea water. Record B_2 was taken from the nerve used for Record B_1 approximately 15 min after replacement of urea in the surrounding sea water with 0.81 M sucrose. The vertical bars represent a change of 6×10^{-5} times the background intensity in Record A and 1.5×10^{-5} times in Record B. Temperature 22°C.

signal produced by an anodal current was not significantly affected by urea. The sign of the optical signal associated with a propagated nerve impulse was changed by the addition of urea in most axons examined (Record B, Figure 4), and this effect of urea was reversible. In other words, 1.5 M urea in the surrounding medium was found to convert the fluorescence behavior of the nerve stained with rhodamine B from the 'pyronin B type' (a decrease in fluorescence intensity during action potential) to the 'acridine orange type' (an increase in fluorescence intensity during action potential). Addition of urea was found to produce a gradual decrease in the amplitude of the action potential, but this effect was also reversible.

Urea is known to be capable of disrupting intra- and intermolecular hydrogen bonds. Such disruption may well be the cause of the reversal of the optical signal by urea. Addition of 1 M sucrose to the medium produced no reversal of optical signals (see Record B2 in Figure 4).

VII SUMMARY

The development of new apparatus and techniques has permitted new approaches to the problem of elucidating the molecular events occurring in the nerve membrane during an action potential. The internal perfusion of squid giant axons has led to a better understanding of the ionic requirements of excitation, and has resulted in the formulation of a theory of nerve excitation in which a conformational change in the nerve membrane macromolecules, brought on by cooperative uni-divalent cation exchange at fixed negative sites within the membrane, is regarded as the primary event in excitation. Changes in turbidity and birefringence of nerves, as well as changes in the intensity of fluorescence of stained nerves, have been detected during excitation. It is suggested that these changes may also reflect a conformational change in nerve membrane macromolecules coincident with the nerve impulse. Finally, the effect of electric current on the various optical properties of nerves, and the effect of urea on the fluorescence change in nerves stained with rhodamine B, may be of significance in determining the site of the conformational change which occurs during the process of nerve excitation.

References

1. HODGKIN, A. L. 1964, *The Conduction of the Nervous Impulse* (Liverpool, England: University Press).
2. TASAKI, I. 1968, *Nerve Excitation* (Springfield, Ill.: Charles C. Thomas).
3. TASAKI, I., LERMAN, L., and WATANABE, A. 1969, *Am. J. Physiol.* **216**, 130–138.
4. COHEN, L. B., KEYNES, R. D., and HILLE, B. 1968, *Nature*, **218**, 438–441.
5. TASAKI, I., WATANABE, A., SANDLIN, R., and CARNAY, L. 1968, *Proc. Nat. Acad. Sci., U.S.*, **61**, 883–888.

6. CHANGEUX, J. P., THIERY, J., TUNG, Y., and KITTEL, C. 1967, *Proc. Nat. Acad. Sci., U.S.*, **57**, 335–341.
7. LEHNINGER, A. L. 1968, *Proc. Nat. Acad. Sci., U.S.*, **60**, 1069–1080.
8. TOBIAS, J. M. 1952, *Cold Spring Harbor Symp. Quant. Biol.*, **17**, 15–25.
9. HILL, D. K. 1950, *J. Physiol.*, **111**, 283–303.
10. INOUE, S., and SATO, H. 1967, *J. Gen. Physiol.*, **50**, 259–292.
11. BORN, M., and WOLF, E. *Principle of Optics*, 3rd edition (Oxford: Pergamon Press, 1964).
12. EDELHOCH, H., and STEINER, R. F. *Electronic Aspects of Biochemistry* (New York: Academic Press, 1964).
13. COLE, K. S. *Membranes, Ions and Impulses* (Los Angeles: Univ. Cal. Press, 1968).
14. WATANABE, A., TASAKI, I., and LERMAN, L. 1968, *Proc. Nat. Acad. Sci. U.S.*, **58**, 2246–2252.
15. TASAKI, I., TAKENAKA T., and YAMAGISHI, S. 1968, *Am. J. Physiol.*, **215**, 152–159.
16. ALBERS, R. W., and KOVAL, G. J. 1962, *Biochim. Biophys. Acta.*, **60**, 359–365.

DISCUSSION

EDELMAN   It may be useful for me to mention some of the basic physical properties of fluorescent probes. Fluorescent probes such as the N-arylnaphthylamine sulfonates do not fluoresce in water but if placed in solvents of low dielectric constant the quantum yield of fluorescence increases and the wave length of maximal emission shifts towards the blue. On the basis of our solvent studies, we have suggested the following explanation of this behavior. The dipole moment in the excited state is greater than that in the ground state by about ten Debye. In polar solvents the interaction with solvent molecules is decreased and therefore there is a greater quantum yield of fluorescence. The red shift in polar solvents is explained by the fact that the energy of the excited state is decreased by an amount proportional to the solvation energy. In any case, whatever the other details of the mechanism, it should be emphasized that fluorescence probes can be very complex and they are susceptible to changes in concentration, temperature and viscosity, and as I mentioned, the dielectric constant. In well-characterized systems they can be used to detect the most subtle conformational changes of proteins, e.g. the chymotrypsinogen to chymotrypsin conversion. In more complex situations it is difficult to interpret changes in fluorescence unless one measures the quantum yield and particularly the wave length of maximal emission. It is this last parameter which is the most reliable index of polarity of the environment, and therefore unless there is a blue shift, one should be quite cautious about interpreting a change in fluorescence as indicative of a conformational change. It could be merely the result of changes in the absorptivity of the medium.

KURSUNOGLU   I would like to ask a question on this black and white communities cooperative phenomenon. Are there collective oscillations of the membrane, and if there are, what is the role of these collective oscillations on the selective diffusion?

TASAKI   When the external calcium concentration is reduced, the membrane potential is known to spontaneously fluctuate (subthreshold oscillation). This oscillatory phenomenon indicates that the ratio of the number of the black elements to that of the white elements varies periodically under these conditions. Qualitatively, the interaction between two kinds of elements discussed before could explain this phenomenon.

KURSUNOGLU   Therefore, different modes should exist depending on the invasion?

TASAKI   Yes. However, this oscillatory phenomenon is rather unstable. When the number of white elements increases, all of the membrane elements tend to undergo transition to the white state simultaneously.

KURSUNOGLU   Then how do you maintain stability of the membrane?

TASAKI   Divalent cations are needed in the external medium to maintain excitability. I believe the Ca-bridges between negatively charged sites, and probably hydrophobic forces, are contributing to the stability of the external layer of the axon membrane. Dr. Ohki's theoretical approach (presented yesterday) may give us a better understanding of this problem. There is evidence that the inner layer of the membrane is stabilized by the electrostatic forces between oppositely charged groups in the macromolecules (see pp. 90–100 in reference 2).

MYSELS   You mentioned that light scattering was independent of the wave length. That would strongly suggest that the elements that scatter are large, close to the wave lengths of light, 1000's angstroms at least. Therefore, I don't know whether that could be due to changes in the protein molecules in the membrane, which are of course small, and I would like to suggest another possibility. That is that these quite large ion fluxes, which you have of course described, produce changes in the refractive index in the surrounding medium over millisecond periods, which could spread out to volumes that would be large compared to the wave lengths of light and could give those effects. Such changes in ionic concentration could also produce changes in fluorescence, causing changes of absorption of the fluorescence molecules which could probably go both ways. I wonder if you have thought about that being an explanation?

TASAKI   In the case of light scattering, there is strong evidence in support of the view Dr. Mysels has just mentioned. The intensity of the light scattered by a nerve (observed at 90 degrees) persists for a long period of time after the end of the action potential. In some instances, a minimum in light intensity is seen at about the end of the action potential. We interpret this late portion of light scattering signal as arising from a gradual change in the ionic composition on both sides of the membrane. The time required for diffusion of ions across a 1μ thick layer is of the order of 1 msec. Therefore, the structure involved in our early signal has to be located within about 0.5μ from the membrane. I imagine that light scattering by a thin membrane is very different from that by small particles.

ADAM   First, I'm not so much familiar with the terms cathodal and anodal, but isn't it true that in the cathodal current the membrane remains in the resting state. If so, it corresponds to a hyperpolarization of the membrane, and if you get a response of the dye in this case, it would speak against a conformational change? Isn't that so?

TASAKI It is true that an anodal (hyperpolarizing) current pulse produces a change in fluorescence. When the current intensity is close to the threshold for excitation, the optical signal produced by a cathodal (depolarizing) current pulse is larger and reaches a steady level more quickly than the signal produced by a current of the same intensity but with the reversed direction. A long anodal (hyperpolarizing) current pulse can raise the divalent cation concentration in the membrane; this may account for the optical signals produced by such pulses.

PAPER 3

Theory of nerve excitation as a cooperative cation exchange in a two-dimensional lattice

GEROLD ADAM†
Institut für Physiologische Chemie und Physikalische Biochemie der Universität München

INTRODUCTION

THE ELUCIDATION of the molecular processes governing the biological all-or-none phenomena represents a great challenge to present biology. All-or-none processes seem to be the basis for proper functioning of many biological structures in sensory-, nerve- and muscle-physiology.

Most investigations of all-or-none behavior have been spent on the non-myelinated giant axon of the squid. The greatest advance in our understanding of this phenomenon was achieved when Cole and collaborators developed the voltage clamp of the squid axon and thereby reduced, by rigorous electrical characterization of the squid axon, the puzzling phenomenon of an action potential to the conceptually most simple aspect of the phenomenon of axon excitation (for reference see Cole, 1968). Subsequently, Hodgkin and Huxley (see Hodgkin, 1964) gave firstly a physicochemical description of the ion fluxes at excitation. Secondly, these authors provided for a phenomenological mathematical description of the phenomena in question. This theoretical work has been of great help over the past years for the classification and discussion of the vast amount of electrophysiological data of many workers in this field. It did not, however, give a quantitative physicochemical mechanism of how the electrical state of the axon membrane is regulated by the electrical field and/or the cation activities. Furthermore, recent observations on internally perfused axons seem to have made necessary,

† Address: Fachbereich Chemie, Universität Konstanz, 775 Konstanz, Postfach 733, W. Germany.

modifications and major revisions of the physicochemical picture of Hodgkin and Huxley. We do not wish, at this point, to enter into the discussion of the phenomena observed on perfused axons, which are still rather controversial.

The aim of the present paper is to try to give a physicochemical mechanism of axon excitation for the case of the squid axon. For this, we had to leave the physicochemical picture of Hodgkin and Huxley, finding it more useful to start from a model of the axon membrane as a cation exchanger, as proposed and experimentally corroborated by Tasaki (1968).

However, the molecular model of the axon membrane, as introduced in a previous paper (Adam, 1968) and presented in the following, is in many respects different from that of Tasaki. We first represent the molecular model and its theoretical formulation, essentially as given in the previous paper (Adam, 1968, in the following denoted as I). Subsequently, we compare the predictions of the theory with the following experiments:

Voltage clamp, depolarizing from the resting state.
Activation of excitation current by previous hyperpolarization.
Steady state negative resistance.
Heat cycle on excitation.

MOLECULAR MODEL

Our proposed molecular model of the axon membrane is based on the Davson–Danielli model, which seems to incorporate into a coherent picture most of what is known about the structural and electrical properties of biological membranes and artificial phospholipid double layers. According to Davson and Danielli (1952), biological membranes consist of two protein layers facing the electrolyte reservoirs on both sides of the membrane and sandwiching a phospholipid double layer. To account for the observed permeabilities to solutes and electrical conductivities of biological membranes, pores of molecular diameters have been postulated, which penetrate the central phospholipid bilayer as hydrophilic protein channels.

Newer results on specific cation carrier molecules like valinomycin or alamethicin (Mueller and Rudin, 1965, 1968) suggest as a possibility that in natural membranes cation transport might be mediated not by pores, but by mobile carriers. However, since no intrinsic mobile carriers for natural membranes have been demonstrated up to now, we shall use the concept of molecular pores in the following. Most aspects of the theory will remain valid if cation transport by mobile carriers is assumed.

In our molecular model of the axon membrane we add three specific postulates to the Davson–Danielli model as presently accepted.

1. Asymmetry of fixed charges of the membrane

The membrane is supposed to be asymmetric insofar as the outer hydrophilic layer (outer proteins plus hydrophilic parts of the phospholipids) carries an excess of positive fixed charges, whereas the inner hydrophilic layer (inner proteins plus hydrophilic parts of the phospholipids) carries an excess of negative fixed charges (see Figure 1). Thus our modified Davson–Danielli

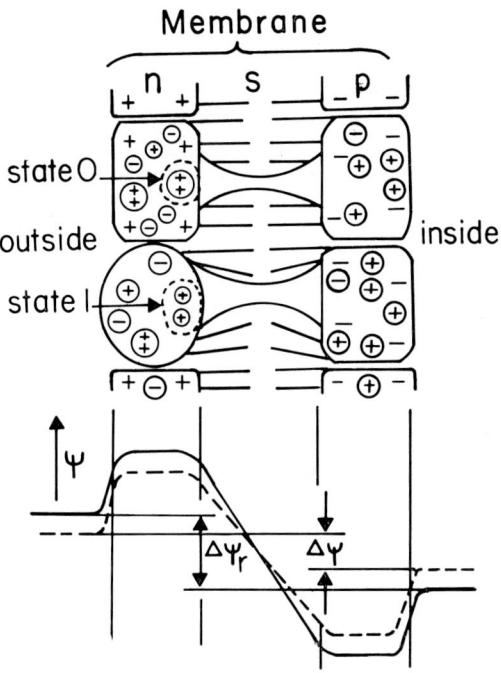

FIGURE 1 Schematic representation of the model of the axon membrane (above) and the profile of the electrical potential. + and − represent fixed charges, + and − encircled represent mobile cations and anions respectively.

$\Delta\psi_r$ = resting membrane potential.
$\Delta\psi$ = depolarized membrane potential.

model resembles a *psn*-junction as used in electron semiconducting devices (Herlet and Spenke, 1955). The charge carriers in the biological case of a *psn*-junction are ions in contrast to electrons and holes in semiconductor devices. A similar asymmetric membrane, a *pn*-junction, has been proposed by Coster, 1965 for plant plasma membranes. Our *psn*-model of the axon membrane exhibits rectification in steady state current flow, and a junction capacity which is in wide limits independent of the density of fixed charges and of the applied voltage (unpublished calculations), as it is observed experimentally for the squid axon membrane (cf. Cole, 1968, p. 157 and p. 136). Figure 1 shows schematically the profile of electrical potential across the ionic *psn*-junction. For a density of excess fixed charges of 1 M for instance, we have a Donnan potential facing the electrolyte reservoir of about 20 mV. Across the *s*-layer the electrical potential drops by an amount which is the sum of the membrane potential and the two Donnan potentials facing the two electrolyte reservoirs. Since the two Donnan potentials at the outer edge of the *n*-layer and the inner edge of the *p*-layer remain fixed, an applied membrane voltage will appear only across the *s*-layer (Coster, 1965). A cation released at the *s*-layer will fall down the potential drop across the *s*-layer, i.e., move inward.

2. *Cooperative binding of cations by the membrane subunits*

The second postulate of our model is a cooperative binding of cations by the membrane subunits.

We think that the membrane is covered by a monolayer of subunits (probably proteins) with the following properties. The subunits constitute the pore proteins and/or the outer (*n*-) layer.

Each subunit has an active center to which either a divalent cation (in the physiological case: Ca^{++}) or two monovalent cations (Na^+ or K^+) can be bound. We need not define more precisely the nature of the cation binding in the active center. It could be by complex formation with uncharged groups, by pair formation with fixed negative charges as considered in electrolyte theory or, less likely, by ionic bonds.

More important is that Ca^{++} bound in the active center confers to the subunit a different configuration (state 0) from that induced by two monovalent cations bound in the active center (state 1). The following consequence of this postulate is most relevant for a mechanism of nerve excitation: The interaction energy between two subunits in state 0, two subunits in state 1, or between one subunit in state 0 and one in state 1, will be different.

If the interaction energy between one unit in state 0 and one in state 1 is lower than between two subunits in equal states, mixed states of the membrane will be unfavorable energetically and the membrane will prefer states with subunits preponderantly in state 0 or state 1.

3. *Cation permeability of membrane subunits*

In addition, we can expect a subunit in state 0 to have a different permeability to cations from a subunit in state 1. Thus the passive conductance of the membrane will be different if the subunits are preponderantly in state 0, compared to the passive conductance if they are mostly in state 1. We can visualize this by the following consideration. From the theory of steady state current across a *psn*-layer (unpublished results; for a *pn*-layer see Coster, 1965), it is evident that the permeability of the membrane is determined mainly by the properties of the *n*-layer and the pores. We have to think that a subunit in state 0, when undergoing the configurational transition to state 1, masks a major part of the excess fixed positive charges of the *n*-layer (for instance, by turning them into a hydrophobic milieu) and thus opens up a direct access of cations to the pore. In addition, there might be a change of the properties of the pore which also leads to a higher cation permeability of the subunit. This would explain the higher conductivity of the membrane in the excited state and the loss of steady state rectification upon excitation.

With these three postulates, our model of the axon membrane is complete and can describe a physicochemical mechanism of membrane excitation.

STEADY STATE OF THE MEMBRANE

In this section, we wish to calculate the steady state of the membrane by a molecular kinetic approach.

We repeat in essence the theory as given in I, but remove certain simplifications to allow for a more detailed description of cation specificity. We determine first the average probabilities W_{01} and W_{10} for the transitions $0 \to 1$ and $1 \to 0$ respectively of a representative subunit. The probability W_{01} consists of the following three multiplicative terms:

1. The partial probability of desorption of Ca^{++} from the active center is proportional to $\exp\{u_2/kT\}$ where u_2 is the energy of binding of Ca^{++} in the active center.*
2. The partial probability for adsorption of the two monovalent cations, exchanging the divalent cation, can be computed in the following way:

    In the physiological case, and in most experimental situations of interest, there is an overwhelming majority of one univalent species of cations in the extracellular medium and of another species in the intracellular medium. For the sake of simplicity we assume that there is only one species l of univalent cations in the extracellular medium (and therefore also in the *n*-layer) and only one species k of univalent cations in the internal medium (and therefore in the *p*-layer).

    We denote the mean electrochemical activity of species l in the *n*-layer by $^n\eta_l$ and of species k in the *p*-layer by $^p\eta_k$.

    Let $^n c_l$ be the rate constant of access of species l from the *n*-layer to the active center and $^p c_k$ that of the species k from the *p*-layer.

---

* Here and in the following an interaction energy is the energy required to move the interacting groups from infinite distance to the distance of interaction. Thus, binding energies are generally negative.

Then, according to the three possibilities of origin of the two univalent cations to be bound in the active center, we can write the adsorption probability as proportional to

$$(^nc_l{}^n\eta_l)^2 + {}^nc_l{}^Pc_k{}^n\eta_l{}^P\eta_k + (^Pc_k{}^P\eta_k)^2 \quad (1)$$

This is the formulation of I, if we set: $c_1 = (^nc_l)^2$; $c_2 = {}^nc_l{}^Pc_k$ and $c_3 = (^Pc_k)^2$.

3. We need further the partial probability that the hindrance energy arising from interactions with neighboring subunits can be overcome.

Let n be the mole fraction of subunits of the membrane in state 1 and $1-n$ that of subunits in state 0.

In order to describe the cooperativity of interaction of the subunits in the membrane we use for a first approximation, the Bragg–Williams formulation. Accordingly, we set the interaction energy of a subunit in state 0 with the surrounding membrane as a linear function of the state of the membrane: $v_{00} - \lambda_0 n$. Here v_{00} is the interaction energy of a subunit in state 0 with the rest of the lattice in state 0 and λ_0 an interaction parameter to be specified later.

Thus the hindrance energy arising from interaction of the representative subunit in state 0 with its environment is $w^* - v_{00} + \lambda_0 n$, where w^* is a critical energy, independent of n and membrane potential, to be attained in the membrane before the transition $0 \to 1$ can occur.

Thus we can write for the total probability of the transition $0 \to 1$ of a subunit:

$$W_{01} = k_{01} e^{u_2/kT} [(^nc_l{}^n\eta_l)^2 + {}^nc_l{}^Pc_k{}^n\eta_l{}^P\eta_k + (^Pc_k{}^P\eta_k)^2] \exp\left\{-\frac{w^* - v_{00} + n\lambda_0}{kT}\right\} \quad (2)$$

where k_{01} is a proportionality constant independent of n and membrane potential.

In a completely analogous procedure, we derive for the average probability of the transition $1 \to 0$:

$$W_{10} = k_{10} e^{2u_1/kT} [^nc_2{}^n\eta_2 + {}^Pc_2{}^P\eta_2] \exp\left\{-\frac{w^* - v_{01} + \lambda_1 n}{kT}\right\} \quad (3)$$

Here k_{10} is a proportionality constant independent of n and membrane potential; $^nc_2$ and $^Pc_2$ are the rate constants of access of Ca^{++} to the active center from the n- and p-layer respectively, $^n\eta_2$ and $^P\eta_2$ the electrochemical activities of Ca^{++} in the n- and p-layer respectively; v_{01} is the interaction energy of a unit in state 0 with the rest of the membrane in state 1; λ_1 an interaction parameter to be specified later; $2u_1$ the energy of binding of the two monovalent cations to be desorbed from the active center.

Equation (3) is virtually identical to I Eq. (2); we have only admitted explicitly the presence of calcium in the *p*-layer and different rate constants for its access to the active center from the *n*-layer or *p*-layer. For all situations of interest in the following, we shall use $^p\eta_2 \ll {}^n\eta_2$ and $^nc_2 \approx {}^pc_2$ as in I.

The fraction n follows the kinetic equation.

$$\frac{dn}{dt} = W_{01}(1-n) - W_{10}n \qquad (4)$$

In the steady state we have $dn/dt = 0$ or with Eqs. (2), (3) and (4):

$$\frac{n}{1-n} e^{wn/kT} = q(T) \left[\frac{({}^nc_l{}^n\eta_l)^2 + {}^nc_l{}^pc_k{}^n\eta_l{}^p\eta_k + ({}^pc_k{}^p\eta_k)^2}{{}^nc_2{}^n\eta_2 + {}^pc_2{}^p\eta_2} \right] \qquad (5)$$

where

$$w = \lambda_0 - \lambda_1 = v_{00} + v_{11} - 2v_{01} \qquad (6)$$

and

$$q(T) = \frac{k_{01}}{k_{10}} \exp\left\{\frac{u_2 - 2u_1 + v_{00} - v_{10}}{kT}\right\} \qquad (7)$$

Relations similar to Eq. (5) have been derived by different authors for interaction of a surface with one reservoir: cooperative adsorption (Hill, 1960), cooperative ion exchange (Barrer and Falconer, 1956), ligand binding to a surface the units of which can isomerize cooperatively (Changeux, et al. 1967). Equation (5) describes the case of cooperative interaction of a two-dimensional system with the electrolyte reservoirs on *both* sides. The electrochemical activities on the right side of Eq. (5) depend on the electrical potential. Thus for fixed cation concentrations, the right side of Eq. (5) depends on the membrane potential $\Delta\psi$. As will be specified later, it increases monotonically with the depolarization of the membrane potential. For brevity, we designate the right side of Eq. (5) as $\ln\{q(T)M(\Delta\psi)\}$. If we plot n versus $\ln\{q(T)M(\Delta\psi)\}$ for a parameter $-w/kT > 4$, an *s*-shaped curve results which is symmetrical about $n = \frac{1}{2}$ (see I, or Hill, 1960, p. 246 f.). We identify the abscissa of symmetry as $\ln\{q(T)M(\Delta\psi_r)\}$, where $\Delta\psi_r$ is the resting potential. This *s*-shaped curve has two stable branches: one, denoted by ${}^0n(\Delta\psi) < \frac{1}{2}$, extends from hyperpolarizing membrane potentials to $\Delta\psi_r$, the other, denoted by ${}^1n(\Delta\psi) > \frac{1}{2}$, extends from $\Delta\psi_r$ to increasingly depolarizing membrane potentials. At the resting potential, n has two stable values $\underline{n} < \frac{1}{2}$ and $\bar{n} > \frac{1}{2}$. All other parts of the curve represent meta- or instable states (see for instance Hill, 1960).

The evaluation of Eq. (5) will be different as to whether $\Delta\psi \geq \Delta\psi_r$ or $\Delta\psi < \Delta\psi_r$.

We first treat the case $\Delta\psi \geq \Delta\psi_r$, which corresponds to the treatment given in I.

If the membrane is in the resting state, specified by n, there is no applied external voltage. As stated in the specification of our model, with subunits predominantly in state 0, the *psn*-structure of the membrane is virtually intact. As shown by Coster (1965) there are Donnan-equilibria at the outer limit of the *n*-layer and at the inner limit of the *p*-layer. With zero applied voltage these equilibria extend in the direction normal to the membrane surface all through the *n*-layer and through the *p*-layer, respectively. Thus we can write

$$^n\eta_l = \eta'_l = a'_l e^{F\psi'/kT}; \qquad ^n\eta_2 = \eta'_2 = a'_2 e^{2F\psi'/kT} \tag{8}$$

$$^p\eta_k = \eta''_k = a''_k e^{F\psi''/kT}; \qquad ^p\eta_2 = \eta''_2 = a''_2 e^{2F\psi''/kT} \tag{9}$$

where η'_i and η''_i are the electrochemical activities of cation species i in the outside and inside medium respectively; a'_i and a''_i are the corresponding chemical activities; ψ' and ψ'' are the electrical potentials in the inside and outside medium respectively. In the stationary state for $\Delta\psi > \Delta\psi_r$, Eq. (9) is again fulfilled, because the Donnan equilibrium of the majority ions (i.e., the cations) in the *p*-layer is not changed by an applied voltage (Coster, 1965). According to what was said in the specification of the model, the relations (8) can also be used for the stationary state with $\Delta\psi > \Delta\psi_r$ because the proteins of the *n*-layer are then in a configuration where the positive fixed charges are masked, and the active centers are exposed directly to the extracellular medium with electrochemical activities of cations η'_1 and η'_2. Thus we can describe the stationary states of the membrane for $\Delta\psi \geq \Delta\psi_r$ by Eq. (5) combined with Eqs. (8) and (9):

$$\frac{n}{1-n} e^{wn/kT}$$

$$= q(T) \left[\frac{(^n c_1 a'_1)^2 e^{2F\psi'/kT} + ^n c_1 {}^p c_k a'_1 a''_k e^{F(\psi'+\psi'')/kT} + (^p c_k a''_k)^2 e^{2F\psi''/kT}}{^n c_2 a'_2 e^{2F\psi'/kT} + ^p c_2 a''_2 e^{2F\psi''/kT}} \right] \tag{10}$$

The derivation of this equation assumed the quantities $^n c_1$, $^p c_k$, $^n c_2$, $^p c_2$ to be independent of the state of the subunit.

If we rearrange and use

$$^n c_2 a'_2 e^{2F\psi'/kT} \gg {}^p c_2 a''_2 e^{2F\psi''/kT} \tag{11}$$

which corresponds to the normal situation of a functional axon, we obtain from Eq. (10)

$$\frac{n}{1-n} e^{wn/kT} = \frac{q(T)}{{}^n c_2 a'_2} \left[({}^n c_l a'_l)^2 + {}^n c_l {}^p c_k a'_l a''_k e^{F\Delta\psi/kT} + ({}^p c_k a''_k)^2 e^{2F\Delta\psi/kT} \right] \quad (12)$$

where $\Delta\psi = \psi'' - \psi'$. This is essentially Eq. (8) of I.

In the steady state at hyperpolarizing membrane potentials $\Delta\psi < \Delta\psi_r$, we have to take into account shifts of concentrations of minority cations as described by Coster (1965). At the junction between the n-layer and the external medium, the Donnan equilibria are still fulfilled:

$$\,{}^n_0\eta_l = \eta'_l; \qquad {}^n_0\eta_2 = \eta'_2 \quad (13)$$

The prefix $_0$ indicates the outer boundary of the n-layer. Using the chemical activities, Eq. (13) can be written as:

$$\,{}^n_0 a_l = a'_l e^{F\psi_D/kT}; \qquad {}^n_0 a_2 = a'_2 e^{2F\psi_D/kT} \quad (14)$$

where ψ_0 is the Donnan potential between external medium and outer boundary of the n-layer.

At the inner boundary of the n-layer, i.e. at the junction to the s-layer, the chemical activities $\,{}^n_i a_l$ and $\,{}^n_i a_2$ are given according to Coster (1965) as:

$$\,{}^n_i a_l = {}^n_0 a_l \exp\{F\Delta_h/kT\} \quad (15)$$

and

$$\,{}^n_i a_2 = {}^n_0 a_2 \exp\{2F\Delta_h/kT\} \quad (16)$$

where $\Delta_h = \Delta\psi - \Delta\psi_r < 0$ is the hyperpolarization of the membrane potential.

In the steady state of hyperpolarization, a linear gradient of chemical activities between outer and inner boundary is established (Coster, 1965).

The average chemical activities of cations in the n-layer, therefore, are:

$$\tfrac{1}{2}({}^n_0 a_l + {}^n_i a_l) = \tfrac{1}{2}({}^n_0 a_l) e^{F\Delta_h/kT} \quad (17)$$

and

$$\tfrac{1}{2}({}^n_0 a_2 + {}^n_i a_2) = \tfrac{1}{2}({}^n_0 a_2) e^{2F\Delta_h/kT} \quad (18)$$

The electrical potential is kept constant across the n-layer by virtue of the negative majority ions.

Thus, we have for the average electrochemical activities of the cations in the n-layer:

$$\,{}^n\eta_l = \tfrac{1}{2}\eta'_l(1 + e^{F\Delta_h/kT}) = \tfrac{1}{2} a'_l e^{F\psi'/kT}(1 + e^{F\Delta_h/kT}) \quad (19)$$

$$\,{}^n\eta_2 = \tfrac{1}{2}\eta'_2(1 + e^{2F\Delta_h/kT}) = \tfrac{1}{2} a'_2 e^{2F\psi'/kT}(1 + e^{2F\Delta_h/kT}) \quad (20)$$

After substitution of Eqs. (9), (11), (19) and (20) into Eq. (5) we obtain after rearranging:

$$\frac{n}{1-n} e^{wn/kT} = q(T) \times$$

$$\left[\frac{(^n c_l a_l')^2 (1+e^{F\Delta_h/kT})^2 + 2 {}^n c_l {}^p c_k a_l' a_k'' e^{F\Delta\psi/kT} (1+e^{F\Delta_h/kT}) + 4(^p c_k a_k'')^2 e^{2F\Delta\psi/kT}}{2a_2'(1+e^{2F\Delta_h/kT})} \right]$$

(21)

The right side of this Eq., valid for the steady state of the membrane at hyperpolarizing membrane potentials, increases monotonically with increasing $\Delta\psi$, as does the right side of Eq. (12).

In the following sections, we shall use the function ϕ, which is the thermodynamical potential for the environmental variables of the membrane system (see I for definition of ϕ). For the parameters used in I, we could evaluate ϕ by a series expansion at the resting state.

A more general formulation for ϕ can be given by using a statistical mechanical argument (to be published in *Z.Naturf.*), which is in essence that given by Hill, 1960, p. 247. Actually the result is the same as his Eq. (14-44).

In our notation, it can be written:

$$\frac{^0\phi - {}^1\phi}{kT} = \ln\left(\frac{1-{}^0n}{1-{}^1n}\right) + \frac{w}{2kT}({}^1n^2 - {}^0n^2) \quad (22)$$

Here, $(^0\phi - {}^1\phi)$ is the difference of the thermodynamic potential of two states of the membrane described by ${}^1n$ and ${}^0n$.

KINETICS OF COOPERATIVE EXCITATION OF THE MEMBRANE

In the following we wish to restate briefly the results derived in I of the kinetic description of membrane excitation after depolarization. In the formulation of the present paper we have removed only the approximation ${}^1n - {}^0n \approx 1$, which could be used for the choice of parameters in I. A more detailed account of the derivation of these modifications is given elsewhere (*Z. Physikal. Chem. (Frankfurt)*, in press). Most of these are obvious.

Nucleation-Growth Theory

Because of the cooperativity of the transitions of the subunits, the membrane undergoes a phase transition upon depolarization, if $-w/2kT > 2$. In order to describe explicitly the kinetics of this phase transition, we have used the nucleation/growth theory, which is well established in the description of phase transitions in solid state physics (see for instance Fine, 1964). According to this theory, the transition of the system from state ^0n to state ^1n takes place by formation of growable nuclei of state ^1n in a matrix of state ^0n. These nuclei subsequently grow until they cover the entire membrane area. As shown in I, the rate of formation of growable nuclei per unit area of untransformed membrane can be given by:

$$I = I_0 \exp\left\{-\frac{b^*(^0\phi - {}^1\phi)}{kT}\right\} \quad (23)$$

Here I_0 is a proportionality constant, which need not be defined in more detail, except as being independent of the membrane potential and the cation activities; b^* is the number of subunits per growable nucleus and given by†

$$b^* = \left[\frac{g\sigma}{2(^0\phi - {}^1\phi)}\right]^2 \quad (24)$$

Here, g is a geometrical factor, which for a circular nucleus is equal to π; $(^0\phi - {}^1\phi)$ is the difference between the thermodynamical potential of the initial and the final state of the transition and can be calculated according to Eq. (22); σ is the extra potential energy necessary to form the nucleus border and can be calculated in the Bragg–Williams approximation as follows.

In this approximation, the interaction energy of a unit in state 1 being in the interior of the nucleus, is:

$$v_{1i} = v_{01} - {}^1n\lambda_1 \quad (25)$$

That of a subunit on the edge of the nucleus is:

$$v_{1G} = \tfrac{1}{2}(v_{01} - {}^1n\lambda_1) + \tfrac{1}{2}(v_{01} - {}^0n\lambda_1) \quad (26)$$

Thus, a subunit in state 1 on the edge of the nucleus contributes the following amount to its energy σ of the border line (as defined in I, Eq. (28)):

$$v_{1G} - v_{1i} = \tfrac{1}{2}\lambda_1(^1n - {}^0n) \quad (27)$$

By the same argument, a subunit in state 0 on the edge of the nucleus contributes to σ:

$$v_{0G} - v_{0i} = \tfrac{1}{2}\lambda_0(^1n - {}^0n) \quad (28)$$

† Eqs. (29) and (30) in I are misprinted; Eq. (29) should be written as the above Eq. (24); Eq. (30) in I is correct only if one takes the reciprocal of the right side.

The fraction of subunits on the edge of the nucleus in state 1 by definition is ^1n; the fraction in state 0 is $(1-{}^1n)$. The average energy σ of border line per subunit is therefore:

$$\sigma = \tfrac{1}{2}\lambda_0({}^1n-{}^0n)(1-{}^1n)+\tfrac{1}{2}\lambda_1({}^1n-{}^0n){}^1n \tag{29}$$

or, using Eq. (6) with $v_{00} = v_{11}$, i.e., $\lambda_0 = w/2$,

$$\sigma = (w/4)({}^1n-{}^0n)(1-2{}^1n) \tag{30}$$

In Eq. (30), the restrictive assumption $^1n-{}^0n \approx 1$ is removed on which I, Eqs. (58) and (60) were based.†

The linear rate u of growth of a growable nucleus is given by I, Eq. (34) as:

$$u = U_0(T)\left\{1-\exp\left[-\left(\frac{{}^0\phi-{}^1\phi}{kT}\right)\right]\right\} \tag{31}$$

$U_0(T)$ is a temperature function.

Next, we give the relations for combined action of nucleation and nucleus growth. Since the transformation starts with the fraction ^0n and ends with ^1n of the subunits of the membrane being in state 1, we have instead of Eqs. (38) and (39) in I:

$$n = {}^1n-({}^1n-{}^0n)\exp\{-ct^3\} \tag{32}$$

and

$$\frac{dn}{dt} = ({}^1n-{}^0n)\,ct^2\exp\{-ct^3\} \tag{33}$$

where

$$c = \tfrac{4}{3}Iu^2 \tag{34}$$

A more detailed argument for the derivation of Eqs. (32) and (33) will be given elsewhere (to be published in Z.Naturf.). The equations do not contain the assumption $^1n-{}^0n = 1$, used in I.

Using Eqs. (34), (23) and (21), the kinetic parameter c can be written as:

$$c = c_0\left\{1-\exp\left[-\left(\frac{{}^0\phi-{}^1\phi}{kT}\right)\right]\right\}^2\exp\left[-b^*\left(\frac{{}^0\phi-{}^1\phi}{kT}\right)\right] \tag{35}$$

where

$$c_0 = 4I_0\,U_0^2/3 \tag{36}$$

is a temperature function, which need not be given more explicitly for the purposes of the present paper.

† Eq (60) in I contains a misprint; it should read $\sigma = -w/4$.

For b^* in Eq. (35) we have from Eqs. (24) and (30) and assuming circular nuclei, i.e., $g = \pi$:

$$b^* = \left[\frac{\pi}{4}\frac{w}{2kT}\frac{(^1n - {}^0n)(1 - 2{}^1n)}{({}^0\phi - {}^1\phi)/kT}\right]^2 \tag{37}$$

With these relations, the dependence of the state of the membrane on the time and on the membrane potential can be calculated from Eq. (33) in the following way. First, we obtain for fixed cation activities from Eq. (12) or Eqs. (12) and (21) the quantities ${}^0n$ and ${}^1n$ for the membrane potentials of the initial and the final state. Then we calculate b^* and $({}^0\phi - {}^1\phi)$ from Eqs. (37) and (22) respectively. These quantities introduced into Eq. (35), give the time course of the state of the membrane for the selected membrane potentials. At this stage we wish to specify more precisely than was done in I the initial state ${}^0\phi = f({}^0n)$ of the membrane transition after a sudden depolarization of the membrane. In I, we had used for the calculation of ${}^0\phi$ the metastable branch $\underline{n} \leq {}^0n \leq n_i$ at the depolarized membrane potential of the voltage clamp. Here n_i is the limit of metastability of the membrane in states $n < \frac{1}{2}$. Within the approximations of I, this was numerically found to be indistinguishable from ${}^0n = \underline{n}$.

To check this approximation for other choices of parameters, we have used the direct molecular-kinetic approach of Eqs. (2), (3) and (4), to estimate the time course of attainment of the metastable states $\underline{n} < {}^0n < n_i$ after a sudden depolarization from the resting state $\underline{n}$ (unpublished results). These calculations show that the time of attainment of the metastable state is not negligible compared to the time of transition from the metastable branch to the stable branch ${}^1\phi = f({}^1n)$.

Thus, for the choices of parameters in the following sections it is necessary to consider as the initial state of the transition the state the membrane occupies while being held at a certain potential prior to the depolarization. Usually in a voltage clamp experiment, this holding potential is hyperpolarizing.

Excitation Current in the Voltage Clamp

In a voltage clamp experiment the concentrations of cations in the intra- and extracellular electrolyte reservoir remain fixed. As described earlier, upon application of the depolarizing voltage step, the binding of Ca^{++} in the active center becomes unstable. A transition of the membrane starts, whereby a fraction $({}^1n - {}^0n)$ of the subunits exchange cooperatively Ca^{++} for univalent cations. This cation exchange is connected with a transient current flow, which in the following will be called excitation current, and is directed inward in most cases of interest, i.e., against the applied depolarizing voltage. In addition to the inward excitation current, there is a passive current flow directed outward for depolarizing membrane potentials. This passive current will depend on the state of the membrane; it will be treated more extensively in a later section.

The net excitation current through the membrane is composed of desorbed Ca^{++}, flowing inward, down the potential drop across the s-layer, and of univalent cations, moving from the n- and p-layers to the active centers. Of course, the current is completed by charges flowing from the external medium into the n-layer and by those flowing from the p-layer to the internal medium. These latter charge transfers can be considered fast compared to the average time of transition of a subunit, thus following the membrane transition without delay.

For all depolarizations in voltage clamp experiments considered in the present paper, the potential drop across the s-layer is inward and large enough so that any outward flow of desorbed Ca^{++} can be neglected. The charge transport due to the exchanging univalent cations is governed by the following three elementary processes:

1. If both univalent cations come from the n-layer, the combined current flow of Ca^{++} and univalent cations contributes per subunit two elementary charges flowing inward. The probability for this elementary process to occur is composed of the average activity of univalent cations

$$\tfrac{1}{2}a'_l(1+\exp\{F\Delta_h/kT\})$$

and the rate constant of access to the active center $^nc_l \exp\{F\psi^1/kT\}$. It is thus proportional to:

$$p_1 = {}^n c_l^2\, a_l'^2 (1+e^{F\Delta_h/kT})^2\, e^{2F\psi'/kT}/4 \tag{38}$$

Here, Δ_h refers to the previous holding potential, whereas ψ' refers to the subsequent depolarized state.

2. If one univalent cation comes from the n- and one from the p-layer, the process contributes only one elementary charge flowing inward. The probability for its occurrence is, in analogy to Eq. (38) proportional, to:

$$p_2 = \tfrac{1}{2}{}^n c_l\, {}^p c_k\, a'_l\, a''_k (1+e^{F\Delta_h/kT})\, e^{F(\psi'+\psi'')/kT} \tag{39}$$

3. If both exchanging univalent cations come from the inside, they will just cancel the current flow due to Ca^{++}. This elementary process does not contribute to the current. Its probability of occurrence is proportional to:

$$p_3 = ({}^p c_k)^2\, a''^2_k\, e^{2F\psi''/kT} \tag{40}$$

Thus, the total excitation current $\mathscr{I}$ per membrane subunit can be written:

$$\mathscr{I} = \frac{\varepsilon}{a_0}\frac{dn}{dt}\left[\frac{2p_1+p_2}{p_1+p_2+p_3}\right] \tag{41}$$

Here, $\varepsilon = 1.602 \times 10^{-19}$ Coul is the elementary charge; a_0 is the membrane area per subunit.

Introducing Eqs. (38) to (40) into Eq. (41) gives:

$$\mathscr{I} = \frac{\varepsilon}{a_0} p(\Delta\psi; \Delta_h)\frac{dn}{dt} \tag{42}$$

where

$$p(\Delta\psi; \Delta_h) = \frac{2 + [2\gamma_a e^{F\Delta\psi/kT}/(1+e^{F\Delta_h/kT})]}{1 + [2\gamma_a e^{F\Delta\psi/kT}/(1+e^{F\Delta_h/kT})] + [2\gamma_a e^{F\Delta_h/kT}/(1+e^{F\Delta_h/kT})]^2} \tag{43}$$

with

$$\gamma_a = {}^Pc_k a_k''/{}^nc_l a_l' \tag{44}$$

In the derivation of Eqs. (42) and (43), a depletion of cations in the n- and p-layers was neglected. In the n-layer, where the cations are minority ions, this assumption might not be valid. For the values of γ_a and $\Delta\psi$ considered in the present paper, the above mentioned approximation is inconsequential however.

APPLICATION OF THE THEORY TO EXPERIMENTS

Voltage clamp: depolarization from the resting potential

In all situations of interest in the present paper the concentrations of univalent species l in the external medium is equal to that of the univalent species k in the intracellular medium. Thus, we use the following approximation:

$$a_l' = a_k'' = a_1 \tag{45}$$

Using Eqs. (44) and (45), we can write for Eq. (12):

$$\frac{n}{1-n}e^{wn/kT} = q(T)\frac{({}^Pc_k a_1)^2}{{}^nc_2 a_2'}\frac{1}{\gamma^2}[1+\gamma e^{F\Delta\psi/kT}+\gamma^2 e^{2F\Delta\psi/kT}] \tag{46}$$

where

$$\gamma = {}^Pc_k/{}^nc_l \tag{47}$$

At $\Delta\psi = \Delta\psi_r$, a formal solution of Eq. (46) is $n = \frac{1}{2}$. Thus, we can write:

$$e^{w/2kT} = \frac{q(T)(^Pc_k a_1)^2}{\gamma^2 \, ^nc_2 a_2'} [1 + \gamma e^{F\Delta\psi_r/kT} + \gamma^2 e^{2F\Delta\psi_r/kT}] \quad (48)$$

After substitution of Eq. (48) into Eq. (46), we have

$$\frac{n}{1-n} e^{-(w/2kT)(1-2n)} = m(\Delta\psi) \quad (49)$$

where

$$m(\Delta\psi) = \frac{1 + \gamma e^{F\Delta\psi/kT} + \gamma^2 e^{2F\Delta\psi/kT}}{1 + \gamma e^{F\Delta\psi_r/kT} + \gamma^2 e^{2F\Delta\psi_r/kT}} \quad (50)$$

We now wish to apply the theory to voltage clamp experiments of Cole and Moore (1960), done at $T = 20°C$ and on an axon with $\Delta\psi_r = -62 \, \text{mV}$.

Using the parameters $w/2kT - 2.4$, $\gamma = 0.6$ and $\Delta\psi_r = -62 \, \text{mV}$, we have from equations (49) and (50): $^0n = \underline{n} = 0.17072$, and for $^1n(\Delta\psi)$ the values listed in Table I.

TABLE I Steady state and kinetic parameters of the theory for: $\gamma = 0.6$; $-w/2kT = 2.4$; $c_0 = 1.083 \times 10^6 \, \text{ms}^{-3}$.

	-40	-35	-30	-20	0
^1n	0.8569	0.8653	0.8744	0.8946	0.9385
$(^0\phi - {}^1\phi)/kT$	0.06469	0.09031	0.1222	0.2117	0.5576
b^*	203.65	112.15	66.00	25.85	5.18
c	0.0080	0.323	4.50	165.00	1.100×10^4

With these numbers we can calculate $(^0\phi - {}^1\phi)/kT$ from Eq. (22) and b^* from Eq. (37). These values are listed in Table I also. As can be seen from these values, the size of the critical nucleus decreases strongly with increasing depolarization. For $\Delta\psi > -30 \, \text{mV}$ the critical nucleus size becomes so small as to render the nucleation growth theory inapplicable because the very concept of a nucleus becomes untenable. This is a disappointing result as it precludes the application of the theory to higher depolarizations. Fortunately, however, a comparison with experimental curves can be made for $\Delta\psi \leq -30 \, \text{mV}$, which includes the region where the experiments show the development of an appreciable inward current. For this comparison with experiments, we have calculated c from Eq. (35), as listed in Table I. Here we chose $c_0 = 1.083 \times 10^6 \, \text{ms}^{-3}$. If we use $a_0 = 23 \times 23 \, \text{Å}^2$ for the area per subunit, we can calculate the theoretical excitation current from Eqs. (33), (42), (43) and (44), where in this case $\Delta_h = 0$. The resulting time curves of the excitation current are plotted in Figure 2, (lower). The curve for $\Delta\psi = -20$ mV, being questionable because of the limitations of the theory discussed above, is drawn in a broken line. The experimental curves in Figure 2 (top) are taken from Cole and Moore (1960), Figure 15. Their records give the combined

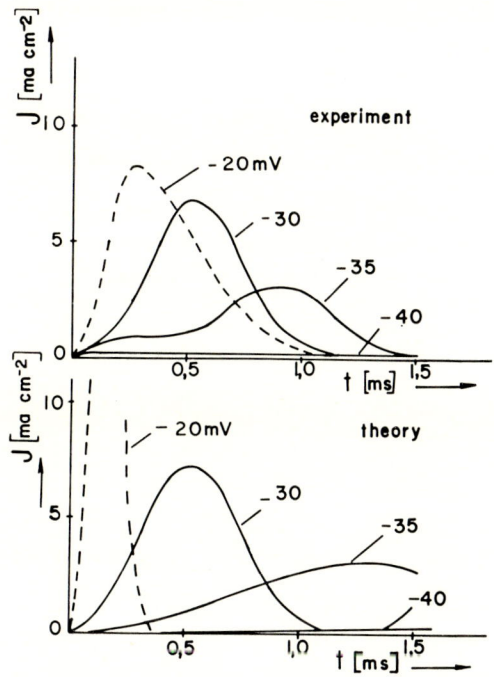

FIGURE 2 Plot of excitation current versus time for different depolarized membrane potentials given by curve parameters in mV. Resting potential at -62 mV.

Experimental curves (above) are taken from Cole and Moore (1960). Theoretical curves are derived as described in the text using $w/2kT = -2.4$; $\gamma = 0.6$; $a_0 = 23 \times 23$ Å^2; $c_0 = 1.083 \times 10^6$ ms^{-3}.

passive outward and excitation inward current. We have not yet calculated the time course of the passive outward current. Since the absolute magnitude of the passive outward current is small for small depolarizations, we introduce only a negligible error if we assume the passive current as established instantaneously. Thus we have subtracted the total observed current from the base line of the asymptotic outward current and plotted this quantity as the excitation current in Figure 2, (top). For $\Delta\psi \leq -30$ mV the probable error of this procedure is only a few percent. For $\Delta\psi = -20\,\text{mV}$ the asymptotic outward current is already appreciable, and the precision of the above separation procedure becomes uncertain. We have therefore indicated this curve by a broken line.

A reasonable comparison between experimental and theoretical curves can thus be made only for $\Delta\psi \leq -30\,\text{mV}$. Within this range, the theory describes well the development of an appreciable inward excitation current in dependence on time and membrane potential.

The parameters used are:

$$a_0 = 23 \times 23\,\text{Å}^2$$

$$-w/2kT = 2.4, \quad \text{i.e. } w = -2.8\,\text{kcal/mole}$$

$$\gamma = 0.6$$

$$c_0 = 1.082 \times 10^6\,\text{ms}^{-3}$$

The area per subunit of $23 \times 23\,\text{Å}^2$ is very reasonable in view of the substructure observed in electron microscopic examinations of the squid axon membrane by Villegas and Villegas (1968). This figure is compatable with the proposed protein nature of the excitable subunits of the axon membrane. Experiments of Tasaki, et al. (1965) have previously given strong evidence for proteins as the excitable structures of the axon. The interaction energy of 2.8 kcal/Mole can be compared with the figure of 1.8 kcal/Mole derived by Wyman (1963) for allosteric oxygen binding of spirographis hemoglobin with a procedure analogous to the Bragg–Williams approximation. Again this figure of 2.8 kcal/Mole is reasonable within the energies to be expected for protein–protein interactions. The parameter of $\gamma = 0.6$ is the ratio of accessability of the active center to species k from inside to that of species l from outside for equal electrochemical activities. This number would say that the access for univalent cation species l from outside is favoured strongly at the membrane potentials considered above. Since on return of the membrane to the resting state the desorbed univalent cations will move inward, in a succession of voltage clamp experiments we should observe a substantial inflow of Na, comparable to the net charge flow. However, in our model the latter is not to be considered as an inflow of Na^+ due to its gradient of electrochemical activity outside to inside, but as a composite flow of Ca^{++} and Na^+ (and some K), following the mechanism formulated in the theoretical sections. As in any application of nucleation-growth theory in physical chemistry, the absolute magnitude of the 'action constant' c_0 cannot be calculated by molecular theory and thus cannot be checked independently.

In this comparison of the results of our model with experimental evidence, we wish to mention some recent results of Cohen, et al. (1968) on light scattering and optical birefringence changes during nerve activity. These experiments indicate that the conformational transition of the membrane during activity, as postulated by our theory (see I, and the theoretical sections above) may well be a reality.

Voltage clamp: depolarization from a hyperpolarizing holding potential

If the membrane is hyperpolarized prior to the depolarizing step in membrane potential, we have to use Eq. (21) instead of Eq. (12) for calculation of ^0n. The quantity ^1n in its dependence on $\Delta\psi$ can be calculated by Eqs. (49) and (50), as before. In papers reporting the effect of previous hyperpolarization (see Hodgkin and Huxley, 1952, and Taylor, 1959), the early peak of the inward current is usually given in dependence on the hyperpolarizing holding potential. For a comparison with experiment, therefore, we have to derive the maximum of the theoretical excitation current in its dependence on time, i.e., we need $(dn/dt)_{max}$. We have from Eq. (33):

$$t_{max} = \left(\frac{2}{3}\right)^{\frac{1}{3}} c^{-\frac{1}{3}} \tag{51}$$

and

$$\left(\frac{dn}{dt}\right)_{max} = (^1n - {^0n})\left(\frac{2}{3e}\right)^{\frac{2}{3}} c^{\frac{1}{3}} \tag{52}$$

The inactivation parameter h, according to theory, can thus be formulated as:

$$h = \frac{\mathscr{I}_{max}(\Delta\psi; \Delta_h)}{\mathscr{I}_{max}(\Delta\psi; 0)} = \frac{p(\Delta\psi; \Delta_h)[^1n(\Delta\psi) - {^0n}(\Delta_h)]}{p(\Delta\psi; 0)[^1n(\Delta\psi) - {^0n}(0)]} \left\{\frac{c(\Delta\psi; \Delta_h)}{c(\Delta\psi; 0)}\right\}^{\frac{1}{3}} \tag{53}$$

Using the same parameters as in the previous section and the same procedure, however taking Eq. (21) instead of (12) for calculation of $^0n(\Delta_h)$, we get, for $\Delta\psi = -30$ mV, the dependence of h on the holding hyperpolarization Δ_h as given by the broken curve of Figure 3. The experimental curve, for $\Delta\psi - \Delta\psi_r = 44$ mV (fully drawn in Figure 3), was taken from Hodgkin and Huxley, 1952, Figure 5. We have not included into Figure 3 the depolarizing branch of inactivation, because at depolarizations >7 mV (see Hodgkin and Huxley, 1952, Figure 4) a substantial early inward current already occurs at the clamping of the holding potential. Although this type of inactivation is not in contradiction with our theory, it seems that its mechanism is a different one from that discussed above. Furthermore, it is not clear to the author how inactivation parameters corresponding to those at hyperpolarizing holding potential can be evaluated from the records given in Figure 4 of the above authors.

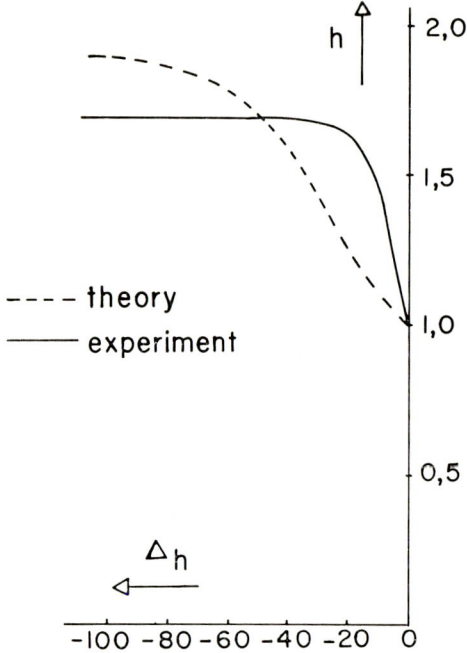

FIGURE 3 Plot of the inactivation parameter versus hyperpolarizing holding potential.
Experimental curve (fully drawn) is taken from Hodgkin and Huxley (1952).
Theoretical curve (broken line) is derived as given in the text, using the parameters given in the legend of FIGURE 2.

As can be seen from Figure 3, the theory predicts a substantial increase of the early current peak by previous hyperpolarization, as it is observed.

The detailed dependence on the holding potential is not reproduced exactly by the theory, which gives a somewhat weaker dependence. However, the finer details are not to be expected from such a crude theory. Furthermore, we have to keep in mind that, due to the limitation of the theory to small depolarizations, we had to use for the theoretical evaluation a depolarization of 32 mV compared to 44 mV used by Hodgkin and Huxley (1952).

Negative steady state resistance

In an external medium of high potassium concentration, Moore (1959) observed a region of negative steady state resistance at hyperpolarizing membrane potentials. On the hyperpolarizing side of this region there is a region of low membrane conductivity α_0, giving a current which in most cases of interest can be written as:

$$\mathcal{I} = \alpha_0 U \tag{54}$$

where U is the applied voltage. In this region the membrane is excitable (Moore, 1959).

On the depolarizing side of the region of negative resistance there is again a region of nearly linear dependence of current on voltage, but with a substantially higher membrane conductivity α_1, giving a current:

$$\mathcal{I} = \alpha_1 U \tag{55}$$

In a typical situation (see Cole, 1968, p. 458, Figure 4:5, curve B) $\alpha_1 \approx 10\alpha_0$. The two branches of the current voltage curve are joined by a region of negative resistance extending over around 25 mV with its central region at about -55 mV. These results are valid for an axon with intact axoplasm maintained in an external solution containing 440 mM K^+, 10 mM Ca^{++} and 50 mM Mg^{++}, and giving a resting potential of 6 mV. We wish to describe this phenomenon in terms of our theory. One basic feature of our model was that a subunit in state 0 has a much lower permeability to ions and thus a lower conductivity increment than a subunit in state 1. Thus, the branch with membrane conductivity α_0 represents a state of the membrane with subunits predominantly in state 0, whereas the branch with membrane conductivity α_1 corresponds to the membrane with subunits predominantly in state 1. Let ρ_0 and ρ_1 be the conductivity increments per subunit in state 0 and state 1 respectively. Let ^0n be the fraction of subunits in state 1 on the hyperpolarizing branch of the region of negative resistance and ^1n the fraction in state 1 on the depolarizing side. Then we have:

$$\frac{\alpha_0}{B} = \rho_0(1 - {}^0n) + \rho_1 {}^0n \tag{56}$$

and

$$\frac{\alpha_1}{B} = \rho_0(1 - {}^1n) + \rho_1 {}^1n \tag{57}$$

where B is the number of subunits per unit area of membrane. In the intermediate region we have

$$\frac{\alpha}{B} = \rho_0(1 - n) + \rho_1 n \tag{58}$$

where n is the quasi steady state value of the fraction of subunits in state 1 in the membrane, corresponding to the value of the applied potential and to the degree of transition of the membrane at the time t_s of measurement. Usually the currents are observed at a time $t_s = 10$ to 40 ms after application of voltage. Due to the strong dependence of the time necessary for the completion of the membrane transition on the clamping voltage, as shown in Figure 2, we expect for fixed t_s non-steady state currents only in a very narrow region of voltage. The fraction of subunits in state 1 at time t_s, according to Eq. (32) is given by:

$$n(t_s) = {}^1n - ({}^1n - {}^0n)\exp\{-ct_s^3\} \qquad (59)$$

where c is given by Eq. (35) and depends on the holding potential and the membrane potential of the measurement. By analogy with the procedure of previous sections, but not necessarily in agreement with the experimental procedure, we think that the holding potential is that of symmetry $\Delta\psi_r'$ in the plot n versus $\ln\{q(T)M(\Delta\psi)\}$, as discussed in the section on the steady state of the membrane. We have to observe, however, that the experimental situation is now different since the species l outside has been changed. In previous sections, we had the normal situation $l = Na^+$, $k = K^+$. Now we have replaced Na^+ outside by K^+. Thus we have to substitute ${}^n c_l$ by ${}^n c_k$, giving $\gamma' = {}^P c_k/{}^n c_k$; all other parameters are unchanged. With this new parameter, γ', the steady state of the membrane is different. At zero applied voltage the state of the membrane with subunits in state 1 is stable. From our theory we can explain this, if we assume that the access of univalent cations from outside is facilitated, i.e., ${}^n c_k > {}^n c_l$, giving $\gamma' < \gamma$. This implies that the potential of symmetry is no longer the potential of zero current, nor the former resting potential. To calculate $\Delta\psi_r'$, we observe that in Eq. (48) the quantity

$$\frac{e^{w/2kT}({}^n c_2\, a_2')}{q(T)({}^P c_k\, a_1')^2}$$

is not affected by a change of the external medium, provided the cation activities are kept constant, which we will assume to be the case. Thus, we have for evaluation of $\Delta\psi_r'$:

$$\frac{1}{\gamma'^2}[1 + \gamma' e^{F\Delta\psi_r'/kT} + (\gamma')^2 e^{2F\Delta\psi_r'/kT}] = \frac{1}{\gamma^2}[1 + \gamma e^{F\Delta\psi_r/kT} + \gamma^2 e^{2F\Delta\psi_r/kT}] \qquad (60)$$

We observe that there is a lower limit $\gamma' \geq 0.58437$ so that a symmetry potential $\Delta\psi_r' > -317$ mV can fulfil Eq. (60). If we use $\gamma' = 0.58437$, and $\gamma = 0.6$; $\Delta\psi_r = -62$ mV as before, we obtain

$$\Delta\psi_r' = -317.2\text{ mV}$$

At this membrane potential once again $^0n = \underline{n} = 0.17072$. If we start from this state and depolarize to different potentials and observe the state of the membrane, then $n(t_s)$ as a function of rising potential, according to Eq. (59), will go from ^0n to ^1n. For the present purposes, we can neglect any dependence of ^1n on the membrane potential. At $n(t_s) = (^1n + {^0n})/2$, the transition of the membrane to the higher conductivity state is half completed. Using $t_s = 10$ ms, we obtain $c_{\frac{1}{2}} = 0.693 \times 10^{-3}$ ms^{-3}. We now use Eqs. (35), (37) and (49) with

$$m(\Delta\psi) = \frac{1/(\gamma')^2[1+\gamma' e^{F\Delta\psi/kT}+(\gamma')^2 e^{2F\Delta\psi/kT}]}{1/\gamma^2[1+\gamma e^{F\Delta\psi_r/kT}+\gamma^2 e^{2F\Delta\psi_r/kT}]} \qquad (61)$$

to obtain $\Delta\psi_{\frac{1}{2}} = -54.2$ mV, corresponding to half completion of the membrane transition. Here we have used $\gamma' = 0.58437$. A choice like $\gamma' = 0.588$ gives $\Delta\psi'_r = -97.7$ mV (i.e., a shift by more than 200 mV in $\Delta\psi'_r$), but again $\Delta\psi_{\frac{1}{2}}$ is around 51 mV, giving a shift in $\Delta\psi_{\frac{1}{2}}$ of less than 5 mV. The range of the membrane transition is about 5 mV. Thus, according to the theory, the range of negative quasi-steady-state resistance is very narrow.

A comparison of these results with experiments is only tentative at this state, because the experiments usually are performed by hyperpolarizing from zero steady state current. As cited before, the half potential is observed at a similar membrane potential as given by the theory; but the width of the region of negative resistance is at least five times broader than given by theory. We think that these results encourage the adaptation of the theory to the description of the experimental procedure and the investigation of the phenomenon in more detail, especially the dependence on the Ca^{++} concentration in the external medium.

Heat cycle on excitation

According to the molecular model, the membrane undergoes a phase transition in the process of excitation by a depolarizing voltage clamp. With this phase transition is connected a latent heat of phase change, which we now want to derive. The free energy of the membrane system can be derived from the partition function given by I, Eq. (12).

Using $s! \approx (s/e)^s$ and $f/kT = -\ln Q$ we have:

$$f/kT = (1-n)\ln(1-n) + n\ln n - \ln q_0 + v_{00}/2kT + n\ln(q_0/q_1)$$
$$-(v_{00}-v_{01})n/kT + wn^2/2kT \qquad (62)$$

For the meaning of the parameters in the Eq. see I. The potential energy u
c

per subunit of the system can be calculated from Eq. (62) by the general thermodynamic relation:

$$u = [\partial(f/T)/\partial(1/T)] \qquad (63)$$

With the aid of I Eqs. (13) and (59), we obtain:

$$u = -(w/2)n(1-n) + n(2u_1 - u_2) + u_2 + v_{00}/2 \qquad (64)$$

If we start from the resting potential with $^0n = \underline{n}$ and depolarize to the state $^1n(\Delta\psi)$, the following amount of energy is released by the membrane:

$$u(^1n) - u(\underline{n}) = -(w/2)[^1n(1-{^1n}) - \underline{(1-\underline{n})}] - (u_2 - 2u_1)(^1n - {^0n}) \qquad (65)$$

Experiments on the caloric effect of a voltage clamp experiment are not known to the author. However, caloric measurements at the action potential of non-myelinated nerve of rabbit have been published by Howarth, et al. (1968).

To compare the theoretical heat with these experiments, we have to go beyond the treatment of the previous sections and characterize the processes underlying an action potential. As outlined in I, in the rising phase of an action potential the membrane goes from the resting state $^0n = \underline{n}$ into the excited state. In contrast to the voltage clamp experiment, the membrane potential is not fixed. Thus, the composite flow of calcium and univalent cations accompanying this cooperative cation exchange, charge the membrane capacity to about $+40\,\text{mV}$. One can estimate easily that only the cation exchange of roughly $\frac{1}{30}$ of the subunits of the membrane suffices to provide the necessary charges.

Near the peak of the action potential, the potential profile across the *psn*-layer will be changed such that no net charge flow occurs through the membrane during continuation of the membrane transition. After this transition is completed, the shifted electrochemical activities in the *p*- and *s*-layers drive the membrane back to the resting state.

Without more detailed analysis of the charge transfers during that part of the rising action potential where there is essentially no net charge transfer, it is difficult to assess the final state of the membrane and the relative proportion of univalent cations bound in the active center. However, we can give reasonable estimates for the terms in Eq. (65).

The term

$$U_c = -(w/2)[^1n(1-{^1n}) - \underline{n}(1-\underline{n})] \qquad (66)$$

arises from the cooperative interaction of this subunits. The initial state in an

action potential is the resting state described by $\underline{n}$ as already assumed in Eqs. (65) and (66). The final state will be characterized by $\bar{n} \leq {}^1n \leq 1$.
Thus we have:

$$0 \geq U_c \geq \frac{w}{2}\underline{n}(1-\underline{n}) \qquad (67)$$

With the parameter $w = -2.8\,\text{kcal/Mole}$, i.e., $\underline{n} = 0.17072$, derived by comparison of theory with voltage clamp experiments, we obtain from Eq. (67):

$$0 \geq U_c \geq -400\,\text{cal/mole} \qquad (68)$$

Negative U_c means an energy released by the system. Thus, we have a contribution from cooperativity of between 0 and 400 cal/Mole released per subunit and rising phase of an action potential.

The term

$$U_b = -(u_2 - 2u_1)({}^1n - \underline{n}) \qquad (69)$$

reflects the release of cation binding energy during the cooperative cation exchange.

We do not know the nature of cation binding in the active center and the relative proportion of Na^+ and K^+ bound in the active center after completion of the rising phase of the action potential. To give an estimate of U_b, we assume pair formation with two negative charges of the active center in the mixed hydrophilic–hydrophobic environment of the membranous proteins of an effective dielectric constant ε_{rel}. If the hydrated cations can interact optimally with both of the hydrated anionic groups, we have:

$$-u_2 = \frac{4\varepsilon^2}{4\pi\varepsilon_0\,\varepsilon_{rel}(r_{Ca} + r_-)} \qquad (70)$$

Here, $\varepsilon_0 = 8.86 \times 10^{-14}\,\text{Coul V}^{-1}\text{cm}^{-1}$, r_- is the hydrated radius of the anionic group, and r_{Ca} is the radius of the hydrated Ca^{++}.

For the interaction energy of the univalent cations we have

$$-2u_1 = \frac{2\varepsilon^2}{4\pi\varepsilon_0\,\varepsilon_{rel}}\left[\frac{1}{r_{Na} + r_-} + \frac{1}{r_K + r_-}\right] \qquad (71)$$

Here, r_{Na} and r_K are the radii of the hydrated Na^+ and K^+ respectively. In Eq. (71) we have assumed that half of the univalent cations bound in the active center is K^+ and half is Na^+. Any other relative proportion will give very similar numerical results in the following estimate.

According to Netter (1959) we have: $r_{Ca} = 4.8\,\text{Å}$; $r_{Na} = 2.56\,\text{Å}$; $r_K = 1.98\,\text{Å}$. We use $r_- = 4\,\text{Å}$ and ${}^1n = \bar{n} = 0.829$.

For $\varepsilon_{rel} = 15$ we obtain per subunit:

$$U_b = -3.7 \text{ kcal/Mole} \tag{72}$$

This number is similar to that of -2.7 kcal/Mole cited by Tasaki (1968), p. 124, for the analogous exchange of Ca^{++} by K^+ in synthetic ion exchangers. For $\varepsilon_{rel} = 80$ we obtain

$$U_b = -0.7 \text{ kcal/Mole} \tag{73}$$

These estimates for U_b are very rough indeed, as is clear from the given arguments.

From Eqs. (68), (72) and (73) we thus have between

$$700 \text{ cal/mol} \leq |u(^1n) - u(n)| \leq 4100 \text{ cal/Mole} \tag{74}$$

of energy released by the membrane in the rising phase of the action potential. Roughly the same amount of energy will be taken up by the membrane in the falling phase of an action potential, where the membrane undergoes the reverse phase transition with a corresponding positive latent heat. These estimates can be compared with results of Howarth, et al. (1968) on the caloric effect of excitation of the non-myelinated vagus nerve of rabbit. These authors measured an amount of 24.5 μcal/impulse of heat released per gram of nerve during the rising phase of the action potential and 22.2 μcal/impulse taken up per gram in its falling phase.

Using their figure for the membrane surface per gram of vagus nerve of $6 \times 10^3 \text{ cm}^2$ and our number for the area per subunit $a_0 = 23 \times 23 \text{ Å}^2$, we obtain from the experimental figure of 22.2 μcal/impulse the following experimental latent heat H per area of a subunit:

$$H = 119 \text{ cal/Mole} \tag{75}$$

Our theoretical estimates, given in Eq. (74), are an order of magnitude higher.

One can expect that a more detailed knowledge of the nature and thus the energy of binding of cations in the active center will provide a more precise theoretical estimate and a more meaningful numerical comparison with experiments. At present the essential point seems to us that the theory predicts a heat cycle upon excitation and a return to the resting state, the sign of which is the same as shown by experiment.

The physicochemical picture of Hodgkin and Huxley does not give an explanation for the re-uptake of heat during the falling phase of an action potential.

CONCLUSION

From the results of the previous sections, we conclude that the theoretical model can describe in reasonable agreement with experiment, several basic observations on excitation of non-myelinated axons. The molecular parameters to be adjusted in this comparison are well in the range to be expected for size and interaction energies of membrane proteins. Unfortunately, the validity of the treatment of kinetic processes by the nucleation and growth theory is limited to relatively small depolarizations. A prerequisite for a theoretical description of an action potential is the calculation of the voltage clamp experiments at higher depolarizations. In order to calculate the excitation current at higher depolarizations, a theoretical method different from the nucleus/growth treatment has to be used. In as yet unpublished calculations we have used the direct molecular kinetic method, i.e., integrated Eqs. (2), (3) and (4) directly. For an almost identical interaction parameter w this theory essentially reproduces the results for small depolarizations, obtained by the quasi thermodynamic nucleation/growth treatment.

For a meaningful comparison of the theory with voltage clamp experiments at higher depolarizations, we also have to calculate the passive current theoretically. This current follows, with a small diffusional delay, the time course of membrane transition. Work on this problem, similar to the treatment of the steady state negative resistance given in the present paper, is in progress.

In spite of the above mentioned incompleteness of the theoretical treatment, the results of the present theory seem to be evidence for the presence in the axon membrane of an excitation process different from that envisioned by Hodgkin and Huxley. We think that our results speak in favour of a homogeneous excitable membrane, where essentially all the protein subunits have similar properties of cation binding, except for isolated patches with specific energy-linked transport enzymes. This picture is in agreement with recent electron microscopic investigations of Villegas and Villegas (1968). It is in contrast, however, to the picture of Hodgkin and Huxley, who imply the existence of isolated pores, the state of which is governed by the electrical potential, and which permit flow of univalent cations according to their gradients of electrochemical activities. We wish to mention that experiments of Moore, et al. (1967) investigating the action of very small amounts of Tetrodotoxin on the axon membrane appear to support the model of few isolated pores in the axon membrane. A clear interpretation of these results is possible, however, only when the mode of action of Tetrodotoxin is elucidated.

In view of the correct description of the phenomena presented in the above sections, however, especially on experiments where the Hodgkin–Huxley theory fails, we consider the present theory a more attractive possibility, as it gives for the first time a physicochemical mechanism of the regulation of the excitation state of the axon membrane caused by external stimulation.

Additional support for our model of regulation of axonal membrane properties by cation-binding seems to come from a survey of electrochemical properties of other biological membranes. We can here only briefly mention some cases which seem to show that the electrical conductivity of biological membranes in general is regulated by binding of Ca^{++} in the membrane and that the mechanism of exchange of Ca^{++} in the membrane by univalent cations is a means by which biological structures perform vastly different vital tasks.

The first example is the release of Ca^{++} by the sarcoplasmic reticulum after excitation of the outer membrane of muscle cells. We consider the sarcoplasmic reticular membrane at rest to contain subunits predominantly in state 0. After depolarization, the Ca^{++} is exchanged by univalent cations and moves 'inward' i.e., eventually to the muscle fibrils where it triggers contraction. For this to happen, the Ca^{++} ions should not be trapped in a p-layer as in the nerve fiber. Thus, the p-layer in the sarcoplasmic reticulum should be absent or different from that of nerve. Furthermore, the cooperativity of the subunits need not be as pronounced as in nerve. Hasselbach and Elfvin (1967) have in fact observed in electron microscopic pictures of vesicles of sarcoplasmic reticulum a marked asymmetry of the osmium tetroxide staining membrane layers, different from what is observed for squid axon (Villegas and Villegas, 1968). Furthermore the membranes of the sarcoplasmic reticulum show a substructure of 25–30 Å diameter in the protein layers. A similar figure of 25–30 Å distance between active sites of Ca^{++} transport was observed by the above authors in inhibitory studies.

A second example of Ca^{++} exchange in membranes performing vital functions for the organism seems to occur in the inner membrane of the mitochondria. Scarpa and Azzone (1968) have shown that in the absence of respiration, Ca^{++} is bound at the outer face of the membrane and is transferred inward with the onset of respiration. This suggests that the stability of binding of Ca^{++} is changed by respiration. In terms of our model we should suggest that it is exchanged from active centers of the membrane by univalent cations (mainly protons) whereupon the conductivity properties of the inner mitochondrial membrane are changed, which should have relevant consequences for the mechanism of oxidative phosphorylation.

The last example of Ca^{++} binding as regulating the conductivity of membranes, which we wish to mention here, is taken from the extremely interesting experiments of Loewenstein (1966). These experiments show that neighboring cells in epithelial tissues have membrane junctions, the electrical resistance of which is negligible compared to the normal membrane resistance. These areas of cell junctions show a tight apposition of the outer faces of the plasmamembranes of the connected cells. In terms of our model we would suggest that the membrane subunits at these junctions are in state 1 with a high increment of conductivity. All other parts of the membrane, being in contact with the relatively high concentration of Ca^{++} of the extracellular medium, possess subunits predominantly in state 0 and thus have low electrical resistance. The intracellular concentration of Ca^{++} is very low. We do not want to comment on how the junctions of cell contact with high conductivity are shielded from the Ca^{++} in the extracellular fluid (see Loewenstein, 1966, for more details). More interesting in this context is the mechanism for the 'sealing' of the junctional membranes, i.e., for their becoming impermeable to ions.

Loewenstein (1966) has reported that Ca^{++} is crucial for this process. For instance, by electrophoretical injection of Ca^{++} into the cells, the junctional membrane conductivity could in essence be removed, although the borders of the junction are still shielded from extracellular Ca^{++}. We can picture in terms of our theory the process of sealing of the junctional membrane in the presence of Ca^{++}, by a transition of this membrane from a state ^1n to ^0n. The kinetics of this process and the related one of establishing electrical cell contact, for instance in wound healing, should also be amenable for theoretical description. In many details this description will be different from that for the axoplasmic membrane in that the asymmetry of the *psn*-structure and the feature of cooperativity seems to be much less pronounced.

In any case, our basic model seems to be useful not only for the understanding of the special case of all-or-none phenomena as observed for non-myelinated axons, but also for regulation of solute permeability and electrical conductivity of biological membranes in general.

ACKNOWLEDGEMENTS

The author wishes to thank Prof. M. Delbrück, Pasadena, for encouragement and stimulating discussion, especially during the early stages of this work which was started in Pasadena.

To Prof. K. S. Cole, Bethesda, the author is grateful for criticism and for making available to him valuable experimental data, published and unpublished.

Thanks are also due to Prof. M. Klingenberg, München, for support of the present work and to the Deutsche Forschungsgemeinschaft for granting a Habilitandenstipendium to the author.

Literature

ADAM, G. 1968, *Z. Naturforsch.*, **23b**, 181.
BARRER, R. M., and FALCONER, J. D. 1956, *Proc. Roy. Soc.* (London), Ser. A **236**, 227.
CHANGEUX, J. P., THIERY, J., TUNG, Y., and KITTEL, C. 1967, *Proc. Natl. Acad. Sci. U.S.A.*, **57**, 335.
COHEN, L. B., KEYNES, R. D., and HILLE, B. 1968, *Nature*, **218**, 438.
COLE, K. S. 1968, Membranes, Ions and Impulses, Berkeley and Los Angeles.
COLE, K. S., and MOORE, J. W. 1960, *J. Gen. Physiol.*, **44**, 123.
COSTER, H. G. L. 1965, *Biophys. J.*, **5**, 669.
DAVSON, H., and DANIELLI, J. F. 1952, The Permeability of Natural Membranes, Cambridge.
FINE, M. E. 1964, Introduction to Phase Transformation in Condensed Systems, New York.
HASSELBACH, W., and ELFVIN, L. G. 1967, *J. Ultrastruct. Research*, **17**, 598.
HERLET, A., and SPENKE, E. 1955, Z. Angew. Physik **7**, 99, 149, 195.
HILL, T. L. 1960, Introduction to Statistical Thermodynamics, Reading.
HODGKIN, A. L. 1964, The Conduction of the Nervous Impulse, Liverpool.
HODGKIN, A. L., and HUXLEY, A. F. 1952, *J. Physiol.*, **116**, 497.
HOWARTH, J. V., KEYNES, R. D., and RITCHIE, J. M. 1968, *J. Physiol.*, **194**, 745.
LOEWENSTEIN, W. R. 1966, *Ann. N. Y. Acad. Sci.*, **137**, 441.
MOORE, J. W. 1959, *Nature*, **183**, 265–266.
MUELLER, P., and RUDIN, D. O. 1968, *Nature*, **217**, 713.
NETTER, H. 1959, Theoretische Biochemie, Berlin-Goettingen-Heidelberg, P. 355.
RUDIN, D. O., and MUELLER, P. 1967, *Biochem. Biophys. Res. Comm.*, **26**, 398.
SCARPA, A., and AZZONE, G. F. 1968, *J. biol. Chem.*, **243**, 5132.
SPYROPOULOS, C. S. 1965, *J. Gen. Physiol.*, **48**, 49 (suppl.).
TASAKI, I. 1968, Nerve Excitation, a Macromolecular Approach, Springfield.
TASAKI, I., SINGER, I., and TAKENAKA, K. 1965, *J. Gen. Physiol.*, **48**, 1095.
TAYLOR, R. E. 1959, *Amer. J. Physiol.*, **196**, 1071.
VILLEGAS, G. M., and VILLEGAS, R. 1968, *J. Gen. Physiol.*, **51**, 445.
WYMAN, J. 1963, *Cold Spring Harbour Symposia Quant. Biol.*, **28**, 483.

DISCUSSION

TASAKI   Why do you have to have anion exchange of property on the surface? If that is the case, can the calcium concentration effect I mentioned occur in this model?

ADAM   I feel that positive charge on the outside is essential for the model of the axon membrane; otherwise rectification wouldn't come about. The axon membrane shows rectification, whereas other biological membranes don't Thus, the rectification properties of the axon membrane probably should not. be explained by a general effect like Dr. Schlögl calculated but as arising from the specific structure of the specific membrane. The proposed *psn*-structure can explain rectification as observed, using your value of the diffusion constant of cations and reasonable densities of fixed charges. This supports the *psn*-structure. Furthermore, the dependence of voltage clamp current on previous hyperpolarization doesn't come out if we have a different structure in our theoretical model. Now, the reasons why you propose outside negative charges come from serious experiments; and I haven't tried to explain these quantatitively. However, there is the possibility that they can be explained by the negative fixed charges in the active center. The active center is that which binds calcium or other divalent cations and the active center is supposed to be on the outside protein layer. Thus, it is tempting to look into the question further of how your effect can be explained by the negative charges within the positive fixed charge layer on the outside.

KEDEM   How can one understand the conductivity changes of the membrane upon excitation?

ADAM   I did not attempt an *a priori* molecular calculation of the α, but I think it can be done. The conductivity of the membrane to cations according to this model is determined mainly by the outer protein layer. This layer, in which the cations are minority carriers, presents the major hindrance for diffusion of cations. Now I think that excitation consists of masking the positive charges in the outside protein layer by maybe turning them into a hydrophobic medium and by way of this the conductivity properties of the subunit are changed. It is a nice result from Dr. Cole's experiments that only in the resting state do you observe rectification; but in the excited state the asymptotic values of the voltage clamp currents are linear with voltage. Thus, the conductivity properties of the membrane are changed in a way which I interpret as meaning that in the excited state the positive fixed charges are masked, and cannot contribute to the rectification properties and to hindrance of diffusion. I think I shall try to work out these considerations quantitatively.

COLE I suggest that, until a satisfactory calculation of the passive current is available, it may be more realistic and logical to accept the Hodgkin–Huxley analysis of the voltage clamp current into the sodium and potassium components corresponding to the excitation and passive currents than to use a linear approximation for the latter. Then experimental curves are available for comparison with theory for depolarizations $\Delta\psi > -30$ mV as well as those shown in Figure 2.

Would you please discuss the transitions on your n vs. $\Delta\psi$ diagram which these excitation currents arise?

ADAM If we are here within the meta-stable branch, the state 0 is thermodynamically unstable, so in the course of time the membrane will move into the higher state. We have only two branches which are stable: $\Delta\psi = -\infty$ to $\Delta\psi = \Delta\psi_r$, where divalent cations are bound in the active center, and $\Delta\psi = \Delta\psi_r$ to $\Delta\psi = +\infty$, where univalent cations are bound in the active center. If we had time enough to wait, then even for a very small displacement Δ of the membrane potential from $\Delta\psi_r$ to $\Delta\psi_r + \Delta$, the membrane will undergo the transition from $n < \frac{1}{2}$ to $n > \frac{1}{2}$. This branch is stable and stays stable at more depolarizing potentials.

I wish to point out a certain specification of the theory as presented today in contrast to what I published in an earlier paper (Z.Naturforsch. **23b**, 181 (1968)). In the earlier paper the initial state of the nucleation/growth theory was assumed to be the metastable state $^0n < \frac{1}{2}$ at depolarizing membrane potentials. Now the question did arise: what is the time necessary for the membrane starting from $n(\Delta\psi_r) < \frac{1}{2}$ to attain the metastable state $^0n(\Delta\psi) < \frac{1}{2}$ after sudden depolarization? In as yet unpublished calculations we could show that this time is not negligible compared to the time of transition of the membrane from the metastable state $^0n(\Delta\psi) < \frac{1}{2}$ to the stable state $^1n(\Delta\psi) > \frac{1}{2}$. Thus, the initial state of the kinetic theory should not be taken as $^0n(\Delta\psi) < \frac{1}{2}$, but rather as $\underline{n}(\Delta\psi_r)$, i.e., as the state of the membrane prior to the depolarization. This point was immaterial in the first paper because of the magnitude of the cooperativity parameter chosen. However, it does matter with the choice of parameters presented today. In addition, the effect of previous hyperpolarization can be described by the theory, if we use $^0n(\Delta\psi_r + \Delta_h) < \frac{1}{2}$ as the initial stage, where $\Delta_h < 0$ is the hyperpolarization from the resting state prior to the depolarization.

COLE Then, as I understand it, for a voltage clamp step to $\Delta\psi > \Delta\psi_r$ n goes from its initial value, near zero, to the appropriate upper stable value, near unity. This gives a total inflow of outside ions which is approximately independent of $\Delta\psi$. However, as is to be expected from the HH equations and as found in Chile, the total sodium inflow rises steadily to a maximum for a voltage clamp of about 20 mV depolarization. Furthermore elimination of the series resistance error will probably move the maximum to a larger depolarization.

Although this difference may not be one of the principal questions about the theory, I wish to note that it can be removed by suitable choice of $-w/2kT < 2$ so that n is a single valued function of $\Delta\psi$.

ADAM Although your suggestion of less cooperativity might lead to an improved description of some particular experiment, the simultaneous description of several experiments with the same set of parameters, as attempted in my paper, would deteriorate. Thus, I feel I should stick with the cooperativity parameter I chose.

PAPER 4

A non-linear term in the ion transport across membranes†

REINHARD SCHLÖGL

THE COMPARISON between a kinetic model of Parlin and Eyring[1] which pictures the membrane as a series of potential barriers to the migration of particles, and the phenomenological equations of Nernst and Planck, gives a rough estimate of the limit of validity of phenomenological laws in small phases. It shows that the transport even across membranes with a thickness of 50 Å or more, may be described by the classical phenomenological equations, as long as the field strengths are below a break-down field of the order of about 10^6 Volt/cm.

To obtain Eqs. as general as possible, we start with linear laws connecting particle fluxes and gradients of electrochemical potentials. These potentials have to be conceived as characterizing the thermodynamical state of small imaginary liquid phases in equilibrium with a corresponding cross section of the membrane parallel to the membrane surfaces. The differential Eqs. of motion can be integrated explicitly under special idealizations.

1. We assume a stationary state.
2. The bulk solutions are conceived as dilute solutions.
3. The ONSAGER-coefficients L_{ii} can roughly be approximated as proportional to c_i, L_{ik} proportional to $c_i c_k \kappa$, where c_i and c_k are the corresponding concentrations of species i and k in the bulk solutions and $\div \kappa$ is a Debye length.
4. The electric field is taken as constant throughout the membrane.
5. Chemical reactions are excluded.

Condition 1 may become a poor assumption in very fast processes like membrane excitation. In the case of slow processes, a description of the over-

† Extract from a paper given at the Jerusalem Symposium, July 1968.

all process would require a consideration of the time dependent storing of species near the membrane surfaces in the outer phases. In this paper however, we shall be concerned with the transport only across the membrane itself.

Condition 4 will be a reasonable approximation in membranes with a thickness comparable or less than a characteristic DEBYE length. If we have to deal with thick membranes, 4 will be fulfilled as long as the membrane potential is in the order of less than 100 mV.

The resulting flux Eqs. for solutes are

$$J_k = (1-\sigma_k)\bar{c}_k \cdot q + p_k z_k \bar{c}_k \cdot \frac{F}{RT}\Delta\phi + p_k \frac{v_k}{tgh\, v_k} \cdot \Delta c_k \tag{1}$$

Here J_k = flux of species k
q = volume flux
$\Delta\phi$ = outer membrane potential
Δc_k = concentration difference between outer phases
σ_k = reflection coefficient of k (defined by Sauer)
$\bar{c}_k$ = arithmetic mean of the outer concentrations of k
p_k = permeability of k
z_k = valency of k (positive, negative or zero).

p_k and σ_k are taken as constant.
v_k is a generalized 'driving force' of the form

$$v_k = \frac{z_k}{2}\frac{F\Delta\phi}{RT} + \frac{1-\sigma_k}{2p_k}q$$

These equations can be made up by an additional equation for the volume flux. This, however, reduces only to a simple form in case of a rather dense membrane (small hydrodynamic permeability)

$$q = L_P(P-\pi_i) + L_{PE} \cdot \Delta\phi - L_P RT\sum_k \frac{\sigma_k v_k}{tgh\, v_k} \cdot \Delta c_k \tag{2}$$

where P = pressure difference between the bulk phases
π_i = osmotic difference of the non-permeating species
L_P = hydrodynamic permeability
L_{PE} = electro-osmotic permeability

The comparison between (1), (2) and the linear laws, given by Kedem and Katchalsky[2] and in more general form, by Sauer[3] indicates a non-linear field dependence, given by the factor $v_k/tgh\, v_k$. Even in the case of rather small concentration differences this term reduces to 1 only if the potential difference $z_k\Delta\phi$ becomes less than about 25 mV.

The non-linear term is due to a rearrangement of the ions within the membrane. This is in strong analogy to the model theory of Teorell. Eqs. (1) and (2) are more general, however.

References

1. PARLIN, R. B., and EYRING, H. 1954, in *Ion Transport Across Membranes* Academic Press N.Y., p. 103.
2. KEDEM, O., and KATCHALSKY, A. 1958, Biochem. at Biophys. Acta, **27**, 229.
3. SAUER, F., to be published.

DISCUSSION

JUNGE   You have discussed a very thin membrane phase (e.g. 50 Å thickness) in a state far from equilibrium. To make this off-equilibrium system accessible to a thermodynamic treatment, you have split up the membrane into subphases, which are close to equilibrium. I wonder how the physical idea of a thermodynamic phase is compatible with a say 1 Å sheet, where fluctuations along the only coordinate of interest may be important. There should be a lower limit for this slicing procedure, otherwise the linear approximation of irreversible thermodynamics would hold for any system, no matter how far from equilibrium it may be.

SCHLÖGL   Well, I would say of course it is not at all evident that you can do so. The fact is the theory shows that if there are more than let us say 15 jumps then you get correct formulas in just assuming a thermodynamic phase in inner equilibrium. Now what I am doing here in splitting it up in small sheets is not a physical process. This is just a mathematical process. All you need is the fact that having more than say 15 or 20 jumps, you get from an integral point of view the same equations as starting from a kinetic theory like Parlin and Eyring have done, or starting from a thermodynamical theory as Nernst and Planck have done.

SNELL   Weren't you doing a bit of trickery there in taking an integral equation, and after dividing it up, differentiate it and go back to integrating it again?

SCHLÖGL   I don't quite understand the point about the integral equation.

SNELL   Well, your original equation was in terms of differences; then you divided it up conceptually into thin sections and thereby got a differential equation which you then turned around and integrated.

SCHLÖGL   Oh, there is a difference. If you start with the description of discontinuous systems, then you assume that the right hand phase and the left hand phase are almost in equilibrium. However, we have the situation where this does not hold. We have larger concentrations on one side than on the otherside. Thus what I am doing is to split up the parts into small sheets, and I claim that between these small sheets there shall exist a state of near equilibrium, not exact equilibrium of course, so that we can again use the thermodynamic equations. And then what we are doing is to integrate from one boundary to the other.

EISENMAN   Does that correspond to anything more than assuming continuity in the electrochemical potential between the membrane at that point and a hypothetical external solution? That is, does it correspond to doing anything more than assuming that the electrochemical potential is a continuous function?

SCHLÖGL   It is of course no more. The only thing that has to be obtained is the constant of integration. And to do this you have to take into account that the electrochemical potentials change from point to point only slightly within the membrane, but rather drastically from one boundary to the other.

TASAKI   Did you derive equations applicable for highly charged membranes under these conditions—that would be interesting for a biologist.

SCHLÖGL   No. The highly charged membrane has complications which could not be taken into account.

COLE   I find it interesting to remember that, about thirty years ago, Torsten Teorell used a similar arrangement to obtain analogue solutions for some of these problems. A number of membranes in series were separated from each other by well stirred layers of electrolyte. Measurements of ion concentrations and electric potentials in the compartments then gave transient as well as steady state solutions to a variety of difficult electrodiffusion problems.

I have repeatedly emphasized the simple problem presented by experiments on several different membranes where the external sodium was replaced by potassium. Here the system was close to equilibrium with a nearly zero membrane potential difference for no current flow. With normal external calcium, the inward current was not only less than proportional to internal negative potentials but as the potential went from about -50 to -100 mV the steady state current actually decreased to give what may be defined as a region of negative, steady state, conductance. I became very discouraged with a simple form of electrodiffusion because in my hands it did not produce such a negative resistance. Does your analysis produce such a behavior? And if so what is the physical basis for it?

SCHLÖGL   I am not sure that I can comment on this question unambiguously without knowing more about the experimental set-up. At first glance, however, I would not expect that the behavior of the system that you mentioned can be explained by my formulae.

SNELL   Would it not come out from the difference between two terms that you change?

SCHLÖGL   Yes, I think you are right. In a steady state condition, this has in fact been tried by Frank in Aachen. He tried to explain those characteristics by interplay between 'water flux' and particle flux.

KEDEM   Time-dependence of coefficients during an impulse may be equivalent to velocity dependent coefficients in stationary states. Assume e.g. that a transport coefficient changes with an average concentration in the membrane —this concentration will change with time in the nonstationary measurement, and it may be a function of the flow rate in a stationary state.

ONSAGER   I am toying with an idea here that the negative characteristics might depend on a carrier of opposite charge to the ion that is actively transported all the way—that is actively exchanged with the aqueous phase. In a strongly polarized membrane this might be sort of piled against one end and not returned.

SCHLÖGL   I think that this is a very important remark since many people, I guess, thought that the cross coefficients should be positive if there is exchange of impulse. However, as Professor Onsager remarks, that must not be the case if there is any sort of a chemical binding. For instance, one can assume the membrane to be a sort of a matrix. This matrix has degrees of freedom itself— thermal motion back and forth—and so a particle may sometimes be bounced to the matrix and with the matrix fluctuate back and forth, then released again, and then be bound again. In such a case, it can happen that those coefficients even become negative. This explains why it is necessary that we have competition between two types of particles. Say there are two solutes, solute 1 and solute 2. If solute 1 has a concentration gradient from left to right, and solute 2 has a concentration gradient the opposite side, and both particles can be bound to the matrix, then it can happen that the movement of species 1 from left to right gives a push to species 2 in the opposite direction. This means that those coefficients can become negative.

ONSAGER   Yes, that is speculation in the same general sense.

MYSELS   I think there is a very simple analogous phenomenon in the permeation of many vapors through plastics. If you had a high gradient of vapor pressure, let's say from vacuum to saturated solution, the permeation may be very slow. If you have them both at close to half saturation, permeation is very fast simply because the plastic, the polymer dries out on the outer side in vacuo and becomes the impervious. Now can your theory account for that? There you definitely have a series of very thin membranes, each of which behaves in a well-known way, and yet the total effect is that you get this kind of a negative conductivity as you increase the driving potential.

SCHLÖGL   This is a very interesting remark, but I don't dare to comment on it.

DODGE   If we are going to continue discussion of this negative resistance characteristic which Dr. Cole brought up—I think it is rather an important specification that on any operating point of that negative conductane limb the high frequency conductance is a straight line that goes straight back to the origin. This should be an important specification for any theory. Is that not right, Dr. Cole?

COLE   Yes.

DODGE   I wonder if anyone, including our chairman, might wish to comment on how cross-coupling between the various ionic species in the theory of irreversible thermodynamics might relate to processes that go on in nerve in a quantitative manner.

SNELL   I'll refer that to Professors Schlögl, Kedem, Onsager, and Hill.

SCHLÖGL   One point I think one should make is that as far as I know today there does not exist a rigorous general theory of the time dependence in transport processes across membranes. This does not mean that there are not a number of very elegant phenomenological approaches but the general theory of the time dependence processes as far as I can tell from a thermodynamic point of view has as yet not been developed.

SNELL   I think Dr. Dodge's inquiry is more directed at the nature of the cross-coupling between ionic species such as sodium and potassium that might be directed more toward the functioning of nerves. Is that correct?

DODGE   Yes. On the space scale of the thickness of the membrane, we might expect that the characteristic time constants for relaxation of ion profiles to be in the order of fractions of a microsecond. We might therefore assume that the slow milisecond long transients of sodium and potassium currents are a sequence of steady states then one can ask the theoretician to consider the many types of experimental evidence in the nerve membrane where the effect of sodium current flow on potassium and vice-versa are not seen. Does this say anything strongly about the cross-coupling coefficients in the nerve membrane? For an example, Frankenhauser on the nodal membrane has performed a separation of current level in which one record was taken in the normal solution and the other in a solution with high external potassium to reduce the driving force on the potassium current. This separation of the sodium and potassium components was just the same as that obtained by the various other indirect manners. This is strong evidence that high concentrations of potassium on the outside does not affect the sodium current—strong evidence also that coupling is very weak. Now, does this say anything about the structure of the membrane?

MULLINS   There isn't any such thing of course as negative resistance in squid axons. You measure something that looks this way because you change the permeability from one ionic battery to another. Of course, in making this change, you get this kind of result. So I would rather see people direct their attention not to the question of why negative resistant regions are apparent but to why this system changes from one permeability state to another.

DODGE   I would like to point out that the separation currents in the Hodgkin–Huxley analysis carries the implication that the structure of the membrane as measured by the permeability coefficient (the conductance depends on the concentration of current carriers) is not dependent on the net current that flows through the membrane. The time-course of the permeability relations for the early transient current which is predominantly sodium current is the same whether current is flowing into the membrane or flowing out of the membrane. The Hodgkin–Huxley analysis is an accurate statement on the functional dependences for the membrane current, and theoreticians should look at this data very carefully with respect to the specifications it puts on any membrane theory. I would like Dr. Cole to discuss more explicitly the time dependence of the negative resistance characteristic in the case where potassium is the only ionic species carrying current. Isn't it the case that if one applies a voltage-clamp pulse from the resting state at zero mv to a value of -40 mv, the current would initially start off at a quite high value and then decline to the steady-state value with a time constant of about a millisecond, and this time constant is strongly dependent on temperature. I believe this slow relaxation must be related to a change in the structure of the membrane itself rather than an ionic relaxation of the simple theoretical kinds we have been discussing.

COLE   The negative resistance is most easily represented formally by a state variable as shown by FitzHugh. But I am not yet prepared to commit myself to any one of the possible mechanisms which have been proposed to explain such a state variable such as change of structure, change of mobility, or change of cross coefficients.

TASAKI   Dr. Cole said that in the absence of calcium ion in the external medium, the voltage–current relation for an axon in a K-rich medium shows only a slight indication of a negative conductance. I think that some Ca-ions are always present in the connective tissue under these conditions, unless a chelating agent (such as EDTA) is used to remove them. When a steady inward current is applied to such a depolarized axon, Ca-ions tend to accumulate in the axon membrane, because the transport number of Ca is far smaller than those of univalent cations. This process of Ca-accumulation in the membrane could produce transitions of the membrane macromolecules from one conformation to the other. I think this process could explain how a 'negative conductance' is created under these conditions.

COLE The idea of a change of transport number at a membrane boundary has appeared a number of times and is an attractive possibility. A crude calculation showed that the calcium pile-up, as described by Dr. Tasaki, should be proportional to the current density. This was not a satisfactory explanation because the negative resistance would require an increasing calcium pile-up as the current decreased. Further, the estimated amount of the pile-up for a nominal current density appeared to be too small by several orders of magnitude. Obviously a better calculation is needed and I hope that Dr. Tasaki will provide it.

DODGE I want to backtrack a little bit. When I use the term structure change that was my ignorance about what's going on in the membrane. It might just as well be a change in the mobility of the ion in the membrane phase. What I wished to do by the discussion was to raise a question whether or not the specifications are sufficient to eliminate over-simplified mechanisms involving relaxation of ion profiles or changes in cross coupling. The point that Dr. Cole made is that this phenomenon is more or less accurately described by the potassium terms in the Hodgkin–Huxley equations whose constants and parameters were deduced under very different situations of current flow. The fact that they obtain in this very different set of circumstances strongly indicates that the coupling is not to the current term. Also, I think there will be direct experiments to show that there is no other current carrier except potassium under these circumstances. Therefore I'd like to ask if there would be some way to save the idea that this characteristic might come around from a change in the coupling between different flows. If so, what other flow might this coupling be coupled to?

COLE Dr. Kedem has suggested that calcium may actually enter the membrane and combine to block potassium conductance and she hopes to build a model to demonstrate this behavior. In another direction, Professor Teorell has shown very interesting membranes in which a coupling between hydraulic and electro-osmotic water flow produces a region of negative resistance between solutions of different conductances. As Dr. Dodge has hinted, it seems highly significant that in the HH formulation of the squid axon membrane behavior there is but a single independent variable, the membrane potential difference. Thus the steady state and transient characteristics of the ion conductances and current flows are completely determined by the membrane potential and by the membrane potential alone. Although I am not sure that this remarkable achievement requires the conduction mechanisms themselves to be only field dependent, it is an important, necessary condition for any theoretical model to approximate.

PAPER 5

Ion distribution equilibria in bulk phases and the ion transport properties of bilayer membranes produced by neutral macrocyclic antibiotics†

GABOR SZABO, GEORGE EISENMAN, and
SERGIO M. CIANI‡

INTRODUCTION

WE HAVE proposed elsewhere (1) that the essential barrier to ion movement across a bilayer lipid membrane is the liquid-like hydrocarbon interior (2) through which ions must move, essentially unaffected by the polar groups at the membrane-solution interfaces. It was shown how the increased ionic permeability of such membranes produced by neutral macrocyclic antibiotics such as Monactin was expected as a result of the ability of these molecules to form complexes with cations and thereby to solubilize them in the membrane. Such molecules were therefore postulated to act as molecular carriers for cations, a view previously proposed by Simon (2a) by Lardy (2b) and by Pressman (2c).

Under the assumption that the membrane could be viewed as a thin liquid-like hydrocarbon phase through which the complexed-cations moved, essentially unaffected by the lipid polar head groups, it was possible to deduce the expected behavior for a bilayer membrane from the Nernst–Planck flux equations (together with the Poisson and Boltzmann equations, where

† This work was supported by National Science Foundation Grant GB 6685 and by USPHS Grant GM 14404-02/03. Ciani is the recipient of a Fulbright Traveling Scholarship and Szabo of Fellowships from the Canadian National Research Council and the University of Chicago.

‡ Permanent address: Institute of Physics, University of Genoa, Genoa, Italy.

appropriate) without making arbitrary assumptions as to electroneutrality or as to profiles of concentration or potential within the membrane. In particular, expressions were developed for the effects of such molecules on the extraction of salt into bulk organic solvents and for the membrane potential and membrane conductance of bilayers in those situations where the concentration of antibiotic was the same on both sides of the membrane.

Measurements of potential and resistance of phospholipid bilayer membranes, as well as studies of salt-distribution equilibria between aqueous solutions and an organic solvent (n-hexane) chosen to represent the interior of the membrane were carried out for Monactin and the cations, Li, Na, K, Rb, and Cs. The theoretically expected identity between the conductance ratio and permeability ratio of bilayer phospholipid membranes, and also between these ratios and the ratio of salt extraction equilibrium constants in n-hexane, was found experimentally; and we therefore concluded that it was likely that the neutral Monactin molecule acts as a molecular carrier for cations across phospholipid bilayer membranes.

In more recent studies (3–9) this work has been considerably extended, with the results summarized here.

THE MODEL AND ITS THEORETICALLY EXPECTED PROPERTIES

This simplest model for the effects of neutral macrocyclic molecules (such as the Macrotetralides of the Nonactin type illustrated in Figure 1) on lipid bilayer membranes is a three phase system in which the membrane is represented by a thin (60 Å) liquid hydrocarbon phase interposed between two aqueous solutions. The macrocyclic molecules are assumed to be preferentially partitioned in the organic phase and to form stoichiometric complexes with cations, thereby solubilizing them in the membrane.† Such a membrane is schematized in Figure 2, omitting for simplicity of presentation, the effects of charged polar head groups of the lipid. The surface charge is expected (7) to be without effect for the antibiotic and ionic concentration ranges characterized here, being important only in the limits of low antibiotic and ionic concentrations, as will be discussed in relation to Figure 19.

The organic phase need not be thin (although this is the only membrane situation we will examine here), nor need it be studied as a membrane. Indeed, for purposes of comparison with the properties of bilayers, the membrane can be considered as expanded into a bulk liquid phase. In this case, the ionic effects of the above neutral macrocyclic molecules on simple distribution

† The partition coefficient of monactin has been measured to be about 5000 to 1 in favor of the membrane phase (8).

Ion distribution equilibria in bulk phases

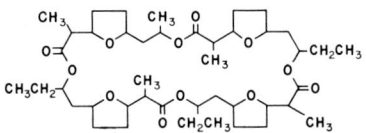

NONACTIN

MONACTIN

DINACTIN

TRINACTIN

FIGURE 1 A. Chemical formulas of the macrolide actin antibiotics. B. Space filling model for nonactin. A CPK model of the nonactin molecule is shown in the configuration which we believe to be likely to exist in a low dielectric constant solvent or in the interior of the membrane, where the molecule is folded around the cation, sequestering it in relation to the four carbonyl oxygens within its interior (in the configuration similar to that of the Nonactin-K^+ complex in crystals (10)). The size of the cavity is seen to be appropriate to accomodate the potassium ion of ionic radius 1.33 Å illustrated below. It can be seen that the overall configuration and external size of the molecule is not expected to vary greatly for different alkali cations within the interior. Note also that the addition of methyl groups to form the more highly methylated members of the series, would not alter greatly the external size of the molecule.

equilibria with a single aqueous phase can be deduced and confronted with appropriate membrane measurements.

If the organic phase is of low dielectric constant, then the concentration of dissociated ions in it will be low (witness the high electric resistance of artificial bilayer membranes in the absence of added materials (11) and the low salt uptake characteristic of pure hydrocarbon solvents (12)). However, in the presence of lipid soluble neutral macrocyclic molecules, both the electric conductance of thin membranes and the salt uptake by bulk phases are increased markedly, as analyzed theoretically and experimentally below.

In this model, three independent types of species are assumed to exist in the aqueous as well as in the organic phases: the free cations I^+, the free anions X^- and the free neutral macrocyclic molecule S. For $\underline{n}$ species of monovalent cations, $\underline{n}$ reactions of the type

$$I^+ + S \underset{K_{is}^+}{\rightleftarrows} IS^+ \tag{1}$$

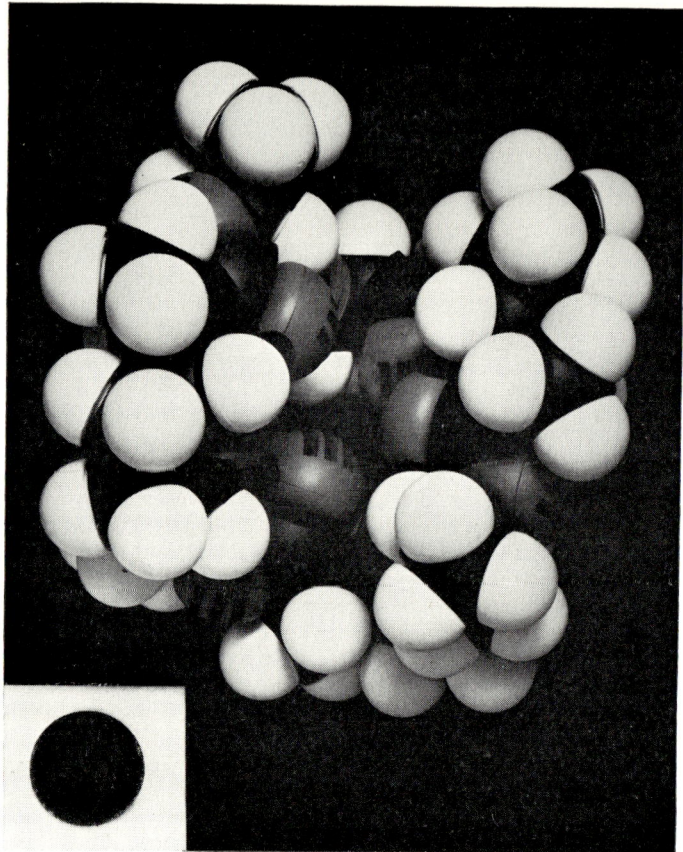

FIGURE 1 B.

are assumed to occur between the cations I^+ and the macrocyclic compound S, leading to the formation of the positively charged lipid-soluble complexes IS^+ as indicated by the equilibrium between the upper three species in Figure 2. Also indicated in Figure 2 is the possibility of further association between the IS^+ species and the anions X^- according to the reaction

$$IS^+ + X^- \underset{K_{isx}}{\rightleftarrows} ISX \qquad (2)$$

The effects of this reaction on bilayer properties are expected to be negligible since the concentration of ISX species should be low for the usual lipid incompatible anions such as chloride, and moreover the ISX species do not directly effect the electrical properties, being neutral. Furthermore, the bulk phase

Ion distribution equilibria in bulk phases

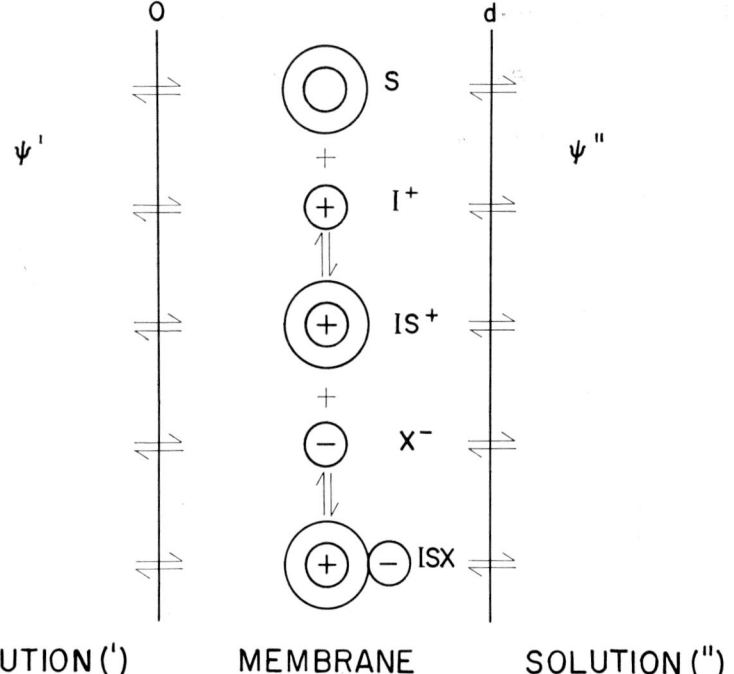

FIGURE 2 Diagram of the membrane. A diagram of the membrane is indicated in which it is seen to be interposed between two aqueous solutions whose electric potentials are designated by ψ' and ψ''. The species I^+, S, IS^+, X^- and ISX refer to the free ion, the neutral molecular carrier, the complexed-cation, the free anion, and the neutralized complex, respectively. Although these species are illustrated within the membrane phase, the arrows at the membrane-solution interfaces indicate that equilibria exist between these species and their counterparts in the aqueous solutions.

extraction experiments carried out using lipid soluble anions such as picrate, have been so designed that reactions (2) occur only to a negligible extent. (The concentration of the ion pairs IX is also so insignificant under the usual experimental conditions that the presence of this species will be entirely neglected in the following considerations.)

When an aqueous solution is in contact either with a bulk organic solvent or with the membrane, viewed simply as a thin membrane-like portion of the solvent, the equilibrium distribution for each species between the two phases can be formally described by heterogeneous reactions of the type

$$S \underset{k_s}{\overset{k_s}{\rightleftarrows}} S^*, \qquad k_s = a_s^*/a_s \qquad (3)$$

for the neutral species S and ISX, and of the type

$$l \underset{k_l}{\rightleftarrows} l^*, \qquad k_l = \frac{a_l^* \exp[(z_l F/RT)\psi^*]}{a_l \exp[(z_l F/RT)\psi]} \tag{4}$$

for the charged species ($l = I^+, IS^+, X^-$), where the partition coefficient k_l can be also viewed as the equilibrium constant of the reaction by which an individual component crosses the boundary between the two phases. In Eq. (4) and throughout the rest of this paper a is the activity in moles/liter, ψ is the electrostatic potential, z_l is the valence of l, and the asterisk (*) will be used to designate any quantity characteristic of the organic phase.

Although the set of Eqs. (1), (3) and (4) is sufficient for a thorough description of the distribution *equilibria* between the aqueous solution and the solvent phase, whether in bulk or as a membrane, the interrelationships between membrane and bulk systems can be seen more clearly by analyzing suitable combinations of these reactions. Namely, subtracting reaction 3 from Eq. (1) and adding Eq. (4) for $l = IS^+$, we get

$$I^+ + S^* \underset{k_s}{\overset{k_{is} K_{is}^+}{\rightleftarrows}} IS^{+*} \tag{5}$$

which is a heterogeneous reaction expressing the formation of the complexed-cation species IS^{+*} within the organic phase in terms of the aqueous concentration of I^+ and the concentration of S^* in the organic phase. By simply adding reaction (4) for $l = X^-$ to Eq. (5) we get

$$I^+ + S^* + X^- \underset{k_s}{\overset{k_{is} K_{is}^+ k_x}{\rightleftarrows}} IS^{+*} + X^{-*} \tag{6}$$

which expresses the extraction of the salt $I^+ X^-$ from the aqueous phase into the organic phase.

When the thickness of the membrane is of the order of the Debye length and the partition coefficients k_{is} of the complexed cations are much higher than those of all the other ions, electroneutrality does not hold in the membrane; and the IS^+ complexes are present in excess as a 'space charge' (3). Since they constitute the predominant charged species in the membrane, its electrical properties can be described solely in terms of the parameters appearing in the equilibrium constant of reaction 5:

$$K_i' = \frac{k_{is} K_{is}^+}{k_s} \tag{7}$$

This constant can normally not be measured directly since its determination requires a knowledge of the electrical potential difference between the two phases.

On the other hand, when studying partition equilibria in bulk phases where the electroneutrality condition governs all chemically detectable concentrations, the equilibrium constant of reaction 6

$$K_i = \frac{k_{is} K_{is}^+ k_x}{k_s} \tag{8}$$

is directly measurable. This constant depends on k_x, the partition coefficient of the anion X^-; but, of course, k_x drops out when the ratios of K_i are compared for two cation species I^+ and J^+ (and the common anion X^-), so that a simple relationship is expected between the constants of reactions 5 and 6. Comparing the ratio of Eq. (7) for two different cations I^+ and J^+ with the corresponding ratio of the equilibrium constants of reaction 6, it can be seen that

$$\frac{K'_j}{K'_i} = \frac{K_j}{K_i} = \frac{k_{js} K_{js}^+}{k_{is} K_{is}^+} \tag{9}$$

showing that the ratio of the equilibrium constants of reaction 6 for two different cations I^+ and J^+ is identical to the corresponding ratio for reactions 5.

Relation 9 is of the utmost importance since it provides a way to test the proposed liquid solvent model for the membrane through the expected identity between the value of K_j/K_i measured in bulk phases and the value of $k_{js} K_{js}^+/k_{is} K_{is}^+$, which will be shown below to be measurable directly from the electrical properties of bilayer membranes.

Evaluation of the equilibrium constant K_i from equilibrium salt extraction into a bulk organic phase

Assuming ideal behavior of the species S, IS^+ and X^- in the organic phase, the equilibrium constant K_i of reaction 6 is given in terms of the measurable salt uptakes for a given set of solution conditions by

$$K_i = \frac{C_{is}^* C_x^*}{a_i a_x C_s^*} \tag{10}$$

where C_{is}^*, C_x^*, and C_s^* are the concentrations of IS^+, X^-, and S in the organic phase and a_i and a_x are the activities of I^+ and X^- in the aqueous phase. Since, in the present experiments, the concentration C_i^* of the uncomplexed cations is expected to be negligible with respect to C_{is}^*, the electroneutrality condition, namely $C_{is}^* = C_x^*$, allows Eq. (10) to be written in either of the two following forms:

$$K_i = \frac{C_{is}^{*2}}{a_i a_x C_s^*} \quad \text{or} \quad K_i = \frac{C_x^{*2}}{a_i a_x C_s^*} \tag{11}$$

When the experimental conditions are so arranged (e.g. at sufficiently low ionic activity a_i) that the association reaction 2 occurs only to a negligible extent in the organic phase, the evaluation of the total concentration of either one of the species IS^{+*} or X^{-*} will give the value of K_i once the activities a_i, a_x and the concentration C_s^* are known. In the experiments described here, a lipid compatible anion having convenient chromophore properties (e.g. picrate or 2-4 dinitrophenolate) has been used and its concentration in the organic phase measured spectrophotometrically in a procedure first used by Pedersen (13).

To express the uptake of the anion X^- in terms of controllable parameters, such as the activity a_i of the cation in the aqueous solution and the initial total concentration of S, C_s^{In*}, Eqs. (10) or (11) have been rearranged in the following way: designating by V and V^* the volumes of the aqueous and organic phases, respectively, and by α_i the ratio of the total anion concentration extracted into the organic phase C_x^{Tot*} to the concentration of the anion initially present in the aqueous phase C_x^{In}, Eq. (11) becomes:

$$\alpha_i = \frac{V^*}{V}\sqrt{\left(\frac{k_i K_i}{k_s(V^*/V)+1+K_{is}^+ a_i}\right)}\sqrt{\left[\left(\frac{V}{V^*}-\alpha_i\right)\left(\frac{C_s^{In*}}{C_x^{In}}-\alpha_i\right)a_i\right]} \quad (12)$$

or, when the partition coefficient k_s is sufficiently large so that $k_s(V^*/V) \gg 1 + K_{is}^+ a_i$,

$$K_i = \frac{V}{V^*} \frac{\alpha_i^2}{[(V/V^*)-\alpha_i][(C_s^{In*}/C_x^{In})-\alpha_i]\,a_i} \quad (13)$$

Membrane Potential. When the concentrations of the neutral macrocyclic molecules are identical on both sides of the membrane and reaction 1 occurs only to a negligible extent in the aqueous solutions, the membrane potential (i.e., the potential difference between solutions ('') and (')) at zero current in mixtures of two cations I^+ and J^+ has been shown (1, 3) to be given in terms of their activities in solutions (') and ('')

$$V_0 = -\frac{RT}{F} \ln \frac{a_i'' + [u_{js}^* k_{js} K_{js}^+/u_{is}^* k_{is} K_{is}^+]a_j''}{a_i' + [u_{js}^* k_{js} K_{js}^+/u_{is}^* k_{is} K_{is}^+]a_j'} \quad (14)$$

where the bracketed combination of parameters, which controls the relative effects of I^+ and J^+ on the membrane potential, differs from the right hand side of Eq. (9) only in that the ratio of the mobilities of the complexed cations in the membrane u_{js}^*/u_{is}^* also appears. Eq. (14) will be recognized as the classical Goldman–Hodgkin–Katz equation (14, 15) and is also characteristic of a

Ion distribution equilibria in bulk phases

variety of Ion Exchange membranes (16). It will be convenient to define the particular combination of parameters in Eq. (14) as the Permeability ratio, P_j/P_i:

$$\left[\frac{u_{js}^* k_{js} K_{js}^+}{u_{is}^* k_{is} K_{is}^+}\right] = \frac{P_j}{P_i} \tag{15}$$

Membrane conductance. Under the same experimental conditions as above, the zero-current electric conductance of a membrane of thickness d interposed between solutions containing identical activities a_i of a single cation I^+ is (3)

$$G_i = \frac{F^2}{d} u_{is}^* k_{is} K_{is}^+ C_s^{\text{Tot}} a_i \tag{16}$$

Taking now the ratio of the membrane conductances G_i and G_j for two different cations I^+ and J^+ at the same concentration level and same carrier composition, and comparing it with Eq. (9) and the permeability ratio defined in Eq. (15), we find the following important relationships between these three independent measurements:

$$\frac{P_j}{P_i} = \frac{G_j}{G_i} = \frac{u_{js}^* K_j}{u_{is}^* K_i} \tag{17}$$

Assuming, however, that the external size and shape of the ion-carrier complex depends very little on the particular species of the sequestered cation (as expected for the Macrolide Actins from examining Figure 1B), the ratio u_{js}^*/u_{is}^* should be very nearly unity; so that Eq. (17) is expected for such molecules to become

$$\frac{P_j}{P_i} = \frac{G_j}{G_i} \cong \frac{K_j}{K_i} \tag{18}$$

Eq. (18) predicts that the permeability ratios and conductance ratios of the membrane can be directly inferred from the ratio of the constants K_j and K_i, deducible from salt uptake measurements in bulk phases. The validity of relation 18 was previously found to hold rather well for Monactin when n-hexane was chosen as the model solvent (1). The experimental studies summarized here will not only verify the applicability of relation 18 for more precise measurements with Monactin in solvents such as dichloromethane and hexane-dichloromethane mixtures, but will extend its validity to Nonactin, Dinactin and Trinactin as well.

THE EXPERIMENTALLY OBSERVED PROPERTIES

1. Equilibrium studies in lipid-free bulk phases

In the theoretical section it was deduced that Eq. (18) should relate such salient electrical properties of a bilayer membrane as the permeability ratio and conductance ratio to the equilibrium constant for salt extraction into an appropriate model solvent. This section describes measurements of the values of this equilibrium constant for Li, Na, K, Rb, Cs, and NH_4 characteristic of Nonactin, Monactin, Dinactin, and Trinactin in dichloromethane and in hexane-dichloromethane mixtures. The values of these constants will be compared in the next section with the appropriate quantities measured from the electrical effects of those molecules on phospholipid bilayer membranes.

The ability of neutral macrocyclic molecules to extract the picrate salts of Li, Na, K, Rb, Cs, and NH_4 into dichloromethane (Dielectric constant $D = 9.08$ (17)) is illustrated in Figures 3 and 4, where data for Trinactin

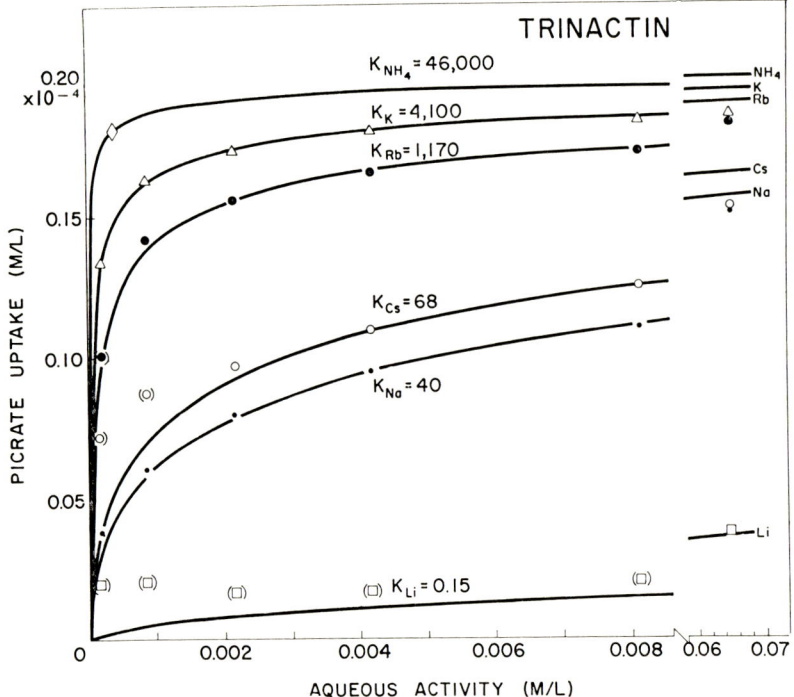

FIGURE 3 Equilibrium uptake of the picrate salts of Li, Na, K, Rb, Cs, and NH_4 into a dichloromethane phase containing 2×10^{-5} Molar Trinactin.

Ion distribution equilibria in bulk phases

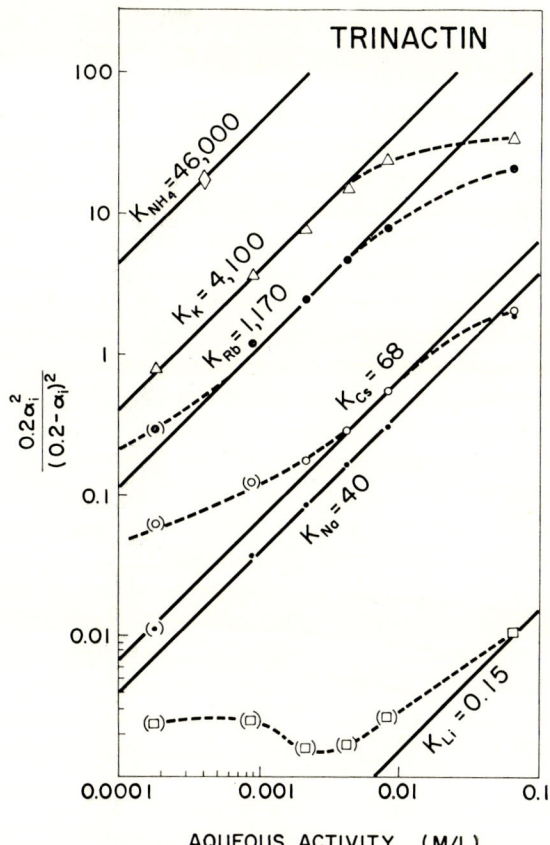

FIGURE 4 Equilibrium uptake of the picrate salts of Li, Na, K, Rb, Cs, and NH_4 into a dichloromethane phase containing 2×10^{-5} Molar Trinactin. The data of Figure 3 have been presented in logarithmic form according to the expectations of Eq. (19).

(but typical of Nonactin, Monactin and Dinactin as well (4)) are presented. The experiment was carried out by equilibrating 10 ml aliquots of 2×10^{-5} Molar Trinactin in CH_2Cl_2 with 2 ml volumes of aqueous solutions containing a constant 10^{-4} Molar picric acid and the various activities of the alkali hydroxides indicated on the abscissa. The picrate extracted by the organic phase was measured from the optical absorbance at 378 mμ using standard spectrophotometric procedures (for full experimental details see reference (4)). The total picrate concentration of the CH_2Cl_2 phase was calculated from this using a precisely measured value for the extinction coefficient of Picrate in CH_2Cl_2 of 18,300 (1000 cm^2/mole).

In the absence of Trinactin, the picrate uptake by the CH_2Cl_2 phase is undetectable on the scale of Figure 3 (less than 10^{-8} M/L even for the highest salt concentrations of Figure 3); but in the presence of 2×10^{-5} Molar Trinactin the uptakes plotted as points in Figure 3 are obtained. It is clear that at sufficiently low cation concentrations there is negligible uptake, whereas at high cation concentrations the uptake becomes substantial, approaching the maximum limit of 2.0×10^{-5} M/L expected for 1:1 stoichiometry. The relative amount of picrate extracted also can be seen to depend markedly on the nature of the cation, being greatest for NH_4^+ and least for Li^+.

The extent to which the observed uptakes are consistent with those expected from the simple equilibrium chemistry postulated for this system can be seen by comparing these experimental points with the solid curves drawn according to Eq. (10) for the indicated values of K_i (obtained from the Y intercepts of Figure 4, as discussed below).

First examine the data for K^+, Rb^+, and Cs^+. The solutions for these measurements were made using pyrex-redistilled distilled water; and in the case of the two lowest Cs^+ concentrations (indicated by parentheses) uptakes of Picrate due to the NH_4^+ present in the water led to deviations from the theoretically expected uptakes.† For K^+ and Rb^+ the effects of NH_4^+ in these solutions was negligible for all concentrations; and the excellent correspondence between experimental observations and theoretical expectations is seen in the agreement between the experimental points and the theoretical curves for K^+ and Rb^+ over the aqueous concentration range 0.0002 Molar to 0.01 Molar.

Because of the greater importance of NH_4^+ as a contaminant in the case of the less strongly bound cations Na^+ and Li^+, solutions of these ions were prepared using Pyrex-redistilled deionized distilled water (referred to hereafter as 'PRDD·H_2O'). The lower NH_4^+ content of this water allowed us to characterize the Na^+ extraction with negligible error over almost the entire concentration range; and the agreement between experimental points and theoretical curves is seen to be excellent. For Li^+, only the highest (0.1 M) concentration point was relatively unaffected by NH_4^+ (note that the preference for NH_4^+ over Li^+ is over 100,000 to 1).

The data of Figure 3 indicate that the Trinactin complexed-cation and the picrate anion in CH_2Cl_2 are completely dissociated under these experimental

---

† Our normal Pyrex-redistilled distilled H_2O was found to contain about 10^{-6} Molar total NH_3. Specially purified distilled H_2O, which was deionized with a mixed-bed ion exchange column to a conductance of less than 0.1 ppM (as NaCl) and then Pyrex-redistilled, had a total NH_3 concentration of less than 3×10^{-7} Molar.

conditions. This has been found to be true also for the other Macrolide Actins (4), and will be further examined in Figure 4.

Although Figure 3 describes the extraction equilibrium in an immediately comprehensible form, it is more convenient to plot the expectations of Eq. (13), which for the experimental conditions of Figure 3, where

$$V/V^* = C_s^{\text{In}*}/C_x^{\text{In}} = 0.2$$

can be written in a simple logarithmic form as:

$$\log \frac{0.2\alpha_i^2}{(0.2-\alpha_i)^2} = \log a_i + \log K_i \qquad (19)$$

where α_i is the ratio of picrate concentration taken up by the organic phase to the initial concentration of picrate in the aqueous phase, and $(0.2-\alpha_i)$ is the ratio of the picrate concentration remaining in the aqueous phase to that initially present.

Equation (19) indicates that a graph of the logarithm of $0.2\alpha_i^2/(0.2-\alpha_i)^2$ as a function of the logarithm of the activity of the hydroxides of the various alkali cations in the aqueous solution should yield a straight line with a slope of unity and a Y-intercept equal to K_i. Such plots of the data of Figure 3 are presented in Figure 4 and were the basis for extracting the values of K_i used to calculate the theoretical curves for Figure 3.

The extent to which the observed data are in agreement with the theoretical expectations can be seen by comparing the data points with the diagonal straight lines expected theoretically from Eq. (19). For K^+, Rb^+, and Na^+, for which effects of the traces of NH_4^+ in the distilled water are negligible, the agreement between the points and the lines is excellent, except for the highest concentration (0.1 M). This indicates not only that the complexed-cation and picrate anion are completely dissociated in CH_2Cl_2 under the present conditions, but also that the elementary chemistry postulated for reaction 6 underlies the behavior of the system over a wide concentration range. Even for Cs^+, the data points in the middle of the concentration range are sufficiently unaffected by the NH_4^+ effects to be usable to extract an accurate value for K_{Cs} with confidence.

The logarithmic presentation of Figure 4 exaggerates the deviations at the highest and lowest uptakes. Those at the highest concentrations are nevertheless real and are probably due to complex formation between Trinactin and the cations in the aqueous phase (4); but since they do not affect our measurement of K_i, they will not be discussed further here.

The remarkably simple behavior of the salt extraction equilibria suggests that ideal behavior (activity = concentration) for most species can be safely assumed in both the aqueous and organic phases. For the organic phase this is

TABLE I Salt extraction equilibria with monactin

Chromo-phore	Solvent	K_i					
		Li	Na	K	Rb	Cs	NH_4
Picrate	CH_2Cl_2	0.070	7.6	780.0	290.0	23.0	16000.0
2-4 DNP	CH_2Cl_2	<0.003	0.092	13.0	3.6	0.20	240.0
Picrate	64% Hexane, (V/V) 36% CH_2Cl_2	0.00000036	0.000054	0.0135	0.003	0.00013	0.59
		K_i/K_K					
Picrate	CH_2Cl_2	0.00009	0.0097	1.0	0.37	0.029	20.5
2-4 DNP	CH_2Cl_2	<0.00023	0.0071	1.0	0.28	0.015	18.5
Picrate	64% Hexane, (V/V) 36% CH_2Cl_2	0.000027	0.0040	1.0	0.22	0.010	43.8

supported by the agreement of the present values of K_i with those measured for ten times higher Monactin concentrations in the CH_2Cl_2 phase (4).

For the aqueous phase, such ideal behavior is in accord with the fact that varying the ionic strength by addition of a poorly extracted salt (e.g. LiCl), as well as the addition of the chlorides of Ca^{2+} Mg^{2+} and Th^{4+}, is without important effects on the values of K_i (4). This suggests that these ions produce no alterations in the configuration of the Macrolide Actin molecules in the aqueous phase. This detail is of relevance because of the marked effects of ionic strength and of these species which will be demonstrated to occur on bilayer properties in Figures 15 and 18.

Independence of the anion species used as a chromophore

By carrying out studies such as those of Figures 3 and 4, using 2-4 dinitrophenolate anion in place of picrate, it has been possible to establish that the relative selectivity of one cation compared to another (i.e., the ratio of K_i to K_j) is independent of the anion used to accompany the cation into the solvent. This is illustrated by the data for Monactin presented in Figure 5, which compares the equilibria for picrates of Li, Na, K, Rb, Cs, and NH_4 at the left, with those for the dinitrophenolates at the right. Although the values of K_i are considerably less in the case of DNP than in the case of picrate, the relative displacements between the individual isotherms are essentially the same in the two cases, indicating that the ratios K_i/K_j are independent of the anion, as can be seen in Table I.

Independence of solvent

Figure 6 characterizes for Monactin the equilibrium extraction of the alkali picrates in the mixed solvent, 64 per cent *n*-hexane and 36 per cent CH_2Cl_2 (V/V), for direct comparison with the equilibria in pure CH_2Cl_2 previously illustrated in the left hand portion of Figure 5. The experiments of Figure 6 were carried out under conditions identical to those of Figure 5 except that the volume of the organic phase was 14 ml instead of the previously used 10 ml and the concentration of Monactin was 1.43×10^{-5} instead of 2×10^{-5} Molar. The decreased dielectric constant of the mixed solvent can be seen to lead to considerably smaller values of K_i corresponding to lower extractions; (the uptake of picrate from 0.1 N KOH by the mixed solvent was less than one hundredth than by CH_2Cl_2).

An interesting detail observable in Figure 6, is the deviation, at the highest concentrations of K^+ and Rb^+, of the data points *above* the theoretical lines. Such deviations would be expected if association between Picrate and the Monactin complexed-cations were becoming significant because of the lower dielectric constant of the mixed solvent. Nevertheless, the agreement of the

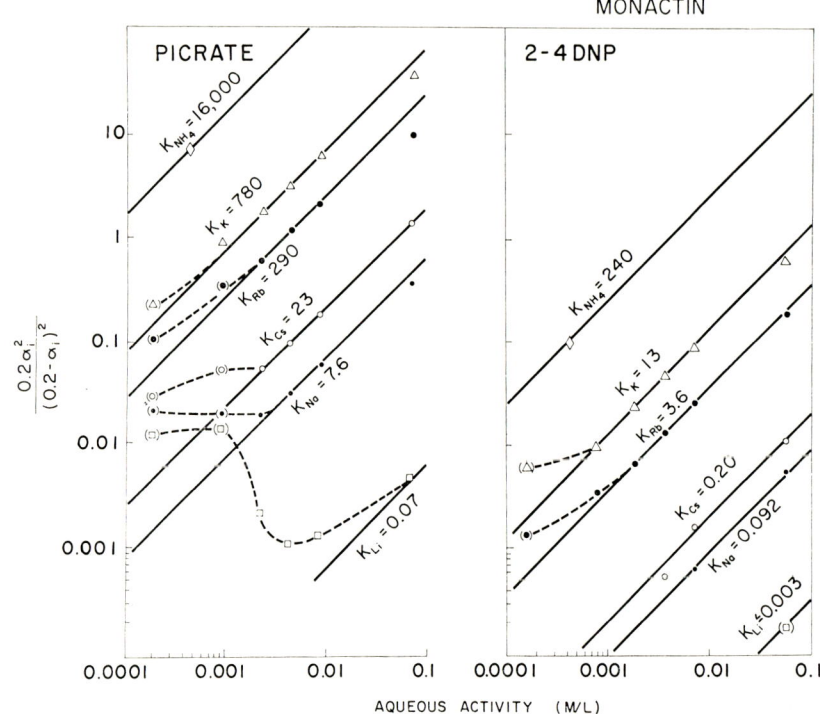

FIG. 5 The effects of varying the chromophore anion on the equilibrium salt extraction by Monactin into dichloromethane.

K^+ and Rb^+ data with the straight lines at intermediate concentrations indicates that the picrate anion is still dissociated from the complexed cation at concentrations as high as 3.86×10^{-7} Molar (corresponding to the concentration calculated for the highest K^+ uptake which still falls on the straight line). It is therefore reasonable to assume that no association with picrate is occurring for the data points measured for the considerably lower Cs, Na, and Li extractions (corresponding to 1.1×10^{-7}, 7×10^{-8}, and 6×10^{-9} Molar concentrations in the organic phase, respectively).

Despite the large difference in magnitudes of K_i between the data of Figure 6 and Figure 5, the displacements between the curves are similar, indicating that the ratios of K_i/K_K are independent of the solvent. This can be seen most clearly for the values of K_i/K_K summarized in Table I. The independence of the *ratio* K_i/K_K of the solvent is indeed a fortunate finding since it renders less critical the particular choice of the solvent to be used as an appropriate model

Ion distribution equilibria in bulk phases

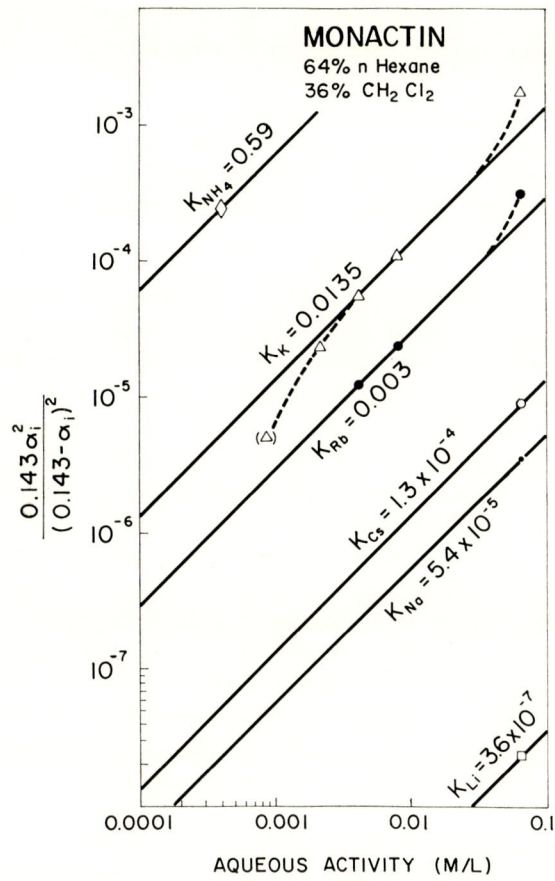

FIGURE 6 Effect of decreasing the dielectric constant of the solvent phase on the equilibrium extraction of alkali picrates by Monactin. This figure is to be compared with the left hand portion of Figure 5.

for the liquid hydrocarbon interior of the bilayer membrane. Such an independence is expected theoretically if the cation is indeed well-sequestered in the interior of the Macrolide Actin molecule as we have assumed (cf. Figure 1B); since in this case the partition coefficient ratio k_{is}/k_{js} of Eq. (9) should be close to unity; so that K_i/K_j is determined essentially by the ratio K_{is}^+/K_{js}^+ of the association constants in the aqueous phase. K_i/K_j is therefore expected to be totally independent of the solvent. An important consequence of this result is that, whereas n-hexane was previously chosen as the most appropriate model solvent for the postulated liquid hydrocarbon interior of a bilayer membrane (1), it is now clear that comparable data can be obtained more easily in CH_2Cl_2 with greater accuracy and reliability.

The above finding has its counterpart in the independence of lipid composition observed for the effects of a given antibiotic on the *ratios* of cation permeabilities and conductances of bilayer membranes (cf. Figures 9 and 17).†
It should therefore be clear why the model solvent need only be a crude approximation to the true interior of the lipid bilayer for it to suffice as a satisfactory model for the bilayer.

The effect of varying the chemical composition of the macrocyclic molecule

Having shown that the ratios of K_i/K_K measured from simple salt extraction into CH_2Cl_2 are independent of the solvent as well as of the chromophore anion, it becomes meaningful to examine how the values of K_i for a convenient solvent such as CH_2Cl_2 depend on the molecular composition of the antibiotic. From measurements similar to those of Figures 3–4 we have characterized the equilibria for Nonactin, Monactin, Dinactin, and Trinactin and obtained the values of K_i and K_i/K_K which are presented in Table II and plotted in Figure 7 as a function of the increasing number of methyl groups for this series of macrocyclic antibiotics.

A systematic increase in K_i is seen to occur with increasing number of methyl groups, as well as a systematic tendency for the ratios of K_i/K_K to change slightly. The increase in value of K_i with increasing methylation reflects an increasing strength of association, which is consistent with the observations for Nonactin and Monactin in methanol (18). Interestingly, the trend in K_i/K_K values with increasing methylation corresponds to that expected, by analogy with the factors controlling selectivity of charged ion exchange sites (19, 20), from the known electron repelling effects of the methyl group, which

---

† If the complexed-cation is deeply sequestered within the interior of the antibiotic molecule and the external structure and overall size of the molecules are approximately independent of the particular cation bound, the ratio of mobilities u^*_{js}/u^*_{is} is expected to be close to unity regardless of the composition of the membrane, even though the individual values of the mobilities are expected to be dependent on the composition of the medium in which the complexed-cations move. The partition coefficient k_{is} depends only on the work to take the charged complex from the aqueous solution into the membrane (or bulk solvent) phase. If the complexes have the same external size, then, since they bear the same charge for the present univalent cations, this work should be independent of the particular species bound; and the ratio of partition coefficients k^+_{is}/k^+_{js} should be unity. Under these circumstances it follows from equations 8, 15, and 16 that the permeability ratio, conductance ratio and salt extraction constant ratio all reduce simply to K_{js}/K_{is}, which depends solely on aqueous phase parameters and is therefore totally independent of the composition of the membrane lipid (or of the bulk solvent).

TABLE II Selectivity parameters of the macrolide actin antibiotics for bulk phases and for bilayer membranes.

		K_i/K_K	P_i/P_K	G_i/G_K	K_i	$G_i \times 2 \times 10^6$
Nonactin	Li	0.00019	0.001	0.0004	0.035	0.074
	Na	0.016	0.0075	0.0065	2.8	1.2
	K	1.0	1.0	1.0	180.0	180.0
	Rb	0.44	0.48	0.46	80.0	84.0
	Cs	0.056	0.037	0.038	10.0	6.8
	NH_4	50.0	—	—	9000.0	720.0
Monactin	Li	0.000090	0.00055	0.00025	0.070	0.22
	Na	0.0097	0.0072	0.0048	7.6	4.2
	K	1.0	1.0	1.0	780.0	880.0
	Rb	0.37	0.49	0.34	290.0	300.0
	Cs	0.029	0.023	0.014	23.0	12.5
	NH_4	20.5	—	—	16000.0	—
Dinactin	Li	0.000056	0.0006	0.00021	0.10	0.46
	Na	0.013	0.0067	0.0081	24.0	18.0
	K	1.0	1.0	1.0	1800.0	2300.0
	Rb	0.43	0.42	0.48	770.0	1100.0
	Cs	0.24	0.014	0.013	44.0	30.0
	NH_4	13.3	—	—	24000.0	—
Trinactin	Li	0.000037	0.00057	0.000033	0.15	0.1
	Na	0.0098	0.009	0.0043	40.0	13.0
	K	1.0	1.0	1.0	4100.0	3100.0
	Rb	0.29	0.32	0.38	1170.0	1170.0
	Cs	0.017	0.015	0.013	68.0	40.0
	NH_4	11.2	—	—	46000.0	—

should increase slightly the negative dipole moment of the ligand oxygens and thereby increase their effective 'anionic field strength'.†

Within the framework of the present theory, which assumes that the Macrolide Actin antibiotics produce their characteristic effects on phospholipid bilayers by acting as neutral molecular carriers of ions, the K_i and K_i/K_K values of Figure 7 and Table II permit a direct prediction for the effects of this series of antibiotics on the electrical properties of phospholipid bilayer

† A recent analysis (20a), extending the previous calculations for charged ligand groups (19) and (20) to the case of neutral dipolar ligand groups has been found to lead to the same selectivity pattern over a wide range of charge and coordination distributions; and it is worth noting that the selectivity sequences so far observed for a wide variety of neutral complexers (21–25) are the same as those of that pattern.

membranes Eqs. (9, 14–18). The following section of this paper tests this expectation and demonstrates the ability of the present theory to predict the properties of the bilayer membrane in remarkable detail from the parameters characterized above.

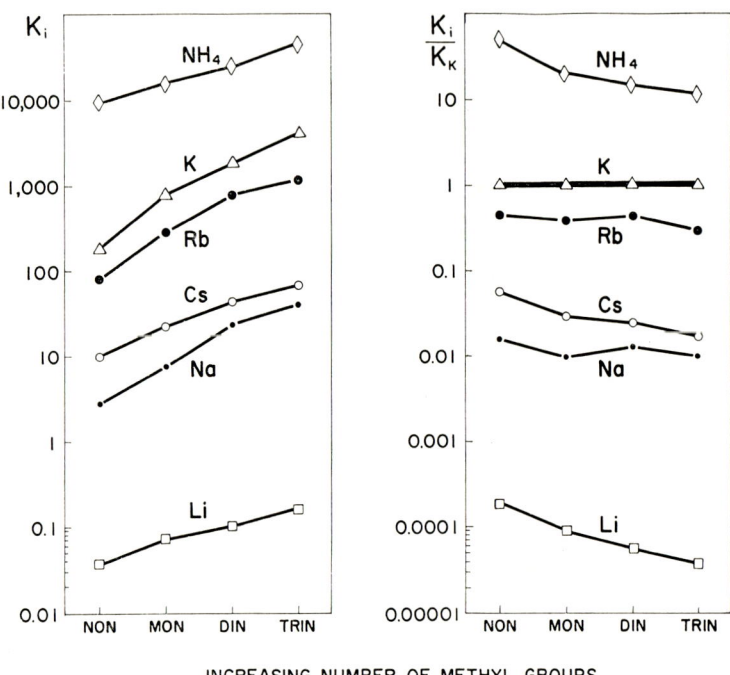

INCREASING NUMBER OF METHYL GROUPS
(INCREASING NEGATIVE "FIELD STRENGTH")

FIGURE 7 Dependence of the equilibrium constant for salt extraction on the composition of the macrolide actin antibiotics. The left hand portion of the figure plots the dependence of K_i for the indicated cations as a function of the increasing number of methyl groups as one proceeds from nonactin through trinactin. The right hand portion plots the ratios of K_i/K_K in the same manner. Note that increasing the number of methyl groups is expected to increase the negative dipole moments of the ligand oxygens through the electron repelling effect of the methyl group. Hence, the abscissa also is a measure of the effective negative electrostatic 'field strength' (19, 20) of the ligand oxygens.

Although the success of this does not unequivocally prove that the antibiotics are molecular carriers of cations, it is certainly a striking expectation of such a mechanism for their action; and it renders such a postulate extremely likely. It, moreover, offers the possibility that the problem of understanding

2. Properties of lipid bilayer membranes

Methods

The experimental methods used were essentially the same as those previously described (1) except for the following modifications. To remove traces of NH_4^+ from our distilled water, this was deionized with a Barnstead mixed-bed ion exchanger column (to a conductivity of less than 0.1 ppm. as NaCl) before being redistilled in a Corning Pyrex Still. In this way the total NH_3 was reduced to less than 3×10^{-7} Molar. This procedure was necessary because the NH_4^+ selectivity of the Macrolide Actin Antibiotics was found to be so large that the residual NH_4^+ concentration in the glass redistilled water used previously was not negligible in dilute solutions of Li^+ and Na^+ (see footnote 3 of reference (1)). Lipid from soy bean lecithin (Asolectin, Associated Concentrates, Inc.) was prepared by dissolving 30 gm of Asolectin in 24 ml of chloroform; followed by the addition of 34 ml methanol, while stirring continuously. The resulting precipitate was discarded, and the supernatent was filtered and poured into 200 ml acetone. This yielded a precipitate which was then washed with acetone, vacuum dried, and dissolved in *n*-decane (Eastman, practical grade) to give a stock solution of about 100 mg lipid per ml decane. This was diluted with *n*-decane before the start of an experiment to give a 'normal lipid' solution of 15 mg lipid per ml of *n*-decane. Since the properties of the membrane were found to be sensitive to the amount of cholesterol present (as will be seen in Figures 12, 14 and 17), control studies of further purification were carried out after several reprecipitations with acetone. No significant changes were detected in the membrane properties with respect to those found after the first precipitation. The same batch of lipid was used for all comparisons presented here among different antibiotic molecules. Full details of the experimental methods are given elsewhere (5).

Samples of Nonactin, Monactin, Dinactin, and Trinactin given to us by Hans Bickel of CIBA as well as Nonactin (SQ 15,859) given to us by Barbara Stearns of SQUIBB were used without further purification. (The nonactins from the two sources were identical in their properties as studied by solvent extraction). Normally, the concentration of antibiotic was kept the same in the solutions on both sides of the membrane except in those experiments where a gradient in the antibiotic concentration was deliberately produced in order to examine the rate limiting step.

Bilayer membrane potential

Typical data for the membrane potential behavior of phospholipid bilayer membranes exposed to 10^{-7} Molar Nonactin, Monactin, and Trinactin are presented in Figures 8 and 9, for which the points are experimentally observed and the curves are those theoretically expected according to Eq. (14) for the values of the permeability ratios given on the figures.

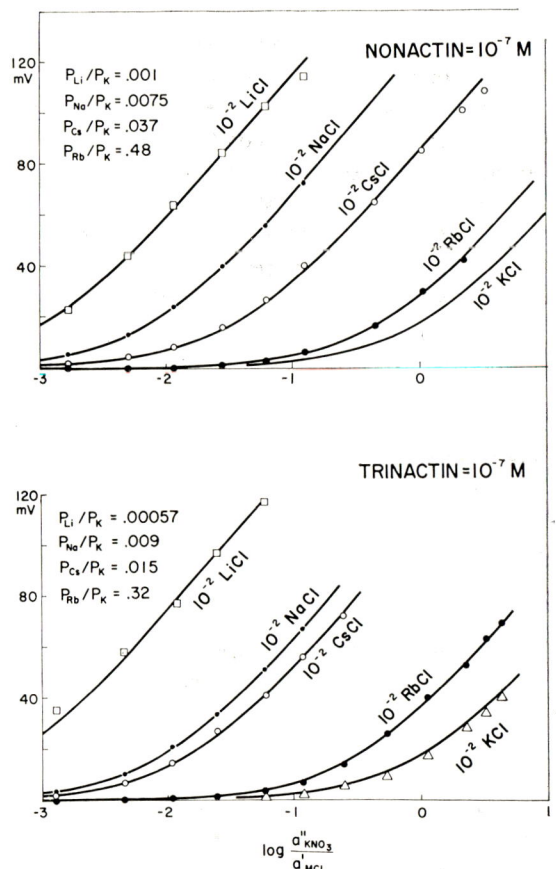

FIGURE 8 Membrane potentials of lecithin bilayers in the presence of Nonactin and Trinactin.

Membranes were formed on the teflon partition of the teflon chamber diagrammed in Figure 10 from approximately three microliters of 'normal lipid' in decane which were introduced initially, using a disposable pipette, into

Ion distribution equilibria in bulk phases

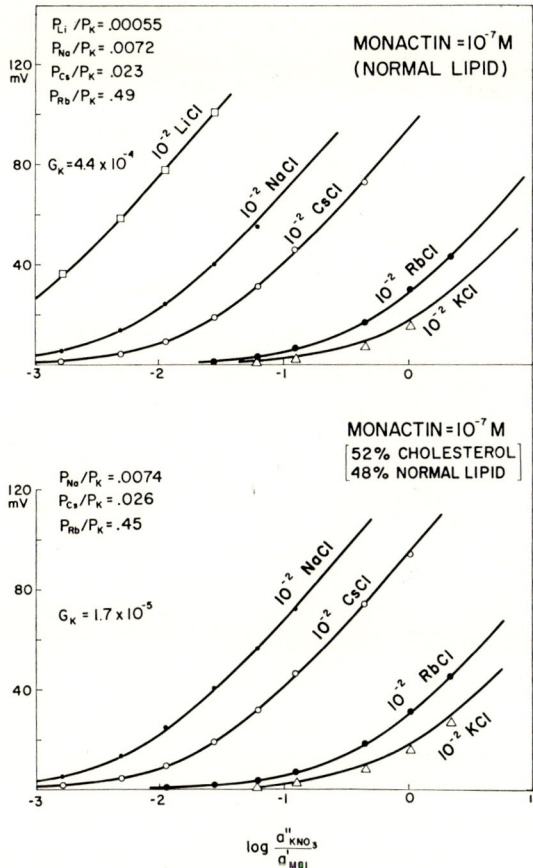

FIGURE 9 The lack of effect of cholesterol on membrane permeability ratios.

the chamber containing a total of 30 ml of 10^{-2} Molar unbuffered (pH 5–6) alkali metal chloride and 10^{-7} Molar antibiotic solution (15 ml on each side of the partition). The potential difference as well as conductance of the system with or without a membrane present in the partition could be measured at this stage. Chlorided silver electrodes were used instead of the usual salt bridges in order to have low electrode resistances for current-voltage measurements as well as to circumvent liquid junction errors and possible KCl contamination or carry-over of macrocyclic compounds from one experiment to another. (In the control measurements with no membrane in the chamber, the observed potential difference gave the assymetry potential between the pairs

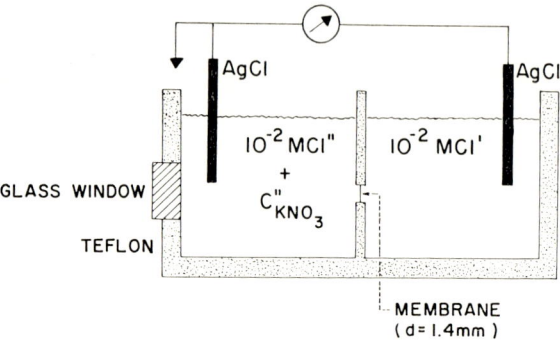

FIGURE 10 Diagram of the chamber. Bilayer membrane potentials as well as conductances were measured in the teflon chamber illustrated, using chlorided silver electrodes. Conductances were calculated from the zero current slope of the steady-state current-voltage relationship obtained by applying small potential differences (less than ± 10 mV) across the electrodes and measuring the corresponding current with a Keithley 601 Electrometer. (The adequacy of this procedure was checked with a voltage clamp measurement using four electrodes). Since the membrane conductances were usually influenced by stirring of the aqueous phases, measurements were made only after intermittent vigorous stirring was found no longer to have a noticeable effect.

of Ag AgCl electrodes, which was never more than ± 1 mV, and which was used as reference for all subsequent measurements of potential differences, while the conductance gave the combined effects of electrolyte and electrode conductances as well as that of electrode polarization, needed to be known in order to insure that their effects would be negligible in membrane conductance measurements.)

Mixtures of K^+ with the various salts were produced by adding small volumes of KNO_3 solution of appropriate concentration to the front compartment of the chamber (designated by ″). After each addition, the solutions were stirred to obtain a uniform concentration; and the potential difference was measured with the front chamber serving as the reference. The steady state value of the potential was attained rapidly (by the time mixing was complete). The values plotted in the figures were measured shortly after stirring was stopped and were time independent over several minutes. All points represent single observations while the curves are drawn according to Eq. (14), which for these experimental conditions simplifies to:

$$V_0 = 58.5 \log\left[1 + \frac{P_K}{P_M}\frac{a''_K}{a'_M}\right] \tag{20}$$

The agreement between the experimental points† and the expectations of Eq. (20) is seen to be excellent, verifying this prediction of our theory.

Figure 8 and the upper portion of Figure 9 illustrate for normal lipid the characteristic differences in selectivity among the alkali metal cations produced by Nonactin, Monactin, and Trinactin. For all three of these antibiotics the decreasing sequence of relative permeability is found to be K, Rb, Cs, Na, Li (i.e. order IV of references (19, 20)), as was the case for solvent extraction.

TABLE III Independence of permeability ratios of the cholesterol content of the lipid.

	Normal lipid	52% Cholesterol 48% Normal lipid
$\dfrac{P_{Rb}}{P_K}$	0.49	0.45
$\dfrac{P_{Cs}}{P_K}$	0.023	0.026
$\dfrac{P_{Na}}{P_K}$	0.0072	0.0074

However, quantitative differences are observable from one antibiotic to another (see for example the relative displacement between the Na^+ and Cs^+ isotherms in the case of Nonactin versus Trinactin in Figure 8), as was also previously noted to be the case in the data obtained from solvent extraction (see Figure 7). These differences will be examined in more detail below.

Figure 9 illustrates for Monactin the effects of adding cholesterol to the normal lipid. Careful comparison of the permeability ratios characteristic of Na^+, K^+, Rb^+ and Cs^+ between the upper and lower portion of Figure 9 indicates that, despite the large reduction in the Monactin-induced membrane conductance which results from adding cholesterol to the bilayer,‡ the effects of Monactin on the potential behavior are completely independent of the presence of cholesterol as can be seen from the fact that the permeability ratios summarized in Table III are virtually unchanged.

† The use of AgCl electrodes necessitated small corrections for changes in the activity coefficient of Cl^- (due to ionic strength changes) in the front (″) chamber when calculating the membrane potential V_0 from the observed potential.

‡ The membrane conductance in 10^{-2} Molar KCl was more than 20 times lower in the cholesterol-containing lipid, being 4.4×10^{-4} mho cm² in the case of normal lipid and 1.7×10^{-5} mho cm² in the case of 52 per cent cholesterol—48 per cent normal lipid.

These results allow us to conclude that although the presence of cholesterol markedly diminishes the effectiveness of the antibiotic on the lipid (as will be examined in more detail in the section on Effects of Cholesterol), it has no influence whatsoever on the relative permeabilities of one cation versus an-

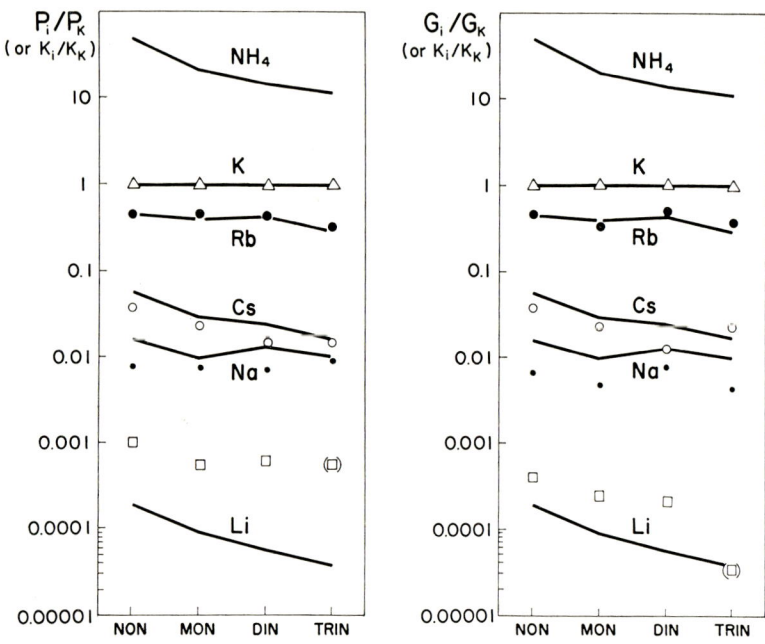

FIGURE 11 Comparison of phospholipid bilayer permeability ratios, bilayer conductance ratios and bulk phase extraction ratios for each of the five alkali metal ions as a function of increasing number of methyl side groups for the Macrolide Actins.

Left: Permeability ratios, represented by points, were obtained from potential measurements on normal phospholipid bilayer membranes following the procedure described in the text. The abscissa ranks the antibiotic molecules in order of increasing number of methyl groups. Solid lines are drawn to connect the K_i/K_K values obtained by bulk solvent extraction measurements of Figure 7.

Right: The curves for K_i/K_K are the same as those at the left, but the points represent the membrane conductance ratios for 0.01 M alkali metal chloride and 10^{-7} M antibiotic. The bracketed points for the LiCl conductance and permeability ratios signify that the membrane was not as black as usual but was rather silverish, indicating a greater thickness than normal.

other, which are solely characteristic of a given antibiotic molecule. Recalling the discussion given in the footnote of page 98, this finding is expected theoretically from Eq. (17). It is the counterpart of the observation previously noted for salt extraction that the relative selectivities of the one cation compared to another are independent of the solvent.

The above results can be immediately confronted with the theoretical expectations of Eq. (18). This is done on the left hand side of Figure 11, where the experimentally observed permeability ratios (plotted as points) are compared with the ratios of the salt extraction equilibrium constants (K_i/K_K) (plotted as curves) traced from the data of Figure 7.

The detailed agreement between the permeability ratios and the solvent extraction equilibrium constants is truly remarkable but expected from the present theory. Only for Li^+ is there a discrepancy between P_i/P_K and K_i/K_K. In this case the difference is almost certainly due to the fact that the effect of dissociated NH_4^+ could be minimized in the solvent extraction experiments for Li^+ at high concentrations and at alkaline pH, whereas for the bilayer measurements at neutral pH the amount of dissociated NH_4^+ was not completely negligible. Even such details as the relative changes in the Cs^+ to Na^+ selectivities and Rb^+ to K^+ selectivities seen in the salt extraction experiments as one goes from Nonactin to Trinactin are also found for the permeability ratios of these ions measured in bilayers.

Membrane Conductance

The right hand side of Figure 11 compares the ratios of zero current membrane conductances (plotted as points) with the previously measured salt extraction selectivity ratios K_i/K_K (plotted as curves). The conductances were measured in 0.01 Molar alkali metal chloride solutions containing 10^{-7} Moles per liter of the antibiotics. The theoretical expectation of Eq. (18) for the conductance ratios G_i/G_k to be identical to the ratios K_i/K_K is satisfactorily fulfilled.

Comparing the Permeability ratios at the left with the Conductance ratios at the right, the agreement required by Eq. (17) for these is also seen. This successful confrontation of theoretical expectations of Eq. (18) with the observed ratios of the salt extraction equilibrium constants, of the permeabilities and of the conductances stands in strong support of the usefulness of the present theory.

In addition to these results for the ratios of relative ionic effects, a correspondence also has been found to exist between the *individual* values of K_i for the salt extraction equilibrium and the measured values of membrane conductance for each ion, G_i. This is illustrated in Figure 12, where the membrane conductances of normal phospholipid bilayer membranes, formed in 10^{-2}

Molar salt and 10^{-7} Molar antibiotic solutions, are plotted as points. For comparison, the values of K_i from the right hand side of Figure 7 are plotted as the solid curves referred to the right hand ordinate. The logarithmic G_i and K_i ordinates have been interrelated by superposing the K^+ data for G_i and K_i

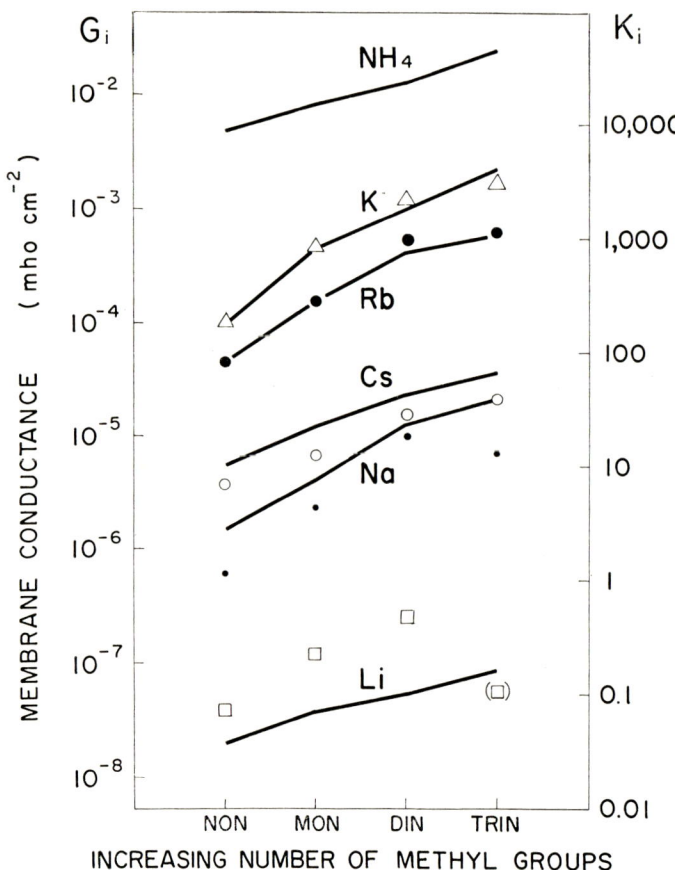

FIGURE 12 Comparison of bulk extraction coefficients and bilayer membrane conductances for each of the five alkali metal ions as a function of increasing number of methyl side groups of the Macrolide Actin antibiotics.

for Nonactin. The close correspondence seen between the G_i and K_i values for the other antibiotics, as well as the other cations (see Table II), indicates that a single proportionality constant (2×10^6) suffices to relate G_i to K_i. This is a useful result which enables one to 'predict' the detailed effects on membrane conductance of the other macrolide antibiotics for each cation directly from the equilibrium extraction data. This behavior would be expected if the addition of a few methyl groups to the large nonactin molecule (cf. Figure 1B) did not significantly alter its mobility or partition coefficient.†

The effects of cholesterol on membrane properties

Since cholesterol is known to immobilize the hydrocarbon tails of lipid (25a–25d), and should thereby decrease the mobilities of the postulated carriers (25d), it is possible to make a prediction for the present model of the expected effect of adding cholesterol to the lipid of the membrane. Namely; adding cholesterol should decrease markedly the conductance of the membrane when exposed to a given concentration of antibiotic and salt without, however, altering the ratios of the conductances or of permeabilities.

The observed effect of adding cholesterol is illustrated in Figures 13 and 14, where it can be seen to decrease the effectiveness of Monactin in increasing membrane conductance by nearly a thousand fold. This is exactly what would be expected if Monactin acts as a mobile carrier for cations, as postulated here; and the principle effect of adding cholesterol is to decrease the mobility of the complexed-cations within the membrane. This effect is not expected *a priori* for the 'tunnel' mechanism of Monactin action and is a strong piece of evidence in favor of the carrier mechanism.

Also notice that the linear dependence expected from equation 6 of membrane conductance on antibiotic concentration which has demonstrated previously for Monactin in solutions of Li, Na, K, Rb, and Cs (cf. Figure 6 of reference (1)), also is shown in Figure 13 to apply to a lipid containing large amounts of cholesterol. Here, conductances measured for membranes formed in 10^{-2} Molar KCl from the cholesterol-free 'normal lipid' and from a mixture containing 80 per cent cholesterol and 20 per cent 'normal lipid' are plotted as a function of increasing concentration of monactin. Figure 14 illustrates the effects of cholesterol content on the increase of membrane conductance produced by Monactin as a function of KCl concentration. For

† This result was not expected *a priori*, since from Eqs. (8) and (16) the relationship between G_i and K_i is

$$G_i = \frac{F^2 C_s^{Tot} a_i}{k_x d} u_s^* k_s^{(s)} \frac{k_{is}^{(m)}}{k_{is}^{(s)}} K_i$$

where the superscripts (s) and (m) refer to the bulk solvent and membrane phase, respectively. The combination of parameters $u_{is}^* k_s^{(s)} k_{is}^{(m)}/k_{is}^{(s)}$ would be independent of the antibiotic only if the additional methyl group had little effect on each of these parameters individually, although some compensation due to opposing effects is conceivable.

the cholesterol-free membrane, the linear dependence of membrane conductance on KCl concentration expected theoretically from Eq. (16) is observed over the range from 10^{-5} to 10^{-3} Molar. With increasing cholesterol content, the region over which this linearity is seen is markedly diminished and this behavior provides a basis for reconciling the differences between our previous

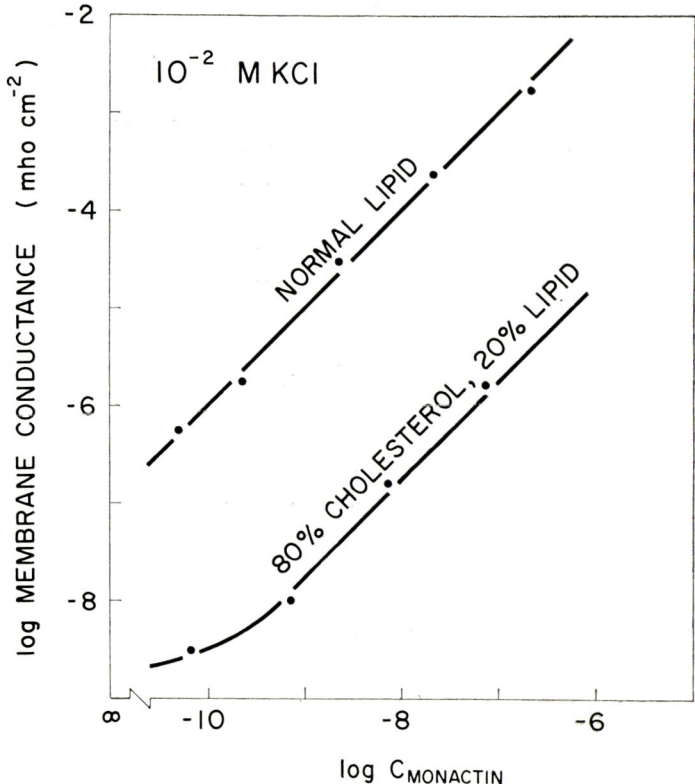

FIGURE 13 The effects of cholesterol on bilayer conductances in 10^{-2} M KCl as function of the monactin concentration. (See legend of Figure 14 and the text for Experimental details).

linear data for cholesterol-free membranes (Figure 7 and footnote 3 of (1)) and the data of Tosteson (Figure 8 of (26)) for cholesterol-containing membranes.

Notice that the conductance increase is less than a factor of ten between zero KCl and 0.1 Molar KCl in the case of a membrane containing 80 per cent cholesterol; so that it is clear that the presence of sufficient cholesterol can

Ion distribution equilibria in bulk phases 109

lead to an apparent lack of affect of Monactin on membrane conductance.†
This is seen strikingly in the data of Figure 13, where the addition of 5×10^{-11} Molar Monactin markedly increases the membrane conductance for the cholesterol-free membrane; but in the presence of cholesterol such an addition produces no significant effect.

The appearance of a 'saturation' in the membrane conductance at high KCl concentrations, which becomes increasingly evident as the cholesterol content of the membrane is increased, might at first glance appear to be inconsistent with the expectations of Eq. (16) of the present theory. However, it will be shown below that this effect is not a true 'saturation' but has another explanation, being due to an ionic strength effect on the physical properties of the membrane.

The effects of ionic strength on membrane properties

Figure 15 presents typical data for the dependence of membrane conductance on salt concentration characteristic of cholesterol-free membranes exposed to the Macrolide Actin antibiotics (5). Figure 15A illustrates for 10^{-7} Nonactin the dependence of membrane conductance on aqueous ionic concentration (note that a higher concentration range is explored than in Figure 14); while Figure 15B displays these data as a function of the calculated aqueous activities. The behavior is seen to be complex in that not only does an apparent 'saturation' occur at concentrations above 10^{-3} Molar, but in the case of LiCl the membrane conductance actually *decreases* with increasing concentration.

The data of Figure 15A looked suspiciously like pure ionic strength effects; and we tested this notion by carrying out the experiment of Figure 15C where LiCl was added to the solutions containing 10^{-3} KCl, 10^{-4} KCl, and 10^{-2} KCl to increase the ionic strength of these while holding their K^+ concentration constant. Since Li^+ has much less effect than K^+ on the membrane conductance in the presence of Nonactin, its effects in Figure 15C should be referable primarily to its effects on the K^+ conductance of the membrane. If

† Comparing the 80 per cent cholesterol curves of Figures 13 and 14, a discrepancy becomes apparent between the conductance values for 10^{-2} Molar KCl and 10^{-7} Monactin in these two figures, the conductance on Figure 13 being larger than that of Figure 14. This came about because, when large amounts of cholesterol are present in the lipid, membranes formed shortly after the lipid was introduced into the chamber (as is the case of Figure 14) have initially lower conductances than those observed in the steady state following 30 minutes of equilibration (as was the case for the curve on Figure 13). Slow dissolving of some cholesterol from the lipid into the aqueous phase is believed to be responsible for this effect. At lower lipid cholesterol content (36 per cent and 52 per cent of Figure 14) such phenomena were almost unnoticeable.

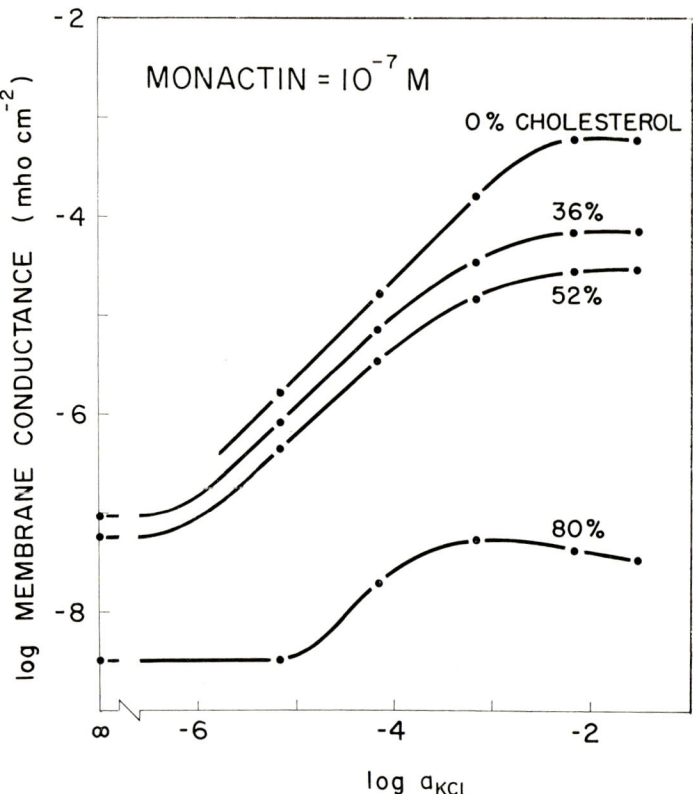

FIGURE 14 The effect of cholesterol on bilayer conductance in 10^{-7} M aqueous Monactin solutions as a function of KCl concentration. The lipids had the following composition:

    0 per cent cholesterol: 15 mg 'normal lipid' per ml decane
    36 per cent cholesterol: 15 mg normal lipid, 8.3 mg cholesterol per ml decane
    52 per cent cholesterol: 15 mg normal lipid, 16 mg cholesterol per ml decane
    80 per cent cholesterol: 4 mg normal lipid, 16 mg cholesterol per ml decane

2 µl of lipid was introduced into the front chamber onto the teflon wall near the opening for the membrane using a disposable Pasteur pipette. Membranes were formed by passing an air bubble from the Pasteur pipette over the opening. This method has the advantage of introducing lipid into the system only at the beginning of each experiment. Possible contaminations are also avoided since a new Pasteur pipette is used for each new experiment. Membranes formed from cholesterol-free normal lipid in pure 10^{-7} M aqueous Monactin solution do not thin spontaneously and a DC potential of about 100 mV (or an AC potential of

this were ionic strength independent, one would expect horizontal curves since the K^+ conductance should be unaltered by the addition of Li^+. Instead, the K^+ conductance is seen to decrease in exactly the manner expected for an effect of ionic strength, pure and simple.† The apparent 'saturation' of the 'pure KCl' curve in C appears to be a consequence of a balance of the opposite effects of the decrease in conductance due to the ionic strength increase with increasing KCl and the increase of conductance with increasing K^+ activity expected from Eq. (16).

That all the peculiarities of conductance in Figures 15A and 15B are due to effects of ionic strength is illustrated by the data of Figure 15D, where the experimental data of Figure 15B have been corrected for ionic strength effects by subtracting the LiCl curve from the conductance value for each ion at a given concentration so as to obtain a horizontal curve for LiCl. Such a correction leads to the data points which can be seen to fall nicely on the solid lines drawn with a slope of unity and corresponds to assuming that the effect of LiCl is purely that of ionic strength (the decrease in conductance seen for 'pure LiCl' actually representing the ionic strength effect of LiCl on the membrane conductance to trace NH_4^+).

The most striking confirmation of this postulate is seen by the experiment illustrated in Figure 16 where the membrane conductance for KCl was measured when the ionic strength was maintained constant with LiCl. Under this condition, the linear dependence of $\log G_i$ on $\log a_K$ unequivocally demonstrates that the conductance is precisely proportional to the activity of

† That this is purely an effect of ionic strength is further indicated by the fact that when CsCl is added instead of LiCl (as can be seen in Figure 18A), the effect is the same for concentrations less than 0.5 Molar (see (5) for a detailed discussion).

---

somewhat higher magnitude) has to be applied to force the thinning and to obtain optically black membranes. In the absence of the applied field, the membrane slowly tends to become silverish, starting from the edges: however this process is sufficiently slow to permit meaningful conductance measurements to be made on black membranes. At higher ionic concentrations the membranes thin more easily, but this effect depends greatly on the ionic species present, the sequence of decreasing effectiveness being NH_4, K, Rb, Cs, Na, Li. Addition of large amounts of cholesterol to the lipid also promotes membrane thinning even in the absence of added salts; so that thinning is spontaneous at high cholesterol content. It was noted at high cholesterol levels that lower membrane conductances were measured in membranes formed shortly after the lipid was introduced into the chamber than when time was allowed for the lipid to equilibrate with the aqueous phase. An equilibration time of about 35 minutes was usually sufficient for measurement to be independent of equilibration. The lipids used in this experiment were prepared and tested on the same day.

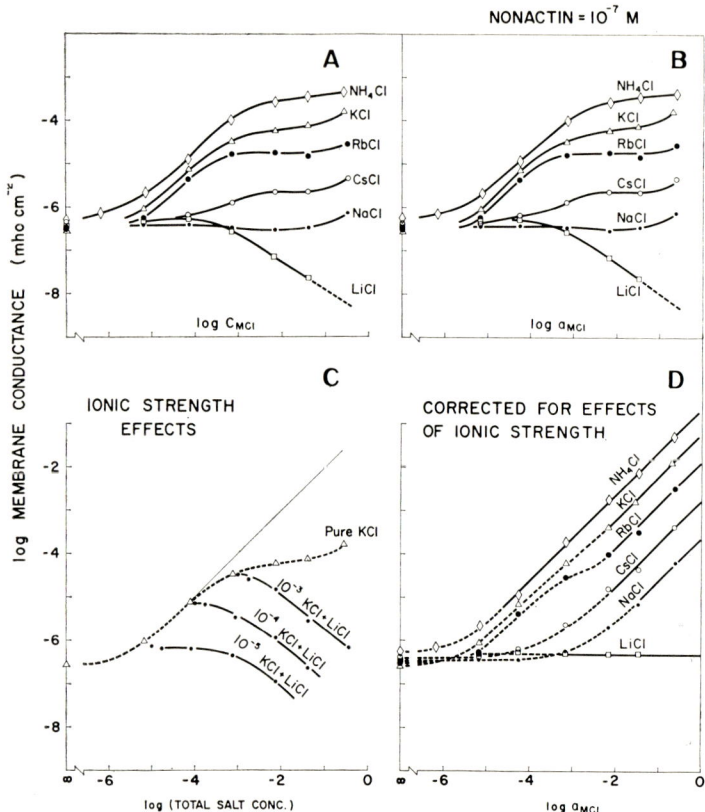

FIGURE 15   Evidence for the ionic strength dependence of the Nonactin-induced membrane conductance. A. Membrane conductances are plotted as a function of salt concentration in the presence of a constant 10^{-7} M Nonactin aqueous concentration. B. Same as A but replotted in terms of activities. C. For membranes formed in 10^{-7} M Nonactin and indicated concentrations of KCl small volumes of LiCl were added to both aqueous compartments to produce the indicated total salt concentrations. At the highest LiCl concentrations, the membrane did not thin spontaneously but could be forced by application of large voltages. In the absence of the applied field, the membrane tended to return slowly to the silverish state, but the process was sufficiently slow to permit measurements to be made on the black films. D. The LiCl curve of B was used to correct all of the curves of B for ionic strength by subtracting the LiCl curve from the others. The solid lines are drawn with a slope of unity; while the dotted lines connect those points that do not fall on this line.

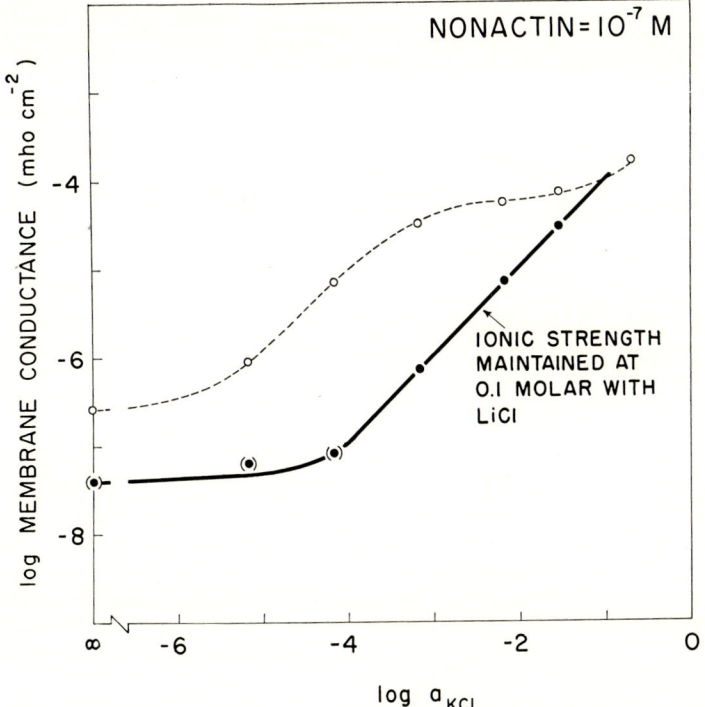

FIGURE 16 The effect of KCl concentration on membrane conductance in 10^{-7} Nonactin in solutions maintained at constant ionic strength by LiCl. The membrane was formed from normal lipid (15 mg/ml decane) in a 10^{-7} Nonactin, 0.1 M LiCl solution. Small volumes of KCl were added (10 μl of 10^{-2} M, 10^{-1} M, 1 M; 100 μl of 1 M, 4 M) thereby changing the total ionic strength only by a negligible amount. Membrane conductances measured following the procedure described on Figure 10 after each addition of KCl are plotted as filled circles. For comparison, the open circles (replotted from Figure 14A) indicate the membrane conductance values obtained in pure KCl solutions. Parenthesized points indicate imperfectly thinned membranes.

K^+ as expected from Eq. (16). It should therefore be clear that the 'saturation' is only an apparent one and not a true saturation—being due to the balance of opposing effects. This result makes it clear now why the relative displacements of the curves of Figure 15A or B can be used so satisfactorily to yield conductance ratios for confrontation with theoretical expectation from equilibrium studies, and also why the relative effects of cations on membrane

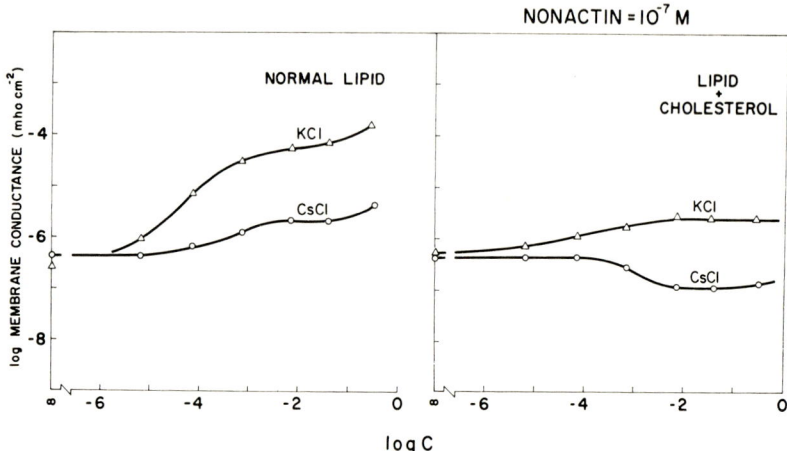

FIGURE 17 The effect of cholesterol on bilayer conductances in presence of 10^{-7} M Nonactin and varying concentrations of KCl and CsCl. The details of the experiment are similar to those in Figure 14. The 'normal lipid' contained 7.5 mg lipid per ml of decane. The cholesterol-containing lipid was prepared by using decane, previously saturated with cholesterol at room temperature, to dissolve the lipid.

conductance are independent of the lipid, as will be illustrated in Figure 17, just as was shown to be the case for permeability ratios in Figure 9.

The conductance ratio between two ions is independent of the lipid composition

Despite the above demonstrated dependence of the effects of a given antibiotic on the composition of membrane lipid, the *ratios* of cation conductances are completely independent of the lipid composition and depend only on the antibiotic molecules. This is illustrated for Nonactin in Figure 17, which compares the K^+ and Cs^+ conductance of a cholesterol-free membrane (left) with those for a cholesterol-rich membrane (right) formed from regular lipid dissolved in decane saturated with cholesterol at 25°C. Despite the obvious differences between the two lipids in the manner in which the individual K^+ and Cs^+ conductances depend on concentration (e.g. for the cholesterol-lipid the conductance is almost independent of the KCl concentration and decreases with increasing CsCl concentration), the conductance *ratio* (i.e. the displacement between the curves) is not only comparable between the two lipids at any given concentration but also constant at concentrations higher than 0.01 Molar. (The decrease in the difference between KCl and CsCl curves at lower concentrations is due to traces of NH_4^+ in our distilled water which are only negligible when the CsCl concentration exceeds 0.001 Molar for normal lipid and 0.01 Molar for cholesterol-rich lipid). A more detailed test of this is shown in Table IV where membrane conductances in 10^{-2} Molar chloride solutions containing 10^{-7} Molar Monactin and their ratios for 52 per cent cholesterol containing lipid are compared against those obtained for normal lipid. Despite large differences in membrane conductance values, the ratios are seen to be independent of the presence of cholesterol.

TABLE IV Effects of cholesterol on bilayer membrane conductances and conductance ratios.

	Normal lipid mho cm^{-2}	52% Cholesterol 48% Normal lipid mho cm^{-2}
G_{Na}	2.1×10^{-6}	9.4×10^{-8}
G_{Cs}	6.2×10^{-6}	2.2×10^{-7}
G_{Rb}	1.5×10^{-4}	7.4×10^{-6}
G_{K}	4.4×10^{-4}	1.7×10^{-5}
G_{Na}/G_{K}	0.0048	0.0055
G_{Cs}/G_{K}	0.014	0.013
G_{Rb}/G_{K}	0.34	0.43

We may therefore conclude that, despite the complexity of the ionic strength dependence of *individual* ionic conductances as well as their dependence on lipid composition, the *ratios* of conductances among the various cations are entirely characteristic of the antibiotic molecules and remarkably independent of wide variations of the lipid composition. This result is consistent with the finding that the selectivity ratios for salt extraction equilibria were independent of the solvent. Indeed, were both of these findings not observed, the correlation we have found in Figure 11 between salt extraction selectivities and the conductance and permeability properties of bilayer membranes would have been extremely unlikely.

This important result is required by our theoretical analysis, as has been previously examined in the footnote on page 109, and is probably the single strongest piece of evidence as to its validity. Questions of validity aside, it should now be apparent that it suffices to characterize cation interactions with the antibiotic molecules in simple solvent equilibria in order to predict their effects on bilayer membrane. The relative cation effects produced in bilayers by the present antibiotics can therefore be regarded as characteristic of the antibiotic molecules themselves which should be understandable directly from the analysis of ion-molecule interactions alone.

On the other hand, whether or not a given antibiotic molecule will produce an effect on a given membrane clearly depends on the particular composition of the membrane lipid. This provides a means by which a variety of membranes exposed to the same aqueous concentrations of molecules such as these antibiotics might exhibit a diversity of responses. The implications of this will be considered in the Discussion.

Some effects of Ca^{2+}, Mg^{2+}, Th^{4+}, H^+, *and* Cs^+ *on the* K^+ *conductance*

Although a detailed analysis will not be possible here, it is worthwhile completing the above description by mentioning the effects of Mg^{2+}, Ca^{2+}, Th^{4+}, Cs^+ and H^+ on the K^+ conductance of the present membranes. Typical data are presented in Figure 18A (which corresponds to Figure 15C);

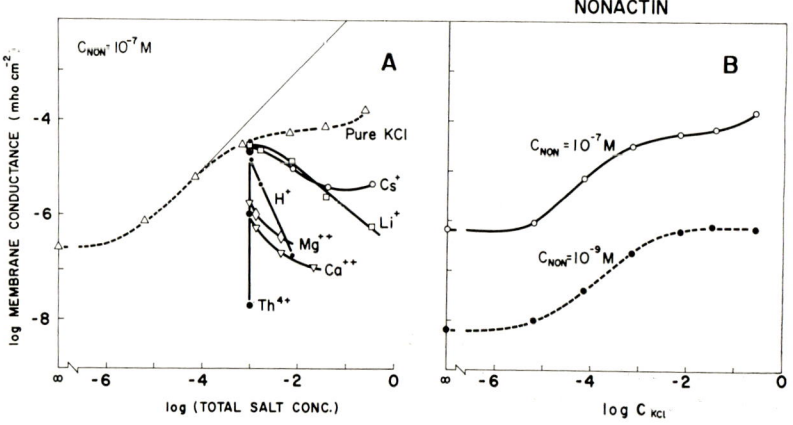

FIGURE 18 A. The effect of addition of some monovalent and multivalent ions on the conductance of bilayer membranes in 10^{-7} Nonactin and 10^{-3} M KCl. B. The influence of Nonactin concentration level on the membrane conductance as function of KCl concentration in the aqueous media.

while Figure 18B illustrates that the conductance dependence on salt concentration is independent of the level of antibiotic concentration. Whereas Cs^+ and Li^+ are seen to be indistinguishable in Figure 18A in their effects for those concentrations (below 0.1 Molar) at which neither of them contribute significantly to the membrane conductance in the presence of 10^{-3} Molar KCl, H^+ is seen to differ from these and therefore has an effect beyond that due to its ionic strength. The even more pronounced effects of Mg^{++}, Ca^{++}, and Th^{4+} cannot be solely attributed to ionic strength. Although these effects are similar to those often attributed to the presence of fixed charges in biological membranes, it should be recalled that they occur here in a membrane for which ion permeation by a *neutral* carrier mechanism seems very likely! Caution should therefore be exercised in attributing effects of ionic strength to the presence of fixed charges.

THE USE OF A GRADIENT IN ANTIBIOTIC CONCENTRATION AS A MEANS FOR EXAMINING THE RATE DETERMINING STEPS FOR ANTIBIOTICS PERMEATION OF BILAYER MEMBRANES

To this point, theory and experiment have been restricted to situations in which the concentration of the antibiotic is the same on both sides of the membrane. Under this circumstance it is neither necessary nor possible to identify the rate determining step of the overall process of diffusion of the antibiotic molecules across the membrane. However, when a gradient of such molecules is maintained across the membrane and equilibrium of reaction 5 is assumed to exist at both interfaces, the concentration of the IS^+ species in the membrane and all the electrical properties will be critically dependent on the concentrations of the neutral species S at each of the interfaces and consequently on the relative rates of the processes by which these species cross the boundaries and diffuse across the membrane (6).

We shall examine here the two limiting cases in which (A) diffusion across the membrane and (B) diffusion across the interfaces are the rate determining processes, respectively, and shall confront the theoretical expectations with the experimental results in order to distinguish between these alternatives.

THEORETICAL CONSIDERATIONS

Membrane potential

From the equilibrium of the heterogeneous reaction 5 at the membrane solution interfaces, the following expression can be derived for the membrane potential at zero current and is valid quite generally:

$$V_0 = -\frac{RT}{F} \ln \frac{\sum_{i=1}^{n} \frac{P_i}{P_j} a_i''}{\sum_{i=1}^{n} \frac{P_i}{P_j} a_i'} - \frac{RT}{F} \ln \frac{C_s^*(d)}{C_s^*(0)} \qquad (21)$$

The first term expresses the dependence of the membrane potential on the activities of the cations in the 'aqueous phases', and corresponds to Eq. (14); while the second term describes the dependence of the potential on the concentrations of the macrocyclic molecules just inside the membrane at the membrane-solution boundaries. As these concentrations are generally not known, an explicit evaluation of $C_s^*(0)$ and $C_s^*(d)$ in terms of the aqueous concentrations C_s' and C_s'' will be necessary in order to obtain an expression for V_0 explicitly in terms of the concentrations in the aqueous solutions.

Using the formalism of reaction rate theory (27) to describe the movement of the carriers across the interfaces, as suggested to us by Norman Davidson, the net flux of carriers J_s across each interface of Figure 2 is given by the difference between inward and outward movements; thus at each interface

$$J_s(0) = \mathbf{k}_i C_s' - \mathbf{k}_o C_s^*(0), \tag{22}$$

$$J_s(d) = \mathbf{k}_o C_s^*(d) - \mathbf{k}_i C_s'', \tag{23}$$

where $\mathbf{k}_i$ and $\mathbf{k}_o$ designate the rate constants of the *inward* (into the membrane) and *outward* (out of the membrane) movements respectively. Within the membrane we have shown elsewhere (6) that the flux of carriers in the steady state is given for the present system by

$$J_s = \frac{RTu_s^*}{d}[C_s^*(0) - C_s^*(d)] \tag{24}$$

where u_s^* is the mobility (assumed constant) of S within the membrane.

In the steady state, conservation of mass allows Eqs. (22), (23), and (24) to be set equal so that the flux of the carrier is

$$J_s = \mathbf{k}_i C_s' - \mathbf{k}_o C_s^*(0) = \frac{RTu_s^*}{d}[C_s^*(0) - C_s^*(d)] = \mathbf{k}_o C_s^*(d) - \mathbf{k}_i C_s'' \tag{25}$$

Solving Eq. (25) with respect to $C_s^*(0)$ and $C_s^*(d)$, and noting that the ratio of rate constants is related to the equilibrium partition coefficient k_s of the neutral carriers by

$$k_s = \frac{\mathbf{k}_i}{\mathbf{k}_o} \tag{26}$$

we obtain

$$C_s^*(0) = k_s \frac{C_s'' + [1 + (\mathbf{k}_o d/RTu_s^*)]C_s'}{2 + (\mathbf{k}_o d/RTu_s^*)} \tag{27}$$

and

$$C_s^*(d) = k_s \frac{C_s' + [1 + (\mathbf{k}_o d/RTu_s^*)]C_s''}{2 + (\mathbf{k}_o d/RTu_s^*)} \tag{28}$$

where C'_s, C''_s are related to the (known) total concentrations of carriers $C_s^{\text{Tot}'}$ and $C_s^{\text{Tot}''}$ in the aqueous solutions by

$$C'_s = \frac{C_s^{\text{Tot}'}}{1 + \sum_{i=1}^{n} K_{is}^{+} a'_i} \quad ; \quad C''_s = \frac{C_s^{\text{Tot}''}}{1 + \sum_{i=1}^{n} K_{is}^{+} a''_i} \tag{29}$$

deducible directly from reaction 1.

Two limiting cases of Eqs. (27) and (28), designated as cases (A) and (B), corresponding to the situations in which the diffusion of the carriers through the membrane is very low or very rapid compared to the process of crossing the interfaces, will be now considered in detail.

Case (A). When the rate at which the neutral carriers cross the interfaces is very high compared with the diffusion across the membrane, we have

$$\frac{k_0 d}{RTu_s^*} \gg 1 \tag{30}$$

Eqs. (27) and (28) then become, approximately

$$C_s^*(0) = k_s C'_s \tag{31}$$

and

$$C_s^*(d) = k_s C''_s \tag{32}$$

showing that, in this case, equilibrium exists for the neutral carriers at the membrane-solution interfaces. By use of Eqs. (31), (32), and (29), Eq. (21) for the membrane potential becomes

$$V_0 = -\frac{RT}{F} \ln \frac{\sum_{i=1}^{n} \frac{P_i}{P_j} a''_i}{\sum_{i=1}^{n} \frac{P_i}{P_j} a'_i} - \frac{RT}{F} \ln \frac{C_s^{\text{Tot}''}}{C_s^{\text{Tot}'}} + \frac{RT}{F} \ln \frac{1 + \sum_{i=1}^{n} K_{is}^{+} a''_i}{1 + \sum_{i=1}^{n} K_{is}^{+} a'_i} \tag{33}$$

The first term of Eq. (33) is recognizable as the same as Eq. (14). The second term indicates that the membrane potential is expected to depend linearly on the logarithm of the ratio of the total antibiotic concentrations in the aqueous solutions. The third term results from the possibility of significant complexation of cations in the aqueous solutions and is significant only when

$$\sum_{i=1}^{n} K_{is}^{+} a_i > 1$$

Since for mechanism (A) all permeant species are in equilibrium at the membrane-solution interfaces, the potential is expected to be expressible in terms of the aqueous solution activities of the predominantly permeant charged species IS^+. That this is so can be shown by noting that the activities $a'_{is}('')$ in the aqueous solutions are given by

$$a'_{is}('') = \frac{K^+_{is} a'_i('') C_s^{\text{Tot}'('')}}{1 + \sum_{i=1}^{n} K^+_{is} a'_i('')} \tag{34}$$

Substituting Eq. (34) in Eq. (33) and recalling the definition of the permeability ratio P_j/P_i, we get

$$V_0 = -\frac{RT}{F} \ln \frac{\sum_{i=1}^{n} u^*_{is} k_{is} a''_{is}}{\sum_{i=1}^{n} u^*_{is} k_{is} a'_{is}} \tag{35}$$

where the simple dependence on the aqueous activities of the complexed-cations is as expected intuitively since the complexed cations are the only permeant species.

Case (B). When the rate of diffusion of the neutral macrocyclic molecules within the membrane is much higher than the rate of crossing the interfaces, we have

$$\frac{k_0 d}{RT u^*_s} \ll 1 \tag{36}$$

Eqs. (27) and (28) can then be approximated by

$$C^*_s(0) = C^*_s(d) = k_s \cdot \frac{C'_s + C''_s}{2} \tag{37}$$

Inserting Eq. (37) in the general expression of the membrane potential, Eq. (21), we find the very simple result:

$$V_0 = -\frac{RT}{F} \ln \frac{\sum_{i=1}^{n} \frac{P_i}{P_j} a''_i}{\sum_{i=1}^{n} \frac{P_i}{P_j} a'_i} \tag{38}$$

This result differs from Eq. (33) by the absence of the second and third terms, the reason being that, in contrast with case (A), the concentrations of the carriers at the interfaces are equalized by diffusion within the membrane, regardless of the different carrier concentrations at the membrane-solution interfaces and of the degree of complexation in the aqueous solutions.

Ion distribution equilibria in bulk phases

Comparison of Eqs. (33) and (38) suggests a clear experiment to establish whether mechanisms (A) or (B) is consistent with the behavior of the system: namely, a measurement of electrical potential in the presence of a gradient of the antibiotic concentration.

Membrane conductance for zero current

When the membrane is interposed between identical solutions, the system is at equilibrium. Since $C'_s = C''_s$, Eqs. (27) and (28) show immediately that the values of $C^*_s(0)$ and $C^*_s(d)$ are not only equal but are independent of $k_0 d/RTu^*_s$, so that no distinction between mechanisms (A) and (B) is possible from such measurements.

In contrast, in the presence of a gradient of the carrier concentration, differences between the expectations for mechanisms (A) and (B) appear.

Let us designate by G^{Asym}_{12} the limiting value of the conductance at zero current when the membrane is interposed between solutions having different concentrations of carriers, say C'_s and C''_s, but identical ionic composition. The following general relationship can be derived (6) between the value of this and the conductances measured for the membrane exposed symmetrically to solutions of carrier concentrations equal to C'_s and C''_s, respectively (defined as G^{Sym}_1 and G^{Sym}_2).

$$G^{\text{Asym}}_{12} = \sqrt{(G^{\text{Sym}}_1 G^{\text{Sym}}_2)} \cdot \frac{1}{k_s} \sqrt{\left(\frac{C^*_s(0)C^*_s(d)}{C'_s C''_s}\right)} \qquad (39)$$

where $C^*_s(0)$ and $C^*_s(d)$ are given by Eqs. (27) and (28) and C'_s, C''_s by Eq. (29).

Let us now particularize Eq. (39) for the cases (A) and (B).

For Mechanism A, Eqs. (27) and (28) can be approximated by Eqs. (31) and (32), so that Eq. (39) becomes simply

$$G^{\text{Asym}}_{12} = \sqrt{(G^{\text{Sym}}_1 G^{\text{Sym}}_2)} \qquad (40)$$

showing that for Mechanism A the *geometric mean* of G^{Sym}_1 and G^{Sym}_2 is expected for the value of the assymmetric conductance.

For Mechanism B, for which Eq. (37) is valid, G^{Asym}_{12} is given by

$$G^{\text{Asym}}_{12} = \tfrac{1}{2}\sqrt{(G^{\text{Sym}}_1 G^{\text{Sym}}_2)}\left(\sqrt{\left(\frac{C'_s}{C''_s}+\frac{C''_s}{C'_s}+2\right)}\right) \qquad (41)$$

From Eqs. (29) and (16) it should be apparent, that, for the case under consideration of identical ionic composition in both solutions

$$\frac{C'_s}{C''_s} = \frac{G^{\text{Sym}}_1}{G^{\text{Sym}}_2} \qquad (42)$$

so that Eq. (41) reduces to

$$G_{12}^{Asym} = \frac{G_1^{Sym} + G_2^{Sym}}{2} \qquad (43)$$

indicating that for Mechanism B the *arithmetic mean* of G_1^{Sym} and G_2^{Sym} is expected for the value of the asymmetric conductance.

Equations (40) and (43) provide a second (and independent) criterion for determining which of the two mechanisms is effectively operating in a given system. As will be shown in the experimental results, both potential and conductance measurements indicate unambiguously that, within the framework of our model, Mechanism B has to be invoked to account for the electrical properties of lecithin-decane bilayer membranes in the presence of the Macrolide Actin Antibiotics.†

EXPERIMENTAL RESULTS

The results of a particularly simple series of experiments, given in Table V, illustrate how the distinction between the two mechanisms is made. In the first experiment of the Table, a lipid bilayer membrane was formed in 10^{-3} Molar KCl in the absence of antibiotic, and its conductance and membrane potential (following sufficient stirring in all cases) measured under this symmetrical condition to be 5×10^{-11} mho and -1 mV, respectively.

Monactin was then added to the rear chamber to produce a concentration of 0.67×10^{-11} Molar. Under this asymmetrical condition the membrane conductance was measured to be 3.4×10^{-10} mho and the potential difference to be 0.5 mV. The solution in the front chamber was then brought to the same antibiotic concentration as the rear, and the membrane conductance and potential were measured under this symmetrical condition to be 6.4×10^{-10} mho and 0 mV, respectively.

Next, the antibiotic concentration was increased 100 fold in the rear chamber to 0.67×10^{-9} Molar to produce an asymmetrical condition with a ratio of 100 to 1 in the antibiotic concentration across the membrane, and the

---

† In the above it has not been necessary to specify whether the rate limiting step is due to a slow surface reaction or to slow diffusion in the unstirred aqueous film adjacent to the membrane. For Monactin film diffusion alone seems to be sufficient for case B to be observed. This can be seen (6) by writing inequality (36) for the case of film diffusion: $\mathbf{k}_0 d/RTu_s^* = u_s d/u_s^* lk_s \ll 1$, where l is the unstirred layer thickness and u_s is the aqueous mobility of S. For $d < 100$ Å° and $k_s = 5000$, unstirred layers as thin as 200 Å° will be important even if the mobility in the membrane is one thousandth that in water.

TABLE V Bilayer membrane conductance

	Monactin concentration		G (Measured)	G (Theoretically expected)			V (Measured)
	Rear chamber	Front chamber		Mechanism 1B $\frac{G_1^{\text{sym}} + G_2^{\text{sym}}}{2}$	Mechanism 1A $\sqrt{(G_1^{\text{sym}} G_2^{\text{sym}})}$	Surface Mechanism $\frac{2}{R_1^{\text{sym}} + R_2^{\text{sym}}}$	

$(10^{-3}$ Molar KCl on both sides of membrane$)$

Sym.	0	0	5.0×10^{-11} mho				−1.0
Asym.	0.67×10^{-11} M	0	2.0×10^{-10}	3.4×10^{-10}	1.8×10^{-10}	9.2×10^{-11}	+0.5
Sym.	0.67×10^{-11}	0.67×10^{-11}	6.4×10^{-10}				0.0
Asym.	0.67×10^{-9}	0.67×10^{-11}	6.9×10^{-9}	6.3×10^{-9}	2.8×10^{-9}	1.2×10^{-9}	+0.1
Sym.	0.67×10^{-9}	0.67×10^{-9}	1.2×10^{-8}				−0.3
Asym.	0.67×10^{-7}	0.67×10^{-9}	5.9×10^{-7}	6.0×10^{-7}	1.2×10^{-7}	2.4×10^{-8}	−0.3
Sym.	0.67×10^{-7}	0.67×10^{-7}	1.2×10^{-6}				−0.3

$(10^{-2}$ Molar KCl on both sides of membrane$)$

Sym.	0	0	1×10^{-10}				−0.2
Asym.	0.67×10^{-11} M	0	1.4×10^{-9}	1.2×10^{-9}	4.8×10^{-10}	1.9×10^{-10}	+0.9
Sym.	0.67×10^{-11}	0.67×10^{-11}	2.3×10^{-9}				—
Asym.	0.67×10^{-9}	0.67×10^{-11}	4.1×10^{-8}	4.1×10^{-8}	1.38×10^{-8}	4.6×10^{-9}	+0.3
Sym.	0.67×10^{-9}	0.67×10^{-9}	8.1×10^{-8}				—
Asym.	0.67×10^{-7}	0.67×10^{-9}	4.2×10^{-6}	5.0×10^{-6}	9.0×10^{-7}	1.65×10^{-7}	0.0
Sym.	0.67×10^{-7}	0.67×10^{-7}	1.0×10^{-5}				0.0

conductances and potentials were measured again. The antibiotic concentration in the front chamber was then increased to a symmetrical 0.67×10^{-9} Molar on both sides of the membrane and the measurements repeated.

This procedure was extended to another 100 to 1 gradient; and, lastly, the antibiotic concentration was raised on both sides of the membrane to 0.67×10^{-7} Molar, as indicated in the remainder of the Table. A total of three such experiments were performed with similar results. Two of these are presented in Table V.

Let us examine the conductance measurements first. The first column under the 'theoretically expected' heading presents the 'arithmetic mean' expectations of Mechanism B, whereas the second column presents the 'geometric mean' expectations of Mechanism A. It is clear that the agreement is excellent between the underlined measured and expected conductances for Mechanism B, except for the least reliable data when the antibiotic concentration is 'zero' in one solution, whereas the conductances expected for Mechanism A differ from the observed results by factors approaching five fold for the highest concentrations.

The present experiment is also pertinent to ruling out an alternative mechanism for the action of the macrocyclic antibiotics proposed by Tosteson (26) in which bilayer membranes are assumed to have 'electric resistance properties characteristic of two monolayers arranged in series'; and the cyclic antibiotics are postulated to increase the potassium permeability of the monolayers 'no matter whether these monolayers are separated by a few Angstrom units as in bilayer membranes of by several millimeters' (26). If this were the mechanism of action of Monactin in the present experiments, the expectation would be for the resistance of a membrane in asymmetrical solutions to be half the sum of the resistances measured in the symmetrical solutions (i.e. the arithmetic mean of the symmetrical *resistances*). The conductance values for this expectation are given in the column labelled 'Surface Mechanism'. The expectations of this mechanism are seen to differ from the observations at all concentrations, and by more than an order of magnitude at the highest concentrations.†

Considering next the membrane potential measurements, we find that in no case was a membrane potential difference larger than 1 mV observed between asymmetrical solutions. This is in accord with the zero potential difference expected for mechanism *B* and in contrast to the potential difference of

† For the large antibiotic gradients of the present experiment, the 'Surface mechanism' requires that the membrane conductance (in asymmetrical solutions) be essentially twice that characteristic of the membrane in symmetrical solutions of the *lowest* antibiotic concentration; whereas mechanism *B* requires that the conductance be essentially halristf that characteic of the *highest* antibiotic concentration.

116 mV expected for mechanism A for a 100 to 1 gradient of antibiotic concentration.

To summarize, in the case of mechanism A (where the diffusion of S is rapid across the interface compared to its diffusion within the membrane) a Nernst potential for the ratio of antibiotic concentration on both sides is expected and the conductance in asymmetrical solutions should be the geometric mean of conductances for the symmetrical solutions. In contrast to this, if mechanism B is operative (for which the diffusion of the S species across the interface is slow compared to its redistribution within the membrane) the expectation is for the absence of any dependence of membrane potential on the gradient of antibiotic and for the asymmetrical membrane conductance to be the arithmetic mean of the symmetrical conductances. The experimental results clearly favor the latter alternative and are sufficiently precise that we can with confidence state that the rate determining step for the antibiotic species is its rate of crossing the membrane-solution interface (or unstirred films), rather than its rate of diffusion within the membrane.

EFFECTS OF SURFACE CHARGES ON THE PROPERTIES OF THE MEMBRANE

In the derivation of the previously given theoretical results the effects of the charged polar head groups of the lipids are explicitly neglected. In particular, continuity of the electric displacement vector was assumed at the membrane-solution interfaces, implying the absence of surface charges other than those due to orientation of dipoles at the interface between the two dielectric media. As, however, a net surface charge, resulting from the dissociable polar head groups of the lipid is likely to be present, it has been felt worthwhile to extend the treatment to estimate to what extent the presence of such fixed charges would affect the expectations of the electrical properties for the membrane. Only those conclusions of this work which bear directly on the present studies will be cited here.

It can be shown (7) that no effects of fixed charges are expected for the membrane potential as long as the presence of anions in the membrane is negligible, whereas their effect on the conductance is only negligible when

$$C_i > \frac{\sigma^2}{32\pi RTD} \tag{44}$$

where C_i is the aqueous concentration of the cation I^+, σ is the surface charge density (assumed to be uniform), and D is the dielectric constant of water. From this it can be seen that for salt concentrations higher than the right hand side of Eq. (44), all of the previously deduced theoretical expressions are valid. Thus, for concentrations sufficiently high that the membrane conductance depends linearly on ionic concentration (i.e. concentrations higher than 0.0001 Molar in Figure 15) the effects of surface charge due to the polar head groups should be negligible.

In contrast, for sufficiently low aqueous ionic concentrations that

$$C_i < \frac{\sigma^2}{32\pi RTD} \qquad (45)$$

the effects of the surface charge should be detectable. These will differ depending on whether the charge is negative or positive.

For negative surface charges the conductance is expected to approach a finite value at low salt concentrations instead of approaching zero (7).

For positive surface charge the membrane conductance is expected to vary as the square of the ionic concentration at low salt concentrations provided that the membrane is still predominantly cation permeable (7).

Although the concentration-independence of conductance at low ionic concentrations seen in Figure 15 is consistent with the expected effects of small negative surface charge, it does not prove that this change is the cause; since the presence of unavoidable traces of NH_4^+ could lead to a similar behavior.†

The effects of surface charge can be emphasized by using a membrane which is expected to have a fixed positive surface charge at neutral pH. Preliminary data for the effects of Monactin on such a membrane made from Oleylamine are presented in Figure 19. In the absence of Monactin, the membrane conductance is seen to be low and largely independent of the salt concentration (the membrane is in fact preferentially permeable to anions in the absence of Monactin). However, in the presence of 10^{-7} Molar Monactin, the membrane becomes selectively permeable to cations and a marked increase of the conductance is observed with increasing KCl concentration. The slope of 2 at the highest concentrations indicates that the membrane conductance depends on the square of the ionic concentration as expected theoretically.

---

† The surface charge density necessary to lead to the deviation seen in Figure 15 from the linear dependence of conductance on ionic concentration for lecithin membranes at concentrations less than 10^{-4} Molar (cf. the K^+ and NH_4^+ data of Figures 15A, B, and D) can be calculated from Eq. (44). For $T = 300\,°K$, $D = 78 \times 1.11 \times 10^{-10}$, and $C_i = 10^{-4}$ moles/liter the calculated density is less than 1 charged group per 10,000 square Angstroms of membrane surface, which is reasonable for our (predominantly lecithin) membrane.

Ion distribution equilibria in bulk phases

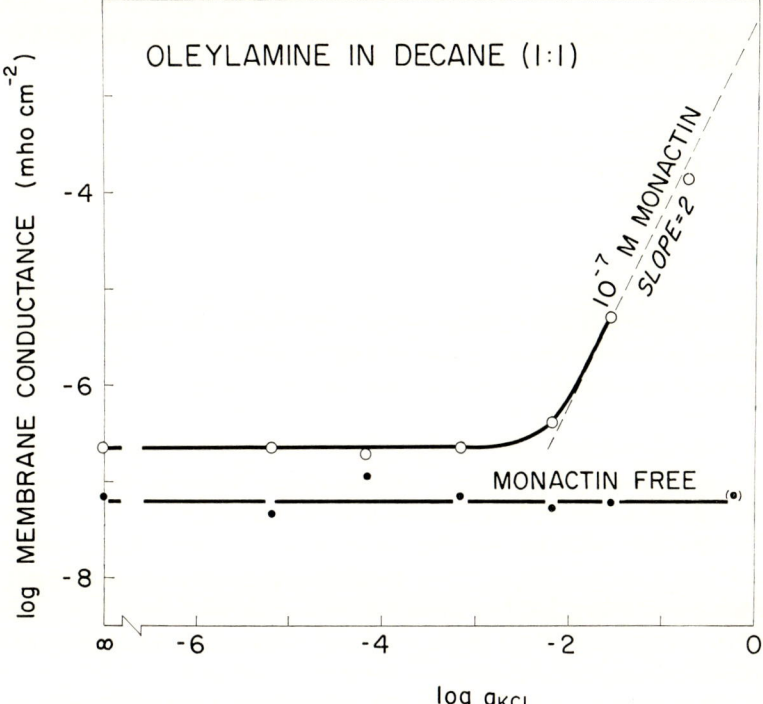

FIGURE 19 Membrane conductance properties of a bilayer membrane having fixed positive charges.

DISCUSSION

Evidence that the Macrolide Actin Antibiotics act as Neutral Carriers of Cations

This paper has postulated that the effects of the Macrolide Actin antibiotics on the electrical properties of phospholipid bilayer membranes are due to the ability of these molecules to solubilize cations in a lipid and thus to act as carriers for cations across the liquid-like membrane interior. Direct evidence for such a cation-solubilizing property as well as for a simple stoichiometry and surprisingly ideal equilibrium behavior has been obtained from studies on salt extraction equilibria into solvents such as dichloromethane and hexane-dichloromethane mixtures. Thus, the simple chemistry postulated has been verified.

The strongest evidence that the Macrolide Actin antibiotics indeed act as the carriers which we have postulated them to be is contained in the successful confrontation of Table II and of Figures 11 and 12 between phospholipid bilayer properties and solvent extraction equilibrium measurements. It is very difficult to conceive how such detailed agreement, which is required by the simple carrier model, could result from any alternative mechanism of action such as the formation of a 'neutral pore' structure.

Another expectation of a neutral carrier model is that, as a consequence of a simple partition equilibrium of the antibiotic between the aqueous solution and the membrane, the conductance of the membrane should depend linearly on the antibiotic concentration in the aqueous solutions. Although the experimental finding of this proportionality *per se* is not sufficient to prove that the present antibiotics act as neutral carriers, the additional finding of an arithmetic mean relationship for the asymmetrical conductances in Table V is only easily understandable for such a neutral carrier mechanism.

The relation of present results to previous work

Solvent extraction experiments similar to those reported here, have been carried out previously by Pedersen for cyclic polyethers (22) and by Pressman for a wide variety of antibiotics (2c). In general, our results are completely in accord with those of these authors. However, it should be pointed out that Pedersen's conclusion that 'the selectivity of a solvent extraction increases as the dielectric constant of the organic solvent is lowered at the same time that the overall efficiency of the solvent extraction drops' should be understood to refer to the total *amounts* taken up in the extraction but not that the selectivity ratios of solvent extraction equilibrium constants vary. Our data indicate that for the present molecules, the ratio is quite constant. A more precise statement would be that the solvent extraction *equilibrium constant* varies with the solvent but not the *ratio* of the equilibrium constants. It should also be noted that the experiments of Pressman have been carried out for much higher concentrations than ours. For this reason, ion pair formation between the anion and complexed cation in the lipid phase cannot be assumed to be negligible in Pressman's experiments; and comparison with our results can only be qualitative in nature.

The present findings are consistent with and extend considerably those reported previously (1), and a number of puzzling phenomena have become understandable in terms of the present results. In particular, the 'peculiar' decrease of conductance previously observed at high concentrations of LiCl (1), as well as the related apparent 'saturation' at high ionic strengths noted by Tosteson (26) are now understandable as due to ionic strength effects. The 'saturation' is not really a saturation, but the outcome of two competing processes leading to the *appearance* of saturation. This phenomenon is markedly enhanced by the presence of cholesterol in the lipid bilayer; and this accounts for the differences between our previous results on cholesterol-free membranes and those of Tosteson on membranes containing cholesterol.

The data of Figures 13 and 14 point out the importance of carefully defining the composition of lipid membranes when attempting to compare results from one laboratory to another. They also point that, although a particular molecule may be found to be ineffective in a given membrane, generalizations from this to all membranes are dangerous.

Lastly, our previous comment (see footnote 4 of reference (1)) that the membrane potential did not show the dependence expected in the presence of a gradient of antibiotic concentration is now understood to be the consequence of the diffusion of antibiotic within the membrane being rapid compared to its rate of crossing the membrane-solution interfaces.

On the likelihood that neutral carrier molecules similar (or even identical) to the present macrocyclic antibiotics constitute a normally occuring mechanism for cation permeation of cell membranes

From the similarities between the electrical properties of artificial bilayers exposed to macrocyclic antibiotics and those of natural membranes (9, 29), it is tempting to speculate that such molecules might provide the basis for normal membrane permeation. Particular similarities to be noted are between the sequences and magnitudes of relative ion permeabilities characteristic of the antibiotics (23, 18, 30) and those characteristic of cell membranes (31). Even more striking is the close correspondence found by Hagiwara recently in Barnacle muscle between cation permeability ratios and conductance ratios (29), a behavior identical to that expected and observed for a neutral carrier mechanism.

If molecules such as the present antibiotics are normally present in all membranes, then they would be expected to exist in finite concentrations in such extracellular fluids as blood and to be in equilibrium with all cells of a given organism. Our finding of the dependence of the effectiveness of such molecules on the lipid composition of the membrane provides a way to understand how a variety of cells having membranes of differing ion permeability might exist all in equilibria with a common extracellular fluid phase in a given organism. There is indeed evidence that the lipid composition of various cell membranes varies from species to species and organ to organ, and it is attractive to view the lipid composition of a given membrane target organ as underlying the specificity of response to antibiotics of that organ.

Perhaps the most suggestive piece of evidence in support of the existance of carrier molecules like the present antibiotics comes from the striking observation made by Osterhout in 1934 (32, 33) and largely ignored since then. Osterhout noted that 'the ability to distinguish electrically between Na^+ and

K^+... appears to exist in *Nitella* only so long as it contains a group of organic substances'... (32), this effect 'can be removed from *Nitella* by leaching in distilled water and *thus removing a substance* (our italics).... This substance can be recovered from the water and applied to the cell and this restores its ability to distinguish electrically between K^+ and Na^+'. To do this, Osterhout extracted the water with petroleum ether and tested the ability of the concentrated residue to restore the normal electrical properties. He concluded that 'the Nitella cell contains a substance which is soluble in petroleum ether and which is responsible for the potassium effect'.

This substance sounds remarkably similar in its properties to the present neutral macrocyclic antibiotics. Osterhout's additional finding that solutions such as blood, saliva, urine, and milk could restore the 'potassium effect' is also consistent with this proposal, since these fluids all share the common characteristics of being aqueous solutions which have been equilibrated with large amounts of membrane material. Were molecules like Monactin present in the normal membrane, then small quantities would be expected to exist in such aqueous solutions. We feel that these observations should be taken seriously and we are presently attempting to extract substances of this type from cell membranes. An alternative to this view would be that antibiotic molecules represent prosthetic groups of larger molecules.

SUMMARY AND CONCLUSIONS

In order to clarify the mechanism by which certain neutral molecules such as the Macrolide Actin antibiotics render phospholipid bilayer membranes selectively permeable to cations, a careful comparison has been carried out, with the guidance of a theory developed for a neutral carrier permeation mechanism, between the ability of these molecules to extract cations selectively into organic solvents and their effects on the membrane potential and conductance of phospholipid bilayer membranes of the Mueller–Rudin type.

Spectrophotometric studies on the equilibrium extraction of picrate, and dinitrophenolate salts of Li, Na, K, Rb, Cs, and NH_4 into dichloromethane and hexane-dichloromethane mixtures over a wide range of ionic concentrations indicates that the simple chemistry postulated for the model is indeed characteristic of the Macrotetralide Actin antibiotics: Nonactin, Monactin, Dinactin, and Trinactin. The equilibrium constant for the extraction of each cation by a given Macrotetralide Actin antibiotic was found to be measurable with sufficient precision for meaningful differences due to the differing number of methyl groups in this series of molecules to be detected. The ratios of selectivities among the various cations were found to be characteristic of a given antibiotic and completely independent of the solvent and chromophore anion.

It has been possible to correlate the effects of the molecular structure of the antibiotic molecules characterizable from the bulk phase partition equilibria with their effects on the electric properties of phospholipid bilayer membranes. Indeed, the properties of the phospholipid bilayer can be 'predicted' in surprising detail from the parameters measured by equilibrium salt extraction. The problem of understanding the effects of macrolides on the ion selectivity of bilayer membranes is therefore reduced, at least formally, to the simpler problem of understanding the equilibrium chemistry of the salt extraction experiments.

By varying the cholesterol content of lecithin membranes we have been able to demonstrate that the effectiveness of a given antibiotic depends markedly on the lipid composition of the membranes. Moreover, the striking ability of cholesterol to decrease the effectiveness of Monactin on the membrane conductance is exactly what would be expected from the expected immobilization of the lipid tails by the cholesterol if Monactin were a mobile carrier. Despite this effect the permeability ratios among cations as well as their conductance ratios are totally independent of the lipid composition, as is expected from the theory proposed here.

Bilayer conductance measurements in presence of the Macrolide Actin antibiotics revealed an apparent 'saturation' at high salt concentrations not expected theoretically. This was found not to be a simple saturation but to be due to a balance between a pure ionic strength effect decreasing membrane conductance in opposition to the theoretically expected effect from increasing the concentration of the permeant cation. When ionic strength was held constant with a relatively impermeant salt, the theoretically expected proportionality between permeant ion concentration and conductance was found.

By carrying out a series of membrane conductance and membrane potential measurements in the presence and absence of a gradient in Monactin concentration, but with the same concentrations of KCl on either side of the membrane, it has been possible to show that the diffusion of the antibiotic molecule is rapid within the membrane compared to its rate of diffusion across the membrane-solution interfaces.

Possible effects of lipid polar head groups on membrane properties have been examined by extending the theoretical considerations to situations in which the surface charge at the lipid-solution interface is no longer neglected. The conditions under which membrane properties are unaffected by surface charges are noted, as well as the conditions under which the effects of these groups become significant. Preliminary experiments on oleylamine membranes are in agreement with theoretical expectations.

All of the above results strongly suggest that the Macrolide Actin antibiotics produce their effects on lipid bilayer membranes by acting as molecular carriers of cations. Since many properties of biological membranes are duplicated by artificial phospholipid membranes exposed to such molecules, the possibility is suggested that neutral carrier molecules like the antibiotics (or even identical to them) are responsible for the electrical properties of biological membranes.

ACKNOWLEDGEMENT

It is a pleasure to acknowledge the generous gifts of Nonactin, Monactin, Dinactin and Tranactin from Hans Bickel of CIBA and of Nonactin from Miss Barbara Stearns of SQUIBB, as well as the perceptive criticisms of Stuart McLaughlin, Max Delbrück and Norman Davidson. We are particularly indebted to the last for clarifying our understanding of the situation in which the diffusion of neutral carriers within the membrane is rapid compared to their crossing the membrane-solution interfaces.

References

1. EISENMAN, G., CIANI, S., and SZABO, G. 1968, *Federation Proceedings*, **27,** 1289.
2. SCHMITT, F. O. 1939, *Physiol. Revs.*, **19,** 270.
2a. PIODA, L. A. R., WACHTER, H. A., DOHNER, R. E., and SIMON, W. 1967, *Helv. Chim. Acta,* **50,** 1373.
2b. GRAVEN, S. N., LARDY, H. A., JOHNSON, D., and RUTTER, A. 1966, *Biochemistry*, **5,** 1729.
2c. PRESSMAN, B. C., HARRIS, E. J., JAEGER, W. S., and JOLINSON, J. H. 1967, Antibiotic-Mediated Transport of Alkali Ions across Lipid Barriers. *Proc. Nat. Acad. Sci. U.S.* **58,** 1949.
3. CIANI, S., SZABO, G., and EISENMAN, G. 1969, A Theory for the Effects of Neutral Carriers such as the Macrotetralide Actin Antibiotics on the Electrical Properties of Bilayer Membranes. I. Neglecting the Polar Head Groups of the Lipid. *J. of Membrane Biology*, **1,** 1.
4. EISENMAN, G., CIANI, S., and SZABO, G. 1969, The Effects of Macrotetralide Actin Antibiotics on Equilibrium Extraction of Alkali Metal Salts into Organic Solvents. *J. of Membrane Biology*, **1,** p. 294.
5. SZABO, G., EISENMAN, G., and CIANI, S. 1969, The Effects of Macrotetralide Actin Antibiotics on the Electrical Properties of Phospholipid Bilayer Membranes. *J. of Membrane Biology*, **1,** p. 346.
6. CIANI, S., SZABO, G., and EISENMAN, G. An Examination of the Rate-determining Step for Ion Permeation of Bilayer Membranes, (in preparation).
7. CIANI, S., SZABO, G., and EISENMAN, G. The Effects of the Charged Polar Head Groups of the Lipid on the Electrical Properties of Bilayer Membranes Exposed to Neutral Carriers, (in preparation).
8. SZABO, G., The Effect of Neutral Molecular Complexers of Cations on the Electrical Properties of Lipid Bilayer Membranes. Ph.D. Thesis, University of Chicago, June, 1969.

Ion distribution equilibria in bulk phases 133

9. CIANI, S., SZABO, G., and EISENMAN, G. 1969, Biophysical Society Symposium on Membrane Selectivity, Feb., 1969, *Biophysical Journal*, **9**, Abstracts, A 81.
10. KILBOURN, B. T., DUNITZ, J. D., PIODA, L. A. R., and SIMON, W. 1967, *J. Mol. Biol.*, **30**, 559.
11. MUELLER, P., RUDIN, D. O., TIEN, H. T., and WESCOTT, W. C. 1962, *Circulation*, **26**, 1167.
12. GEMANT, A. Ions in Hydrocarbons (Interscience, New York, 1962).
13. PEDERSEN, C. J. 1968, *Federation Proc.*, **27**, 1305.
14. GOLDMAN, D. E. 1943, *J. Gen. Physiol.*, **27**, 37.
15. HODGKIN, A. L., and KATZ, B. 1949, *J. Physiol., London*, **108**, 37.
16. SANDBLOM, J. P., and EISENMAN, G. 1967, *Biophys. J.*, **7**, 217.
17. WEISSBERGER, A., PROSKAUER, E. S., RIDDICK, J. A., and TOOPS, E. E. JR. *Technique of Organic Chemistry*, edited by A. Weissberger. New York: Interscience, 1955, vol. VII.
18. PIODA, L. A. R., WACHTER, H. A., DOHNER, R. E., and SIMON, W. 1967, *Helv. Chim. Acta*, **50**, 1373.
19. EISENMAN, G. 1961, *In Symposium on Membrane Transport and Metabolism.* A. Kleinzeller and A. Kotyk, eds. Academic Press, New York, 163.
20. EISENMAN, G. 1962, *Biophysical Journal*, **2**, part 2, 259.
20a. EISENMAN, G., Theory of Membrane Electrode Potentials. In Symposium on Ion-Selective Electrodes of the National Bureau of Standards, R. Durst, editor, U.S. Government Printing Office, in press, 1969.
21. PRESSMAN, B. C. 1968, *Federation Proc.*, **27**, 1283.
22. PEDERSEN, C. J. 1967, *J. Am. Chem. Soc.*, **89**, 7017.
23. MUELLER, P., and RUDIN, D. O. 1967, *Biochem. Biophys. Res. Commun.*, **26**, 398.
24. GRAVEN, S. N., LARDY, H. A., JOHNSON, D., and RUTTER, A. 1966, *Biochemistry*, **5**, 1729.
25. DIAMOND, J. M., and WRIGHT, E. M. 1969, *Ann. Rev. Physiol.*, **31**, 581.
25a. LEHNINGER, A. L. 1968, The Neuronal Membrane. In Neurosciences Research Program Bulletin, Vol. 6, Supplement, Symposium on Frontiers of Molecular Neurobiology, cf. page 17.
25b. CHAPMAN, D., and PENKETT, S. A. 1966, *Nature*, **211**, 1304.
25c. RAND, R. P., and LUZZATTI, V. 1968, *Biophys. J.*, **8**, 125.
25d. FINKELSTEIN, A., and CAFF, A. 1968, *J. Gen. Physiol.*, **52** (2), 145f.
26. TOSTESON, D. C. 1968, *Federation Proc.*, **27**, 1269.
27. GLASSTONE, S., LAIDLER, K. J., and EYRING, H. 1941, The Theory of Rate Processes. McGraw-Hill, New York.
28. PRESSMAN, B. C. 1968, *Federation Proc.*, **27**, 1283.
29. HAGIWARA, S., HAYASHI, H., and TOYAMA, K. 1969, *J. Gen. Physiol.*, in press.
30. LEV, A. A., and BUZHINSKY, P. E. 1967, *Tsitologiya*, **9**, 102.
31. EISENMAN, G. 1965, *Proceedings, XXIIIrd International Congress of Physiological Sciences*, No. **87**, 489.
32. OSTERHOUT, W. J. V. 1940, *J. Gen. Physiol.*, **23**, 429.
33. OSTERHOUT, W. J. V. 1944, *J. Gen. Physiol.*, **27**, 91.

DISCUSSION

GREEN I would like to ask about a problem that you didn't deal with; namely, what are the chemical features of the anion that would determine its effectiveness in forming the *ISX* complex. For example, you compared picrate and 2-4 dinitrophenol. Can you say more about the nature of the anion in relation to the complex formation?

EISENMAN In our experiments on solvent extraction, picrate and dinitrophenolate are substituted for chloride or hydroxide because the work of transferring them from water into a low dielectric constant solvent is less. We performed these experiments in the concentration range where these anions are still dissociated from the complexed-cations so that we were independent of the chemical nature of the anion. In hexane, the reaction between anion and complexed cation occurs to a significant extent, and in that case the parameters depend on the anion and the solvent. Since these are different from those characteristic of the biological or artificial membrane, the correlation between bulk and membrane properties would not be immediate. You will notice, from the values of K_i in Table I, that the partition coefficient of dinitrophenolate between dichloromethane and water is lower than the partition coefficient of trinitrophenolate (picrate). A dependence of the absolute values of K_i on the anion is expected. These differ essentially just because the work of taking the dinitrophenolate anion versus picrate from water into the dichloromethane is different.

GREEN I was wondering about the mechanism by which charges are screened. How are charges in the *ISX* complex screened to make that complex soluble in the organic solvent?

EISENMAN Oh, it's a neutral molecule. Essentially the picture is that you have all cations well screened by the polar oxygens in the center of the complex surrounded by a thick hydrophobic region just as Dr. Onsager presented. The energetic surround of the cation in the antibiotic is, in first approximation, similar to the energetic surround of the cation in water; but the exterior of the complex is hydrophobic, and thus is preferentially partitioned in favor of the organic phase.

GREEN What about the anion? That's my problem. When you add the anion as well and make the complex—

EISENMAN But the anion is not in the complex in the range where reaction 2 is negligible. When the experiments are carried out in conditions for which reaction (2) occurs to a significant extent, we find that there are effects of the anions. Only at higher concentrations in low dielectric constant solvents (e.g.

hexane) does reaction (2) to the completely neutralized complex occur significantly. We specifically used dichloromethane to study the dissociated range where reaction (2) was negligible. The important point is that our solvent extraction experiments were designed to extract the particular combination of parameters which is most meaningfully confronted with the properties of the bilayer, which requires that we measure K_i for reaction (1), uncomplicated by reaction (2). Maybe what I should say is that the bilayer is selectively permeable to cations when chloride is the anion because what we have done with the antibiotic is to solubilize only the cation species preferentially to the anion. If, in fact, the anion in the aqueous-phase had been dinitrophenolate instead of chloride, which of course is one of the ones used as 'poisons' then some dinitrophenolate would be expected to enter the membrane; and you should form some of the neutral ISX species. This would, of course, affect fluxes markedly because the neutral species can shuttle ions across the membrane, but because it has no net charge, it would not contribute to the electrical properties of the membrane in a first approximation. The principal species affecting in the electrical properties would be the excess of the IS^+ species. The anion would markedly affect the fluxes but might still be electrically silent. Is that the answer you were looking for?

MYSELS I wonder whether in your calculation of the average membrane conductivity in the unsymmetrical case, you took into account that the resistance of the entrance of the membrane may not be just at the boundary, but may be in the diffusion layer in the water on both sides. And when you work with such low concentrations and favorable partition coefficients, I would expect that this is where most of it would be.

EISENMAN (see Footnote, p. 124).

IZATT We have recently used a calorimetric titration procedure to measure some stability constants for the interaction of metal ions with the crown 31 compound discussed by Dr. Eisenman. We find it interesting that the variations of these stability constants for the alkali metal ions follow the same order as his permeability ratio measurements; namely, $K^+ > Rb^+ > Cs^+$. Furthermore, we find no heat liberated in the reaction of sodium and lithium with the crown compound. These are aqueous solution measurements. We've also made measurements with the alkaline earth metal ions, and we find that barium interacts much more strongly than potassium. There is no interaction for either magnesium or calcium with the crown 31 compound.

PAPER 6

Possible mechanisms of ion transit

LARS ONSAGER

TODAY I CAN do little more than repeat what I have said elsewhere; but it might be new to some members of this audience.

Like many others I have been wondering how ions get through the walls of cells. This is a controlled process; inside and outside concentrations of a simple ion like Na or K often differ by an order of magnitude or more. As Dr. Kedem has pointed out, water does get through, not quite readily but perhaps more like the way it can penetrate through a thin layer of plastic.

We have more or less accepted the idea, supported by many observations, that the main barrier to the passage of ions is a lipid bi-layer. The hydrocarbon tails of the lipid molecules provide a non-polar layer some 50 Å thick; the polar heads of the molecules remain in contact with each other and with the aqueous phases on both sides of the membrane.

Solubility studies support the notion that such a membrane would serve as an effective barrier for ions. The non-polar interior of the bilayer would be much like a homogeneous solvent of dielectric constant around 2.5, with no polar groups. Water dissolves sparingly in such liquids, and most salts of small ions are practically insoluble. Some salts of very big ions like tetrabutyl ammonium picrate will dissolve and ionize just slightly in benzene or dioxane. The nitrates and the thiocyanates of similar cations behave similarly— one big ion is often enough.

Nevertheless, the permeabilities of membranes to common ions are much greater than we should expect from our general experience with non-polar solvents; so we suspect special facilities for ion transport. This suspicion is very strong, because membranes often discriminate sharply between fairly similar ions, and selective inhibition of ion transport has been achieved by remarkably small quantities of specific poisons like ouabain or tetrodotoxin.

What kinds of natural facilities mediate ion transport? The guesses fall into two categories. For one, there might be mobile carriers, ionic or neutral, which can pass in one direction loaded with one kind of ion and return either empty or with a different cargo. A finite electrical conductivity indicates that charged entities of some sort can pass in at least one direction.

The other class of theories postulates fixed facilities, which might function like pores through the membrane containing an aqueous solvent or something like that. Here an important constraint on our speculation is that the structures must be stable.

Whether we deal with mobile carriers or with fixed installations, the simple considerations of electrostatics severely limit the scope for conjecture.

For a neutral cation carrier we need a compact structure with a central cavity big enough to contain the ion. In order to attract the ion, the electrostatic potential must be negative inside, which calls for a set of dipoles, reasonably symmetrical, all directed outwards. In such an arrangement the interaction of the dipoles is repulsive, and a little thought shows that this is a required characteristic for a chelating agent which will function in a non-polar environment; some electrical stress must be relieved by the incorporation of the ion, or it will not be attracted. Thus we may conclude that a set of separate polar molecules or even polar groups attached to an open, flexible chain will not be very effective; because in either case the dipoles can turn so as to relieve the electrical stress. Such natural carriers as have been found (nonactin, valinomycin) are in fact cyclic compounds (one a polyether-ester, the other a polypeptide), whose structures do not permit much relief of the electrical stresses by changes of conformation.

These carriers can obviously mediate uncontrolled passive transport. It is hard to see how they could have any other capability, and no physiological function has been demonstrated; for mammals they are rather potent poisons.

In order to sustain an electric current, a mobile carrier must obviously traverse the membrane in one direction or the other with a full elementary charge aboard, a neutral carrier of cations in the loaded condition, an anionic carrier when it returns empty.

If we are dealing with fixed facilities for ion transport, then I prefer to look for the narrowest serviceable 'pore'. Its essential feature would be a chain of dipoles or at most a very small set of chains. Such a chain has two configurations of low energy: all positive ends inward or all outward. The most important electrical characteristic of such a chain is its 'end charge', which equals the dipole moment per unit length. For a chain of hydroxyl groups we might expect end charges of about 10^{-10} e.s.u. Apart from very local stray fields, the electrostatic field produced by a chain is just the sum of the fields due to the end charges, and if only the ends are located in regions of high dielectric constant, the entire field energy is quite small.

If a cation moves along such a chain accompanied by a pair of negative end charges of 10^{-10} e.s.u., the net charge present and travelling is not a full 4.8×10^{-10} e.s.u. but only 2.8×10^{-10} e.s.u. and the electrostatic energy is reduced accordingly—by a factor of about 3. We must remember that only the net charge of 2.8×10^{-10} e.s.u. moves with the ion itself, and that it reverses the polarization of the chain in the process. In order to sustain a steady current the chain must be capable of spontaneous repolarization by the passage of a bonding defect, which carries the remaining 2×10^{-10} e.s.u. (two end charges). A hydrogen ion can traverse a chain of hydroxyl groups quite readily by successive proton jumps along the bonds. Some information about the possible properties of dipole chains is available from the studies of protonic semiconductors, particularly ice and other disordered solid phases of water; a measure of general disorder is needed for any mechanism of conduction which involves the rotation of dipoles. Formic acid is clearly a protonic semiconductor; it could yield important information about chains of carboxyl groups. Unfortunately, we know just enough about it to crave more. KH_2PO_4 and its analogs have been studied quite thoroughly; phosphate groups are unlikely candidates for transport facilities in membranes; but they play most important roles in bioenergetics. By the way, in KH_2PO_4 the ions carry about 20 per cent of the electric current and the bonding defects carry 80 per cent. Of course, the former determine the dielectric relaxation time (in the nanosecond range) and the high frequency conductivity; the bonding defects determine the steady conductivity, which is smaller by about 8 orders of magnitude.

It seems quite possible that dipole chains could be formed from polar groups in the side chains of amino-acids—serine and threonine for hydroxyl groups, aspartic and glutamic acids for carboxyl groups, with other kinds just possibly participating. The polar groups of the peptide chains themselves are rather more doubtful candidates; they cannot turn easily, and they may be fully committed if they have to maintain some rigid supporting structure like our α-helix.

DISCUSSION

ADAM   I should like to ask: If you have two of these chains, both aligned in the same direction, and one in this direction and one in the opposite, what would be the difference of chain-interaction energy between the two cases assuming a certain distance of two of these chains?

ONSAGER   As far as I can see, it could be either side of zero. It could be zero, it could be positive, it could be negative. It can be cooperative; it can be anti-cooperative. It is hard to say.

ADAM   I mean, what would be the order of magnitude if you have two of these chains, for instance, within a distance of 20 Angstroms or so?

ONSAGER   20 Angstroms? Very close to zero.

KURSUNOGLU   This polarization you mentioned—does the resultant electric field accelerate the ion?

ONSAGER   The possibility of polarizability would be high because the dipole moment is a quarter of an elementary charge, multiplied by the thickness, so that the dipole moment is 50 Debye. So, it doesn't take very many of them to add something to the observed dielectric polarizability of a membrane.

ADAM   Have you estimated how long it takes, for instance, a cation to move along such a chain? I would imagine that it is different for different cations. Maybe you can tell us how long it takes for a proton; probably that is better known.

ONSAGER   The proton might do it very fast.

ADAM   What do you mean by 'very fast'? In a millisecond or a microsecond?

ONSAGER   You get down to something of the order of 10^{-10} second.

ADAM   What about another cation, say potassium or sodium?

ONSAGER   That's hard. I think it would be slower though than the corresponding passage of potassium ion in water, say.

CHRISTENSEN   Does the model have any provision for the recognition of sodium ions from potassium ions?

ONSAGER   It should be quite selective.

CHRISTENSEN   And how is this arranged?

ONSAGER   Because there is less complete solvation than you find in the liquid, which means, that if you just take a fraction of the solvation energy you have

Possible mechanisms of ion transit

in water and even a fraction of it is many kT, a factor of several orders of magnitude is not out of the question for such a facility.

KEDEM   The proton zipping mechanism could be tested by a comparison of proton conductance and diffusion of tritiated water. A Grotthuss chain would constitute what biologists term a 'single file' mechanism and the isotope permeability should be considerably smaller than the value expected from the conductance and a simple Nernst equation.

ONSAGER   Provided you have proton conduction. Proton conduction here is a possibility. It is not a necessary property of every facility. It is a likely property, however, of a carboxylic acid chain. For that sort of a chain, I would be surprised if you don't have it, as one possible capability.

BLANK   Is it possible to estimate the temperature dependence of such a process?

ONSAGER   Well, experience shows that bridging defects have an activation energy of $13\frac{1}{4}$ kcal. It happens to be the same for ice and for a phosphate. But that is in a solid, three-dimensional network and, for a chain like this, it is not out of the question that it might be slightly more. 15 or 16 kcal would not seem unreasonable for such a model.

ADAM   May I ask another question? How would you think it is possible to detect such a structure? Would you think that one can observe this highly ordered structure of dipoles maybe by infrared spectroscopy; or what are the specific features which would enable us to detect them? Maybe nuclear magnetic resonance? Have you thought about this?

ONSAGER   I'm not the only one that has thought about it. I think Dr. Wallach might tell us something tomorrow. There is some evidence, from rotatory dispersion, of the presence of helical protein in membrane. It may be hard to decide whether it sticks this way or that way. But it is not out of the question that such evidence may become firmer.

PAPER 7

Physical principles in monolayer and membrane permeation†

MARTIN BLANK‡ and JOHN S. BRITTEN
Department of Physiology, College of Physicians and Surgeons, Columbia University, 630 West 168 Street, New York, New York 10032

I INTRODUCTION

IN A conference devoted to 'Physical Principles of Biological Membranes' one should, perhaps, review a group of related studies on biological membranes in terms of the physical principles that apply. This course of action is essential, but it is also very difficult because of the varied properties that are found in biological systems. In addition, one inevitably compares the biological systems to the common physical systems normally studied (e.g. ion exchange membranes), which are usually very different in both composition and structure. An alternative approach is to find a simple physical system that is related to the biological membrane (in size, structure or composition), determine the physical principles that apply and attempt to extrapolate to the case of the biological membrane. This type of approach, using model systems, is somewhat easier in terms of determining physical principles, but the difficulties arise when extrapolating the findings to the case of biological membranes.

The second approach, utilizing monolayers to study the permeation process in very thin films, has been employed in this paper. The physical principles involved in the monolayer process are discussed at some length and this

---

† The work described in this paper has been supported by Research Grants from the U.S. Public Health Service (GM-10101), the Office of Saline Water, U.S. Department of the Interior (14-01-0001-1797) and the National Science Foundation (GB-6846).

‡ Supported by Research Career Development Award (K3-GM-8158) of the U.S. Public Health Service.

is followed by considerable speculation about the expected properties of biological membranes. These speculations lead to somewhat different conceptions about the properties of biological membranes, but they have the advantage of being based on the known properties of physical systems.

II GENERAL CHARACTERISTICS OF MONOLAYER PERMEATION

The study of monolayer permeation is in essence a study of the diffusion process at the molecular level. The size of the permeant is comparable to the size of the monolayer, and the assumption of a multi-collision process with consequent averaging, which applies to macroscopic systems, cannot be made in this case. Because of this fundamental difference one may expect notable differences between the permeation of monolayers and of macroscopic phases. Such differences have been observed.

The permeability of monolayers to various substances has been measured for monolayers composed of various polar and non-polar groups, at different surface pressures and at several temperatures. The variety of measurements provide evidence for some generalizations about the process of monolayer permeation (1, 2). For example, the least permeable monolayers, formed from the saturated, straight chain fatty acids, alcohols, etc. of C_{16} chain length and longer, are of the close-packed incompressible type. In a homologous series, the permeability is related to the length of the hydrocarbon chain, and in a series of compounds having the same chain length, the permeability depends on the size of the polar group. These effects correlate the permeability inversely with the compactness of the film and directly with the amount of free space in it. As is expected, the monolayers that are almost freely permeable form relatively open structures such as those of protein, cholesterol or oleic acid films. In keeping with our ideas about the effect of monolayer structure upon diffusion rates, the differences between the permeabilities of a monolayer to different gases are best explained by the effect of molecular size of the permeating species. (The permeation rate appears to vary exponentially with the molecular cross-sectional area of the permeant.) This last observation means that monolayers have functional pores, although there is no reason to expect the presence of fixed structural pores.

Monolayer permeation appears to resemble the process of diffusion in macroscopic solid phases, where the permeability depends upon the size and number of holes available in a lattice. Further support for this parallel comes from experiments which show interference effects between a permeant and another non-diffusing component of the system. Studies of monolayer permeability to gases (3, 4) have indicated that water vapor is operationally present in the monolayer as an additional resistance. Since the subphase is in equilibrium with the gas phase, water molecules must pass back and forth through the monolayer and they apparently act as a resistance. The movement of water can be reduced by dissolving a salt, such as LiCl, and the total vapor pressure can be raised by dissolving a volatile solute such as methanol. These experiments, as well as those in which the vapor pressure of water is varied by changing the temperature, all support the view that the vapor molecules from the subphase interfere with the passage of other gas molecules. Two substances can usually interdiffuse independently in an inert bulk liquid phase, but in a solid-like phase, if the holes normally available for diffusion are occupied, a lower permeation rate would be expected.

The comparison of monolayer permeation to the permeation of solids is quite reasonable in the light of the above observations. In addition, the magnitudes of the permeability and of the activation energy, although somewhat lower, are nevertheless comparable to analogous values for solids. However, unlike the bulk phase process, the monolayer process exhibits a variation of the activation energy with the monolayer thickness. This observation implies that the permeation occurs as a one-step discontinuous process. Further support for this idea comes from the observation that there are no interfacial partition effects, i.e., interactions of the permeant with the monolayer, during permeation. The permeabilities to the various gases show no correlation with partition coefficients, as estimated by assuming that the monolayer shows the same solubility as hexadecane (1). These results point to a qualitatively different kind of process when dealing with monolayers. (Operationally, we find that Fick's law is not obeyed in monolayer permeation, since the permeation rate is proportional to the concentration difference and area, but not inversely proportional to barrier thickness. This was easily demonstrated by using monolayers of different chain lengths.)

In summary, the monolayer can be thought of as a solid 'phase' for the purposes of understanding the permeation process, but one in which the number of events involved is not sufficient to allow for averaging. Thus the permeation of a monolayer involves a one-step jump across the layer, just as in a solid lattice there is a jump of permeant to a hole. This process should be of considerable interest to the theoretician who now has an opportunity to test some of his assumptions about physical processes at the molecular level. The theoretician also has an additional constraint upon his efforts since any description of a monolayer process must, upon extrapolation to thicker systems, eventually give rise to the familiar properties of diffusion in bulk systems. Monolayer permeation should be of particular interest in connection with the study of natural membranes, since the two systems have a number of similar properties, and monolayer systems can serve as convenient models for the study of membrane processes.

III THEORIES OF MONOLAYER PERMEATION

Monolayer permeability has been described in terms of an energy barrier at the surface where the monolayer prevents the molecules at the lower end of the kinetic energy distribution from passing through, and therefore acts as an energy sieve. The theory of an activation energy barrier to the evaporation of water through monolayers was first suggested by Langmuir and Langmuir (5), and developed further by Langmuir and Schaefer (6). A Boltzmann expression gives the fraction, f, of molecules which have an energy in excess of E, the energy needed to permeate a monolayer:

$$f = \exp(-E/kT) \qquad (1)$$

where k = the Boltzmann constant and T = the absolute temperature and E is the energy needed to compress the monolayer to make a large enough hole. (Therefore $E = \pi a_0$, where π = surface pressure and a_0 = the cross-sectional area of permeant.) Since the rate of permeation should vary directly as the fraction of molecules having the required energy for penetration, f is proportional to the permeability.

Archer and LaMer (7) considered E in terms of the energy required to form a vacant site in a monolayer lattice, and related E to the vaporization energy of the monolayer. They determined the magnitude of the activation energy for the penetration of monolayer and of its individual CH_2 groups. The energy barrier for fatty acids in the liquid-condensed (LC) surface phase is about 15 kcal/mole depending upon chain length, and the contribution for each CH_2 group, about 300 cal/mole, generally varies with the lateral compression of a monolayer (8).

The energy barrier theory has proved useful in giving a quantitative description of a permeation process that depends on the surface phase, the length and size of the hydrocarbon chain, the size and nature of the polar group, the surface pressure, and the cross-sectional area of the permeant. However, other properties intrinsic to the monolayer such as compressibility (9), free surface area (1) and the 'line tension' (10) are also needed to describe permeation, and add to the complexity of the molecular model.

A different approach to a theory of monolayer permeation has been attempted (11), utilizing some of the more recent studies of monolayer properties summarized above. The new model suggests that free spaces in the monolayer become available for permeation from the natural free area in a lattice and from the equilibrium fluctuations in monolayer density at a (gas molecule-monolayer) collision site. From the entropy change associated with an expansion of a monolayer, it is possible to estimate the probability of a given expansion. The monolayer resistance can then be derived if it is assumed that all local expansions which yield an area equal to or greater than the cross-sectional area of the permeant result in passage through the monolayer.

This approach seems quite reasonable in principle, although the earliest work along these lines was deficient in a number of respects. Some of these deficiencies have been overcome in a more recent paper on the transport properties of monolayers (12), and in this paper we hope to summarize our current thinking on this problem.

IV FLUCTUATIONS IN MONOLAYER DENSITY

Let us consider an insoluble monolayer at an air-water interface as a hexagonal close-packed lattice (condensed phase) of molecules at equilibrium. The molecules are assumed to be perpendicular to the surface, and the interactions between the monolayer molecules themselves and between the molecules and the subphase are reflected by the surface isotherm. It will also be assumed that there are no additional effects due to the water that is bound to the monolayer and that the monolayer molecules are mobile.

The area per molecule, A_m, is equal to the total area of the surface divided by the number of molecules present. The cross-sectional area of the molecule, A_x, is less than A_m because of the inability to utilize all the space available for packing. The difference between these two areas, $A_m - A_x$, represents the free area per molecule, A_f. Therefore, A_f represents the area of two trigonal holes in the assumed lattice, and on the basis of experimental evidence is equal to about 10 per cent A_m.

The monolayer is at constant area, but as a result of normal molecular motions there are fluctuations in monolayer density in small regions of the monolayer. To estimate the probability of fluctuations, we assume a change in the area (A) of a small region of the monolayer while the rest of the monolayer acts as a reservoir for this process. For this system, which is at constant temperature (T) and composition, the variation in the entropy due to a reversible expansion (which we assume to be valid for a local expansion also) is

$$\Delta S = \int \frac{dE'}{T} - \int \frac{\gamma \, dA}{T} \tag{2}$$

E' = the surface energy, γ = the surface tension and Eq. (2) is an expression of the application of the First Law to the surface.

Since the surface energy per unit area is

$$E = \gamma - T\frac{d\gamma}{dT} = \frac{E'}{A} \tag{3}$$

then at constant T

$$\Delta S = \int \frac{A d\gamma}{T} - \int \left(\frac{d\gamma}{dT}\right) dA - \int A d\left(\frac{d\gamma}{dT}\right) \quad (4)$$

The last term on the right hand side of Eq. (4) can be set equal to zero according to arguments advanced earlier (12). Because of the presence of a condensed monolayer

$$\gamma = k_1 A_m - k_2 \quad (5)$$

where k_1 is related to the monolayer compressibility and k_2 is the surface tension intercept at $A_m = 0$. The surface pressure, π, is equal to $\gamma_0 - \gamma$, where γ_0 is the surface tension of pure water, [Eq. (5) is sometimes given in terms of π.] Evaluating $d\gamma$ from Eq. (5), substituting into Eq. (4), and integrating over a change in molecular area yields

$$\Delta S = \frac{I \Delta A}{T} \quad (6)$$

Here $I = k_1 A_m - T(d\gamma/dT) = E + k_2$ has the units of energy per unit area, and is relatively insensitive to variations in surface pressure. The magnitude of I is about $10^2 \frac{\text{ergs}}{\text{cm}^2}$ for condensed fatty acid and fatty alcohol monolayers.

The probability, W, of observing a spontaneous decrease in entropy is

$$W = W_0 \exp\left(-\frac{\Delta S}{k}\right) d(\Delta S) \quad (7)$$

Here W_0 is the probability of the equilibrium state ($\Delta S = 0$), and k is the Boltzmann constant. If we substitute for ΔS due to an expansion, we get

$$W = W_0 \exp\left(-\frac{I \Delta A}{kT}\right) d(\Delta A) \quad (8)$$

The frequency distribution of fluctuations around the equilibrium area is not symmetric. However, it is reasonable to assume that the expansions are exactly balanced by contractions so that A_m is the median of the distribution. Therefore, we can evaluate W_0 by considering only the expansions, integrating Eq. (8) over all values from $\Delta A = 0$ to $\Delta A = \infty$ and setting the integral equal to one-half. The result is

$$W_0 = \frac{I}{2kT}$$

Therefore, the probability that a molecule will occupy an area $A_m + \Delta A \pm d(\Delta A/2)$ for $\Delta A \geq 0$ is

$$W(\Delta A) = \frac{I}{2kT} \exp\left(-\frac{I\Delta A}{kT}\right) d(\Delta A) \tag{9}$$

This method of dealing with density fluctuations in a monolayer utilizes macroscopic equations to evaluate the processes that occur with single molecules. Since one is dealing with a large number of molecules and an even larger number of events (expansions and contractions), this approach is statistically valid.

Eq. (9) gives the probability of a given fluctuation, i.e., the relative frequency of a fluctuation. If we let v be the number of fluctuations of any size per molecule per unit time, Eq. (9), multiplied by v, gives the absolute frequency of the fluctuations in area. The following method has been used to evaluate v. π, which is equivalent to the two-dimensional spreading pressure per unit length, can be calculated on the basis of the rate of change of momentum at a boundary due to the fluctuations. Since the expression for π contains the frequency, we can obtain v in terms of the parameters related to the properties of the surface isotherm.

Let us assume that energy is conserved on the submicroscopic level during a fluctuation, i.e., the kinetic energy with which the molecules fill a 'hole' equals the potential energy of the 'hole' given by Eq. (3). The mass that is shifted during this fluctuation is equal to $(\Delta A/A_m)m$, where m is the mass of a monolayer molecule. Therefore, the momentum change that accompanies this shift can be calculated in these terms (12), and for each fluctuation $\Delta A > 0$, is given by

$$\Delta \text{ momentum} = 2\sqrt{\left(\frac{2Em}{A_m}\right)} \Delta A \tag{10}$$

If A_m is the area per molecule, A_m^{-1} is the number of molecules per unit area, and there are on the average $\sqrt{A_m^{-1}}$ molecules per unit length in any random direction.

The momentum change at the boundary (or at any arbitrary line in the monolayer) per unit length per unit time is equal to the surface pressure.

$$\Pi = 2\sqrt{\left(\frac{2Em}{A_m}\right)} \cdot \sqrt{A_m^{-1}} \cdot v \cdot \frac{I}{2kT} \int_0^\infty \Delta A \exp\left(-\frac{I\Delta A}{kT}\right) d\Delta A$$

Therefore,

$$\Pi = v\sqrt{(2Em)} \frac{kT}{IA_m} \tag{11}$$

If we solve for the frequency, we get

$$v = \frac{1}{\sqrt{(2Em)}} \frac{\Pi A_m}{kT} \qquad (12)$$

Substituting values into Eq. (12) we can calculate v. If we use the isotherm data on stearic acid or octadecanol for $\pi = 20$ dynes/cm., $v \approx 10^{11}$ sec.$^{-1}$. This means that there is a very short lifetime for any position of a monolayer molecule. For the sake of comparison, the mean vibration frequency of carbon in a diamond lattice is on the order of 10^{13} sec.$^{-1}$, so the above value is within the range of observed frequencies.

Eqs. (9) and (12) have been used to derive the self-diffusion coefficient, the viscosity and the thermal conductivity of a monolayer in terms of properties that are related to the surface isotherm (12). The results provide some interesting insights into monolayer properties. The permeability of a monolayer can be treated along the same lines, and this will be done in the following section.

V THE PERMEABILITY OF A MONOLAYER

We assume that for a gas molecule to pass through a monolayer, the two-dimensional lattice must contain a free space at least as large as the permeant. The free space may arise from the free area in the lattice, A_f being the free area per molecule, and it can also arise as a result of the fluctuations in monolayer density. [In a previous paper (11) it was assumed that the penetrating molecule can utilize its kinetic energy to do work against the surface forces in the monolayer, but this mixing of molecular properties in bulk phases with average surface energies is not valid.] The probability of permeating is, therefore, equivalent to the probability of the permeant's finding a hole of sufficient magnitude. This assumes that the permeation process is all-or-none, i.e., either a gas molecule gets through in one shot when it collides or else it is reflected.

From the point of view just outlined the free area per molecule is $A_f + \Delta A$, and the probability of hitting a free area is

$$W_f = \frac{A_f + \Delta A}{A_m} \qquad (13)$$

The number of molecules that strike 1 cm^2 of surface per second, N_1, is given by kinetic theory as

$$N_1 = n_s \sqrt{\left(\frac{RT}{6.28M}\right)} \qquad (14)$$

where n_s = number of molecules per cm³ of gas, M = molecular weight and R = gas constant. N_1 divided by A_m gives the number of gas molecules hitting the average area of a monolayer molecule per second. Since $A_f \approx 0.1\, A_m$, the number of gas molecules hitting a free area per second (on a per monolayer molecule basis) is

$$N_f = \left(\frac{A_m + 10\Delta A}{10 A_m^2}\right) N_1 \tag{15}$$

The absolute probability of observing a fluctuation of magnitude ΔA is equal to $v\,W(\Delta A)$, or Eq. (9) times Eq. (12). We can then determine the number of molecules that get through the monolayer,

$$N = \int_{a_0 - A_f}^{\infty} N_f\, v W(\Delta A)\, d(\Delta A) \tag{16}$$

Substituting Eqs. (9), (12) and (15) into Eq. (16) and integrating over the range indicated, we get

$$N = \frac{\Pi N_1}{2\sqrt{(2Em)A_m}} \left[\frac{I a_0}{kT} + 1\right] \exp\left(-\frac{I(a_0 - A_f)}{kT}\right) \tag{17}$$

To get the permeability we can divide Eq. (17) by N_1 and

$$P = \frac{\Pi}{2\sqrt{(2Em)A_m}} \left[\frac{I a_0}{kT} + 1\right] \exp\left(-\frac{I(a_0 - A_f)}{kT}\right) \tag{18}$$

Equation (18) has been derived in such a way as to emphasize the qualitative aspects of the thinking that produced the final expression. For example, no attempt has been made to deal with the problem that the actual free area that a permeant finds is produced by the combined effects of several adjacent monolayer molecules. Nor has there been a consideration of the related problem of the possible coupling effects of the expansion of one molecule on the compression of an adjacent one. These considerations are bound to affect the final results of any derived expression, but they should not alter the general picture of a monolayer that has been discussed here. (We are currently attempting to overcome these problems by dealing with the distribution of holes directly, rather than through the distribution of molecular areas.)

It should be noted that Eq. (18) has the same form as Eq. (1) and it appears to have many of the qualitative characteristics required in a description of monolayer permeation. Although the variation of P with surface compression is complex (π, A_m, A_f, E, I, varying simultaneously), it appears that P should vary (approximately) directly with π as has been generally observed (9, 10).

The compressibility factor (i.e. k_1 of Eq. (5)) appears in I and should have a pronounced effect upon P as has been mentioned. Also the dependence of P on the cross-sectional area of the permeant is approximately the same as has been observed (1, 13). These qualitative properties of monolayer systems that are incorporated in Eq. (18), support this approach to monolayer properties and also give a molecular basis for the observed macroscopic energy barrier. However, the calculated energy is much too low and in general there is no quantitative agreement.

Part of the lack of quantitative agreement must be due to known factors that have not been taken into consideration. For example, the gas that permeates the monolayer and subsequently dissolves in the aqueous subphase is subject to interference from water vapor molecules. Water is always present as a support for a monolayer, is in equilibrium across the monolayer, and water vapor molecules are constantly exchanging between the gas and liquid phases. The holes that become available in the monolayer may be used by the water molecules which then block their use by the permeating gas. This effect has been observed in monolayer permeation (4), and its existence should increase the effective energy barrier over that calculated by Eq. (18).

Another factor that has not been taken into consideration yet is the finite thickness of the monolayer. It has been assumed that all colliding molecules that can fit into a hole actually do pass through. However, the permeant must avoid a collision with the walls of a vacant site in a monolayer if it is not to be reflected, and this imposes a great restriction upon the angle of approach of a permeant to a vacant site that will result in permeation. The permeability can be derived on the basis of angular selection (11), and the result is a much lower permeability with the same qualitative dependence on monolayer parameters. However, this type of calculation ignores another, perhaps even greater, complicating factor which has to do with the short average lifetime ($\approx 10^{-11}$ sec) of a fluctuation. If a molecule travels with an average speed of about 400 meters/sec. (CO_2 at room temperature), then it takes about 5×10^{-12} sec. to traverse a distance of 20 Å. This means that it takes almost as long to cross the monolayer as the lifetime of a fluctuation, and that a permeant is more likely to collide with a monolayer molecule and be reflected. This additional factor means that only the more rapidly moving gas molecules stand much of a chance to permeate, thereby reducing the effective N_1 of Eq. (14) and lowering the permeability considerably.

These problems complicate the derivation of any expression of monolayer permeability, but they do not in any way weaken the qualitative base that has been outlined for considering monolayer processes. It is for this reason that it appears worthwhile to extend this approach to a qualitative consideration of the factors involved in membrane processes.

VI BILAYERS AND NATURAL MEMBRANES

The previous discussion on the properties of monolayers, which has been developed on the basis of fluctuations, utilizes well known theoretical and empirical relationships in the field of surface studies. However, the crucial factor in this development is the relatively small number of molecules involved in a monolayer process, which leads to relatively large fluctuations. Therefore, it is monolayer size (rather than the detailed structure and energetics) which leads to the special properties.

A number of years ago, Schroedinger (14) considered the problem of fluctuations in biological systems, largely in terms of the validity of the statistical laws of physics as applied to the problems of heredity. He thought that since systems must be relatively large in order to minimize the deviations (fluctuations) from these laws, genes were too small to be reliable in a statistical sense. Therefore, nature had to utilize a different mechanism, now well known to us, to guarantee order. Membranes are also too small in the statistical sense, but here their size may provide an important clue to an understanding of their mode of action.

Without going into a long discussion on the pro's and con's of various detailed models of the natural membrane, most investigators agree that the membrane is only about two to three times larger than the monolayers considered earlier. It is therefore reasonable to expect that the molecules composing the membrane undergo similar fluctuations of roughly comparable magnitude and frequency, and that these effects give rise to the passive permeability properties of natural membranes. Therefore, it should be possible to account for the observed functional porosity of membranes, interference effects during permeation (i.e., the observed 'drag' effects), etc., directly as a consequence of membrane structure. The observed (i.e., fixed) permeabilities result from the effects of the fluctuations averaged over the large membrane areas in any group of cells studied.

The view of membrane structure where the actual permeability barrier is the lipid bilayer is considered by many to be a reasonable model for discussing membrane properties. Specifically, this model offers a way of approaching the problem of permeation based on fluctuations as a direct extension of the monolayer process. However, in this case, no simple relations between area and a coordinate intensive property of the bilayer are known. The energetics of bilayers have thus far been studied only in the case of equilibrium with bulk phase reservoirs (15), which complicate the system.

There are additional complications when extrapolating to the case of bilayers and membranes. The larger numbers of molecules involved imply smaller fluctuations, but since the energetics and the degree of coupling between the two layers are not known, it is difficult to estimate the effect on permeation. The molecules forming bilayers are generally longer, decreasing the permeability because of the increased cohesion and the angular selection discussed earlier, but they are also of greater cross-sectional area which leads to greater free areas in any lattice. Furthermore, there are several possibilities that are not present in monolayers. The presence of permeating molecules between the two layers in the event of permeation of one monolayer at a time, the adsorbed protein monolayers on the polar surfaces, the presence of aqueous phases on both surfaces of the bilayer leading to hydration effects (16), etc. all add to the difficulties of applying these ideas to membranes directly.

However, several interesting ideas emerge from a consideration of fluctuations in membrane bilayers as a mechanism of permeation. The existence of functional pores in the absence of structural pores, the interference effects observed during counter transport, etc. should follow directly from this model. The fluctuation model also leads to ideas about membrane structure. For example, if fluctuations are to result in the formation of transient pores, then an oriented structure, as in a lipid bilayer, should be far more efficient than a random arrangement of molecules as in a bulk liquid. Finally, the possibility of water being present between the layers (16), allows the introduction of permeant solubility or partition between the aqueous phases and the membrane, and an extrapolation to the properties of bulk systems.

VII 'CARRIERS' AND ACTIVE TRANSPORT

The approach to membrane properties via fluctuations has certain advantages in attempting to characterize passive permeation properties (1, 2). It has an additional advantage in providing a framework for speculation about the more specialized aspects of natural membranes, such as membrane 'carriers' and active transport.

Let us return for a moment to the case of monolayers and consider the conventional view of an oriented array of molecules with hydrophilic groups toward the water and hydrophobic groups away from the water. This static view is very useful for considering many monolayer properties, but it is considerably oversimplified. The constant random movements of the molecules in the monolayer result in variations in the average molecular areas and must occasionally result in the transient localized inversion of molecules. Although the probability and the rate of these overturnings must be small, there have been reports of such effects in monolayers. For example, Langmuir [see reference (18)] and Gaines (19) more recently, have observed effects in the case of monolayers deposited on solid surfaces that can only be explained by the overturning of molecules. Similarly, the transfer of monolayer molecules from one solid plate to another on contact (19) implies an overturning in the process, and it is interesting to note that the transfer between plates is accelerated in a vacuum. The process of 'monolayer penetration' [see reference (18)] is another phenomenon which probably involves the overturning of molecules adsorbed from the subphase, and this process depends on monolayer charge and packing. Still another observation that may be due to the overturning of molecules is the evaporation of monolayer molecules from a liquid surface (20). All of these instances emphasize the considerable freedom of movement of the molecules in a monolayer, and support the model that has been presented.

Let us now consider the monolayer model in this context. Despite the problems involved in estimating the forces governing the orientation of monolayer molecules, the model predicts that a finite number of monolayer molecules will be overturning at any time. Since we have expressions for the frequency and probability of density fluctuations, it is possible to calculate the number of times per second that a molecule has enough space available for turning around. For example, for stearic acid or octadecanol of cross-sectional diameter 5 Å, the area must expand to about five times the average area in order for the molecule to have enough space to lie on its side. (This, of course, does not take into account the shape factor.) Integrating Eq. (9) over all values of ΔA between $5A_m$ and infinity, we obtain a value of about 10^{-9}, or one chance in a billion. If we multiply this value by the estimated frequency of fluctuations, $\sim 10^{11}$ per sec., we see that there are about 100 chances per second (per unit area) that a molecule has enough space available for overturning. (This order-of-magnitude calculation ignores the fact that the distribution $W(\Delta A)$ may not be valid at the extremes, that a finite time is required for turning around, that molecules from the vapor phase may block the holes that open up, etc., but it does indicate that a possibility exists for the overturning of molecules in a monolayer.)

What factor will determine the distribution of molecules between the normal and the overturned state? Haydon and Taylor (17) have estimated the solubility of a polar group in a lipid environment and determined a factor of about one in 500 as a reasonable distribution based on free energy considerations. If one assumes that the surface potential, which is generally about 400 millivolts, determines the distribution, the factor would be about one in 100,000. In either case, the local expansion of the monolayer that leads to the overturning would cause a transient change in favor of the overturned state, because the expanded state has a lower surface potential, etc.

On the basis of the above discussion and especially in terms of what has been observed in monolayer systems, it is reasonable to suppose that overturning and the inter-layer transfer of molecules occurs in bilayers and in membranes as a result of the same kinds of forces as described above. [Molecular transfers have already been measured between the two halves of a stearate bilayer (21).] Without having to postulate special properties of bilayers or membranes, the normal fluctuations to be expected in small groups of molecules provide us with a possible mechanism of action of a 'carrier' molecule. Carriers are believed to bind and carry small ions and molecules across a membrane, and molecules on one side of a bilayer can bind permeants, overturn as a kinetic unit, move into the second layer of molecules and equilibrate their bound species with the second aqueous phase. (This mechanism does not rule out the presence of non-carrier mediated permeation in membranes.) Phospholipids, which are found in abundance in most membranes and which are known to bind many simple permeants (22, 23), may perform a 'carrier' function by this mechanism.

In the case of bilayers and membranes it is obvious that any scheme incorporating the idea of fluctuations and overturning leading to 'carrier' transport will have to take into consideration the different binding constants for various substances, the fact that the outer of the layers of a membrane contains more molecules than the inner one, and the membrane factor that determines the distribution of molecules between the normal and overturned states. (If this latter factor is related to the membrane potential, then it is possible to see how the 'carrier' transport will be regulated by the membrane potential.) All of these factors would be involved in determining the rates of transport and the steady state distribution of permeants. Despite the complexity of the problem, one can nevertheless see that such a scheme is possible and that it may be worth being considered further as a mechanism of 'carrier' transport.

The conception of a 'carrier' presented here prompts some comments about the recent observations of the specialized 'carrier' properties of several cyclic antibiotics (24). The proposed mechanism of 'carrier' transport implies that spherical molecules ought to be the most efficient carriers, since they are able to overturn without requiring any additional area. Of course, the 'carriers' have to bind (e.g. ions) and be somewhat surface active, but the shape factor is most important, because of the rapid fall-off of $W(\Delta A)$ with larger values of ΔA. The physical properties of these cyclic compounds fit the expected pattern for 'carriers' and lend support to the ideas presented here.

In line with the above discussion it is possible to extend this scheme to the case of active ion transport across a membrane. Let us suppose that a phospholipid molecule (PL) acts as a 'carrier' that transports Na^+ and K^+ across a membrane. If PL is attached by its phosphate group to a protein which is present at the surface of the bilayer, the 'carrier' is effectively immobilized in its normal position. However, if PL is released from the protein by the action of ATP, then it can perform its 'carrier' function. (This may be a possible explanation for the non-active ATP dependent Na^+ exchange diffusion seen by Garrahan and Glynn (25) in red blood cells.) The release of the 'carrier', which we assume to be catalyzed by Na^+, will cause *PL* to bind and carry Na^+ ions to the outside. If the 'carrier' returns with an Na^+ ion it will not cause any change, but when the 'carriers' that bind K^+ on the outside return they will catalyze the splitting of the ATP, the release of K^+ to the inside and the reattachment of *PL* to the protein (see Figure 1).

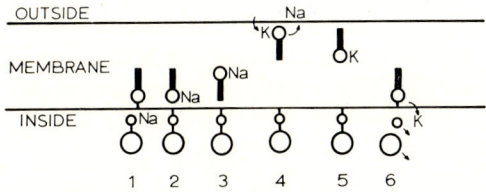

FIGURE 1 A possible mechanism of carrier-mediated active ion transport across a membrane. The process which is discussed in the text proceeds from step 1 to step 6, starting with the approach of Na and ATP on the inside to an enzymatic site where a phospholipid (PL) molecule is attached. The result of the process is an exchange of an Na for a K and the splitting of an ATP, but the actual stoichiometry should depend on the number of ions bound, the probability of splitting an ATP upon detaching, etc.

One way in which this scheme will work is for *PL* to have a much greater affinity for K^+ ions. This would result in a relatively rare splitting of the *PL*-protein bond on the inside, but a heavily favored ion exchange process on the outer surface leading to a binding of K^+ for the return trip. However, one is still left with the problem of why the Na^+ bound to the *PL* does not exchange with K^+ on the inside where K^+ is very abundant. Presumably, the action of the protein at this site on the binding of the ATP produces this situation. This, of course, still leaves us with the problem of how active transport, or the enzyme associated with active transport, works. But it does provide a new way of considering this problem in a framework that is compatible with some of the known properties of interfacial layers.

References

1. BLANK, M. 1962, *J. Physical Chem.*, **66**, 1911.
2. BLANK, M. 1968, *J. General Physiology*, **52**, 191S.
3. BLANK, M. In 'Retardation of Evaporation by Monolayers', V. K. LaMer, editor, Academic Press, New York, 1962, p. 75.
4. BLANK, M. 1961, *J. Physical Chem.*, **65**, 1698.
5. LANGMUIR, I., and LANGMUIR, D. B. 1927, *J. Physical Chem.*, **31**, 1719.

6. LANGMUIR, I., and SCHAEFER, V. J. 1943, *J. Franklin Inst.*, **235,** 119.
7. ARCHER, R. J., and LAMER, V. K. 1955, *J. Physical Chem.*, **59,** 200.
8. BLANK, M., and LAMER, V. K. In 'Retardation of Evaporation by Monolayers', V. K. LaMer, editor, Academic Press, New York, 1962, p. 59.
9. ROSANO, H. L., and LAMER, V. K. 1955, *J. Physical Chem.*, **60,** 348.
10. MILLER, I. R., and BLANK, M. 1968, *J. Colloid and Interface Sci.*, **26,** 34.
11. BLANK, M. 1964, *J. Physical Chem.*, **68,** 2793.
12. BLANK, M., and BRITTEN, J. S. 1965, *J. Colloid Sci.*, **20,** 789.
13. PRINCEN, H. M., and MASON, S. G. 1965, *J. Colloid Sci.*, **20,** 353.
14. SCHROEDINGER, E., 'What is Life?', Doubleday, New York, 1956.
15. TIEN, H. TI. 1967, 1968, *J. Physical Chem.*, **71,** 3395; **72,** 2793.
16. LEVINE, Y. K., BAILEY, A. I., and WILKINS, M. H. F. 1968, *Nature*, **220,** 577.
17. HAYDON, D. A., and TAYLOR, J. 1963, *J. Theoret. Biol.*, **4,** 281.
18. GAINES, G. R., JR. 'Insoluble Monolayers at Liquid-Gas Interfaces', Interscience, New York, 1966.
19. GAINES, G. L., JR. 1960, 1959, *Nature*, **186,** 384; **183,,** 1109.
20. BROOKS, J. H., and ALEXANDER, A. E. International Congress on Surface Activity, Cologne, 1961. Paper B/1/3, number 33.
21. DEAMER, D. W., and BRANTON, D. 1967, *Science*, **158,** 655.
22. SCHULMAN, J. H., and ROSANO, H. L. In 'Retardation of Evaporation by Monolayers', V. K. LaMer, editor, Academic Press, New York, 1962, p. 97.
23. MOORE, T. J. 1968, *J. Lipid Research*, **9,** 642.
24. Physiology Society Symposium. 1968, 'Biological and Artificial Membranes', Federation Proceedings, **27,** 1249.
25. GARRAHAN, P. J., and GLYNN, I. M. 1965, *Nature*, **207,** 1098.

DISCUSSION

KURSUNOGLU   I want to ask a few questions; questions that experts might not like to ask in public. For example, why is the membrane there? That's one question, and the other is: What is the nature of selectivity of the ion diffusion? The third one is related to the first one. What binds the membrane together? What is the general physics of the membrane as a stable structure? Can you say something about this for the benefit of physicists? Maybe the biologists know the answer. We haven't heard.

BLANK   I think probably the only answer one can give at this stage is that the membrane is a stable entity, and that it is there to separate the inside of the cell from the surroundings. Once this permeability barrier is destroyed, then life is no longer possible.

The question dealing with the forces that hold the membrane together might be better addressed to Dr. Ohki, who will be the next speaker, because he has spent much time considering this problem. In dealing with the forces that stabilize a membrane he assumes a bilayer structure, but the same kind of thinking is undoubtedly important regardless of the assumed structure.

With regard to your question on specificity, there are many components within the membrane that can contribute to specificity of action. For example, phospholipids and proteins show a considerable amount of specificity and one can isolate many enzymes from the membrane system which show specificity. Specificity seems to be the rule rather than the exception in the membrane state.

KURSUNOGLU   No, I was asking about selectivity.

BLANK   Selectivity and specificity, I would say, are manifestations of the same forces.

GREEN   I would like to make a few comments about your model and the relation of this model to biological membranes. First let me say that we can recognize two kinds of permeability in the mitochondrial membranes—permeability through the membrane (into the interior space) and permeability into the membrane. Certain salts like sodium chloride readily penetrate the inner mitochondrial membrane via the aqueous spaces or channels which separate repeating units but are incapable of penetrating into the membrane. By contrast, sodium acetate readily penetrates into the membrane but only relatively slowly penetrates into the interior spaces. The point I want to drive home is that there are different kinds of permeation and ion movement and that we oversimplify the problem of permeability by throwing together in one pot phenomena which belong to entirely different categories. My second point is that evidence accumulated in our laboratory by George Blondin strongly suggests that the penetration of ion pairs into the mitochondrial inner membrane does not depend upon the entry of charged molecules into hydrophobic domains, but rather upon the presence in the phospholipid phase of facilitating molecules. These naturally occurring molecules are capable of complexing with ion pairs in such a way that the ion pairs in the form of a complex with the facilitating molecule can penetrate into the hydrophobic areas of the membrane. The difference in the penetrability of the mitochondrial membrane by sodium salts rather than potassium salts is not a problem of the atomic sizes of Na^+ versus K^+ but rather a problem of the availability of facilitating molecules. There is a facilitating molecule for sodium salts in abundance but the facilitating molecule for potassium salts is much less abundant. The natural facilitating agents make possible the formation of complexes with ions that can penetrate into the hydrophobic interior of the membrane—an interior into which the unscreened ion cannot penetrate. My third comment is that there is an alternative to the bilayer model that can account satisfactorily for the permeability properties of biological membranes. I have already considered such a model in a previous session of this conference. The orientation of phospholipid molecules in the membrane at right angles to the plane of the membrane can be achieved in models other than the unit membrane. It may be wise to consider these alternative models and not insist on a particular model of the membrane that is incompatible with a large body of experimental evidence.

BLANK Your first question was in connection with entrance into the membrane as opposed to passage through the membrane. One of the advantages of the approach I presented is that one does not have to consider entrance into the membrane. When a water molecule or an ion passes through a bilayer or monolayer it passes through a hole, and one can neglect interaction. Permeation occurs when the permeant finds an open space and there is a driving force (the chemical potential difference) across the membrane. This is one way to get around the problem of why ions move through systems with very low dielectric constant. Similarly one can account for the large flow of water through membranes despite the high concentration of lipid within the membrane. To recapitulate, this approach avoids the problem of how the permeant gets into the membrane. The membrane is a boundary, and the permeant does not interact with that boundary in normal passive permeation.

I would like to stress my reference to passive transport, because in active transport, this approach is not adequate. When you get enzymatic reactions in membrane systems, and this is primarily what you are dealing with in the case of the mitochondrion, there are undoubtedly many more complications.

Your point about the difference between sodium and potassium, which you ascribe to different binding, can also be explained in other ways. All one needs is a mechanism to get the sodium out of the cell. This can be accomplished by a carrier mechanism, as you proposed, or by an oriented reaction in the membrane. Since the potassium can move through a membrane more easily, its hydrated radius being much smaller, it can redistribute itself passively.

GREEN I accept your thesis that in a monolayer statistical fluctuations can provide holes or pores through which water molecules and ions can move freely but I would like to stress that this concept of statistical pores may be inapplicable to biological membranes. The evidence is clear that ions can penetrate into the hydrophobic domain of the repeating units and that this penetration is intrinsic to the conformational cycle. This special feature of the membrane problem is one which is not present in the monolayer and in this important respect the monolayer is not a satisfactory model of the membrane.

BLANK There is really no difficulty in accounting for water permeation in simple membranes, because water is very small. It is present at a very high concentration, constantly bombarding the membranes, so one would expect it to readjust differences in chemical potential very quickly.

I think that some of this discussion is like the story of the five blind Indians describing an elephant. The fact is that we are talking about different membranes, i.e., a simple membrane as opposed to a membrane system. I think the passive properties of simple membranes can be encompassed within the bilayer model, without making additional assumptions. However, this model does not lead to a complete description of a cell membrane or membrane system. One needs some detailed kind of structure to give you such specialized properties as a pump mechanism, and this does not reside in the simple picture of a membrane that has been presented.

CHRISTENSEN I wonder if it's a good idea to make so sharp a distinction between active and passive movements. It turns out that the movements that are with the gradient are occasioning membrane responses including the co-migration of other substances against their gradient that indicate there is not that much difference. We are probably dealing with the same receptor sites and perhaps the delivery of energy simply in the other direction, when the movement is in the direction of the electrochemical potential. So, therefore, I personally question the desirability of dividing the subject along those lines.

BLANK Until the permeant is greater than a three carbon length, at which size there is usually a tremendous drop-off in normal passive permeation, we can frequently regard the process as passive. The movements of such small solutes can probably be explained without having to invoke specialized mechanisms. After all, most of the movement across natural membranes is due to substances like O_2, CO_2, water, bicarbonate, chloride, lactate, etc., which fall into this category. It is also possible to use the same approach to account for a number of basic observations on excitable membranes without having to invoke special properties (*J. Colloid Science* 20:933, 1965).

CHRISTENSEN   Then the question becomes: Just how commonly does an entirely passive, indifferent passage occur? To how large a degree does the passage of biological membranes have a chemically unstructured nature?

BLANK   I think that the point you made is correct. It is frequently difficult to separate active from passive processes. However, in dealing with the passive movements of small solutes that do not rely on any 'carrier mediation', the term passive transport can be defined in a satisfactory way.

KEDEM   The movement of water is not so very fast in biological membranes, as compared to various synthetic films. Volume change of cells is fast simply because the area per volume is very large and the cell membrane is thin. Thin films of hydrated polymers or liquid organic materials which dissolve a small amount of water will readily give flow rates comparable to those in cell membranes and there is no need to assume the existence of channels.

BLANK   I was referring to instances like the chloride shift in the red cells. This happens twice in every cycle of the blood and involves a significant change in volume of the cells, i.e., a large and rapid movement of water across the membranes.

ADAM   I have a question concerning the problem as to how far your model of solute transport through the membrane is useful for explanation of ion transport through biological membranes. We know from experiments on phospholipid bilayers that without any addition of proteins or of carriers, the membrane resistance is very high, lets say a million, ohms/cm^2, whereas biological membranes in the resting state have a resistance of 10^3 ohms/cm^2. Now the question: Is your mechanism of fluctuation of the structure of the bilayer able to explain things like this? Did you do any quantitative calculations of showing how often it happens, that the phospholipid molecule gets dissolved into the bilayer and turns up on the other side, because if you have any number of how often this could occur, then this would give a lower limit to the numbers of sodium ions which can be transported by such a mechanism.

BLANK   First let me say that it is not possible as yet to make any calculations on the bilayer, even though the monolayer calculations can be made. The

Physical principles in monolayer and membrane permeation

bilayer calculations are limited by the fact that one does not know the energetics. (The relation between the expansion of a bilayer and the surface tension is complicated by the fact that there is a torus of material which contains a large reservoir of solvent and phospholipid.) But the question about the interchange of molecules between two halves of a bilayer has been looked into experimentally by Deamer and Branton (*Science* **158**:655, 1967), who measured the half time of exchange of stearate from one layer to an adjacent one. The half times were on the order of twenty minutes, which means that there was an appreciable exchange between layers.

ADAM Can these numbers be compared with the passive transport of ions through bilayer membranes. If you propose this model, then you should estimate of the order of magnitude of them.

BLANK The experimental observations were on multilayers, which are comparable to bilayers. However, the deposition ratio in multilayers is never equal to one, i.e., there is a lot more free area, and one would expect to get more exchange in these multilayer systems than in an ordinary bilayer. At this stage, it is fair to say that the effects described in the paper are possible, in principle. The effects in multilayers are certainly large enough to account for the ionic permeation rate through bilayers.

PAPER 8

Current rectification and action potentials across thin lipid membranes

D. R. KALKWARF

Pacific Northwest Laboratory, Battelle Memorial Institute, Richland, Washington 99352

and

D. L. FRASCO and W. H. BRATTAIN

Whitman College, Walla Walla, Washington 99362

THE RESEARCH to be described is a further attempt to understand the behavior of biological membranes by determining the range of chemically and electrically induced reactions open to simple biochemical membranes synthesized between two aqueous compartments. Numerous investigators have now shown that such membranes can be formed by painting a solution of lipids across a hole in a sheet of polyethylene or teflon separating the compartments. The use of lipids which are known to concentrate at living cell interfaces and the spontaneous thinning of the membrane to the thickness of the double-layered, osmiophillic zone observed in electron micrographs of certain cell boundaries, gives these structures at least a superficial resemblance to living cell surfaces. Far stronger resemblance, however, is seen in the electrical rectification properties of these synthetic membranes and the action potentials produced when the bare lipid surfaces interact with proteins from the aqueous phases. Rudin and Mueller (1) were the first to report these latter phenomena, and it was a first point of order to see whether their interesting results could be duplicated. If so, these simple systems would appear to be excellent models for determining the basic principles of membrane action and control in living organisms.

The basic membranes used in the present work were prepared from a solution composed of 1.25 per cent stearoyl sphingomyelin dissolved in

D-α-tocopherol: chloroform: methanol in the weight ratio 5:3:2. Each of these constituents was purified chromatographically before mixing. Membranes could be readily prepared by painting this solution across an approximately 1-mm² diameter hole in polyethylene submerged in water or aqueous salt solutions. Technically, the ease of membrane construction was facilitated by painting a solution of sphingomyelin in chloroform-methanol around the edge of the hole and allowing it to dry before immersing it in the solution and preparing the membrane. The stability of the final structure was also improved by boiling the aqueous solution to remove dissolved air, cooling it rapidly and then painting the membrane. Troublesome air bubbles could thus be eliminated from forming on its surface.

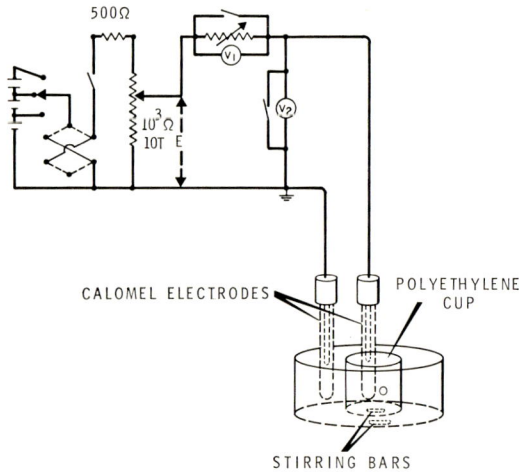

FIGURE 1   Apparatus for measuring membrane potentials.

The experimental arrangement is shown schematically in Figure 1. Electrometer V_1 was a Keithley Model 600 A while V_2 was a Keithley Model 200 B. Both were connected to recorders for simultaneous recording of current and voltage across the membrane. The open-circuit potential-difference across the membrane could be measured by reading the applied potential difference E required to null the circuit current. The surrounding solutions were maintained at 40 ± 3 °C with a hot plate-magnetic stirrer which also served to thoroughly agitate the contents of both compartments. When solutions were added to one compartment an equivalent volume of solution was added to the other to maintain equal pressure on both sides of the membrane.

A lipid layer gradually thinned until its surface took a gray metallic sheen and its electrical conductance increased to $3-5 \times 10^{-9}$ mho. cm^{-2}. The electrical capacitance of the completed membrane was also routinely evaluated by measuring the time constant for charging it with an applied potential difference. Values ranged from 0.4–0.8 μF. cm^{-2}. In this state, the membrane was stable for hours and could withstand temperatures of up to 60° C. The final compositions of the thinned membranes are unknown, but they would be expected to contain both sphingomyelin and tocopherol and hence were termed *S-T* membranes.

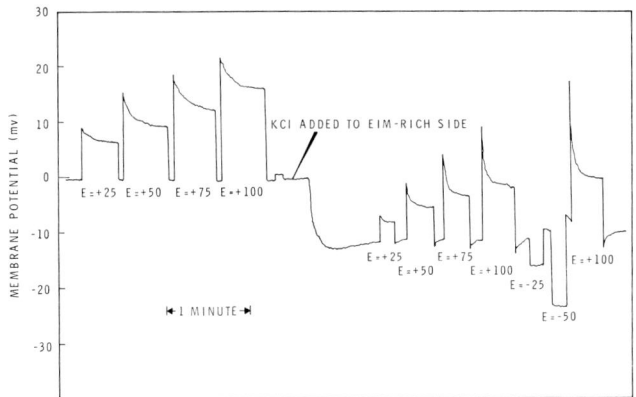

FIGURE 2 Rectification of electrical current through an EIM-activated sphingomyelin-tocopherol membrane. Both sides buffered at 6.9 with 0.005 M histidine-NaCl. *E*-values indicate potential in millivolts applied across membrane and series resistance R = 10^7 ohms. Addition of KCl brought [K$^+$] on EIM-rich side to 0.088 M.

Investigation was first directed toward verifying the report of Rudin and Mueller (1) that these membranes would rectify current and show action potentials when activated by a proteinaceous material called EIM ('excitation-inducing material'). Samples of EIM were obtained from these investigators and many of their results were indeed verified. Addition of aqueous EIM solutions to the inner compartment caused a large increase in conductance to 3–5 $\times$ 10^{-5} mho. cm^{-2}. In this state, addition of KCl to either compartment produced a resting voltage across the membrane with a negative charge on the side which received the excess KCl. The membrane showed strong rectification with positive current passing more easily from the EIM-rich side to the other. This is illustrated in Figure 2. This recording of the membrane potential as a function of time showed that rectification also occurred by the peculiar switching process termed 'delayed rectification' by Rudin and Mueller. This suggests a molecular rearrangement within the membrane from a non-conducting resting structure to a conducting,

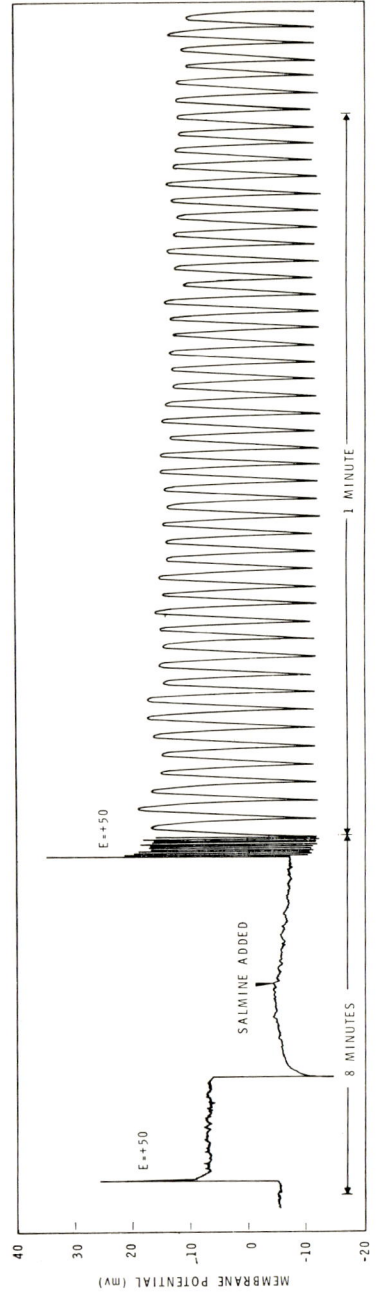

FIGURE 3 Rhythmic action potentials shown by an EIM-activated sphingomyelin-tocopherol membrane following addition of 0.015 ml, 10-mg/ml salmine to EIM-rich side. Both sides buffered at pH 6.9 with 0.005 M histidine—NaCl. EIM-rich side contains 0.040 M Cs$^+$. E-values indicate potential in millivolts applied across membrane and series resistance R = 10^7 ohms.

electrically-induced structure with a time constant of a few seconds. Finally, addition of purified protamines such as salmine and clupeine to the EIM-rich side of the membrane produced rhythmic action potentials as illustrated in Figure 3. In this example with CsCl providing a -8.5 mV resting potential on the EIM-rich side, the membrane altered its polarity at the rate of 0.7 cycle per second. In other examples, the response frequencies varied from 0.1 to 2 cps. These action potentials could occasionally be initiated chemically by adding just the right proportions of protamine and EIM. Generally they could be initiated by a small positive-current pulse through the membrane from the EIM-rich side. Again in verification of Rudin and Mueller's work the frequency of the rhythmic potentials increased with temperature, their intensity could be sustained by addition of more protamine, and their appearance could be terminated by addition of Ca^{2+} to the EIM-rich side.

While the observations described above suggest the creation of more elaborate membranes to duplicate life-like phenomena *in vitro*, it is our intention to first concentrate on obtaining sufficient data concerning the properties of these simple biochemical systems to develop an understanding of the basic principles involved. For instance, one would like to know the identity of the current carriers in the membranes. Tracer experiments using ^{137}Cs failed to reveal this in bare *S-T* membranes due to the low current possible; however, in EIM-activated *S-T* membranes, $^{137}Cs^+$-flow accounted for the full current in the high-conductance direction. Further indication that cations were the principle charge carriers was shown by the spontaneous formation of membrane potentials when any of the ions: Li^+, Na^+, K^+, Cs^+, NH_4^+, Ca^{2+}, Sr^{2+}, or Ba^{2+} were added in excess to one side of the EIM-activated membrane. In each case, the sign of the potential was such as to indicate that the cations carried the major portion of the current through the membrane structure.

In order to understand the rectification properties of the EIM-lipid configuration more fully, current-voltage plots were made of the recorded data. Here, considerable care was required to obtain reproducible data indicative of a stable association. It was realized early in our work that potentials applied across EIM-activated membranes could irreversibly change their electrical characteristics. For example, application of a positive potential to the EIM-rich side accelerated interaction, indicated by the formation of the high-conductance structure. This would be expected if EIM were a positively charged protein. Conversely, negative potential applied to the EIM-rich side caused the membrane conductance to gradually decrease to its 'bare-lipid' value. Interaction between the constituents was considered complete only when the order of successive applied potentials did not alter the current-voltage data. In this state, the membranes generally broke if a large negative potential was applied to the AIM-side in an effort to dissociate the configuration. We would like to stress that the emphasis of our work has been on obtaining data which is reproducible on a single membrane and typical of many membranes.

Typical current-voltage curves for EIM-activated S-T membranes are shown in Figure 4. Rectification and 'delayed rectification' occurred with or without salt-concentration potentials of either sign but were always of such a nature that positive charges could flow more easily away from the EIM-rich side than vice versa.

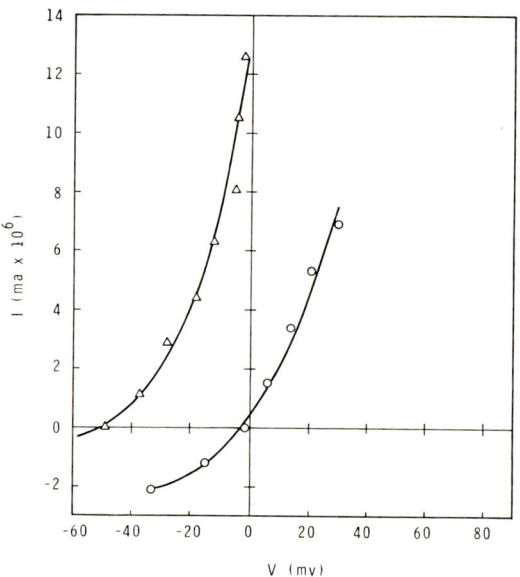

FIGURE 4 Current passing through an EIM-activated sphingomyelin-tocopherol membrane as a function of membrane potential. Both sides buffered at pH 6.9 with 0.005 M histidine-NaCl. ○ Data taken before, and △ data taken after addition of KCl to EIM-rich side to give $[K^+] = 0.038$ M. Current directed away from the EIM-rich side is considered positive.

Observation of current rectification by these biochemical structures prompted an examination of the data in terms of the classical equation for rectifiers, namely:

$$I = I_0 [\exp(neV/kT) - 1]$$

In the above expression, I is the current in the high-conductance direction, I_0 is the saturation current in the opposite direction, n is the charge of the current carriers, e is the electronic charge, V is the voltage across the membrane, k is Boltzmann's constant and T is the absolute temperature. Both cuprous oxide rectifiers and properly made *p-n* junctions obey this equation with $n = 1$. One finds that by properly choosing I_0, all the experimental data fall approximately on a straight line if one plots $\log[(I/I_0)+1]$ against applied potential, V. When this is done, one finds the data to be consistent with $n = 1$ in the above equation indicating that the process involves the transport of singly charged ions at 40° C. This is shown in Figure 5 for data illustrated

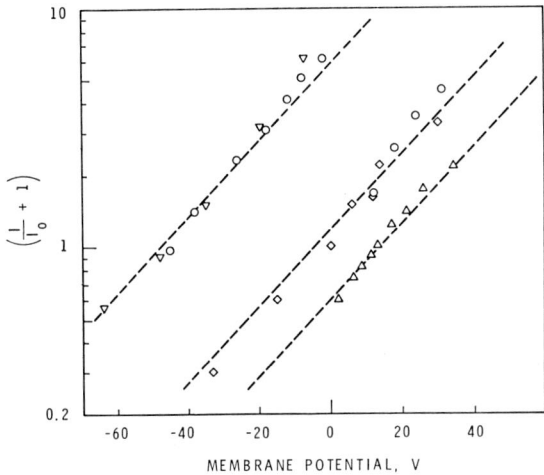

FIGURE 5 Test of equation $I = I_0 [\exp(neV/kT) - 1]$ for EIM-activated sphingomyelin-tocopherol membranes with and without salt-induced bias voltage. Dotted lines indicate theoretical slope for $n = 1$.

previously in Figure 4 as well as other data sets. This graph also shows that if the membrane is subjected to a salt-induced bias voltage, the above equation is still valid provided V is regarded as the algebraic sum of the applied voltage and the salt-induced bias. We have found, however, that there are cases for which the above analysis indicates values of 2, 3 and even 4. These results seemed to be correlated with a very high conductance state of the membrane.

While we are continuing to examine the properties of EIM-activated *S-T* membranes in more detail, we would also like to mention another class of compounds that interact with sphingomyelin-tocopherol membranes so as to increase their conductance. These are the diaminotriphenylmethane dye-anions which have a particular arrangement of sulfonate groups in their molecular configuration. The general structure of this class is shown in Figure 6.

Two members of this class are Patent Blue and Patent Blue A. Addition of either of these dyes to one side of an *S-T* membrane increased the conductance to $3-7 \times 10^{-5}$ mho/cm^2. Numerous other compounds, of course, have been reported to increase the conductance of lipid membranes (2). What makes the action of these compounds particularly interesting is that analogs of these dyes, e.g. Food Greens No. 1, No. 2, No. 3, Food Violet No. 1, Food Blue No. 2, and Alkali Fast Green 10GA, that do not contain sulfonate groups in the configuration shown in Figure 6 are ineffective.

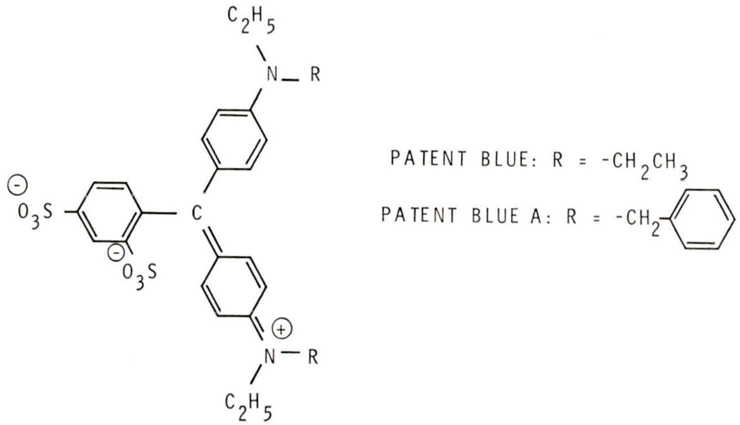

FIGURE 6 Structure of diaminotriphenylmethane dyes capable of increasing conductance of sphingomyelin-tocopherol membranes.

As expected from their negative charge, the Patent Blue and Patent Blue A dye-anions interacted very rapidly with the lipid film if a negative potential was applied to the dye-rich side. Conversely, a negative potential applied to the opposite side of the film slowly broke up the dye-lipid association as indicated by a decreased conductance in either direction. A typical current-voltage curve for these dye-activated structures is shown in Figure 7. The membranes rectify current but in a direction opposite to that of EIM-activated membranes. To this extent, they are mirror images of the latter system. It is to be noted in this case that the sign of rectification appears to depend on the sign of the charge on the organic activator. No 'delayed rectification', however, was observed. Moreover the addition of Li$^+$, Na$^+$, K$^+$, and Cs$^+$ as the chlorides to either side of the membrane produced no bias voltage. By contrast, addition of divalent cations such as Mg^{2+}, Ca^{2+}, Sr^{2+}, and Ba^{2+} as the chlorides to the dye-side broke the membrane immediately. No interaction between these

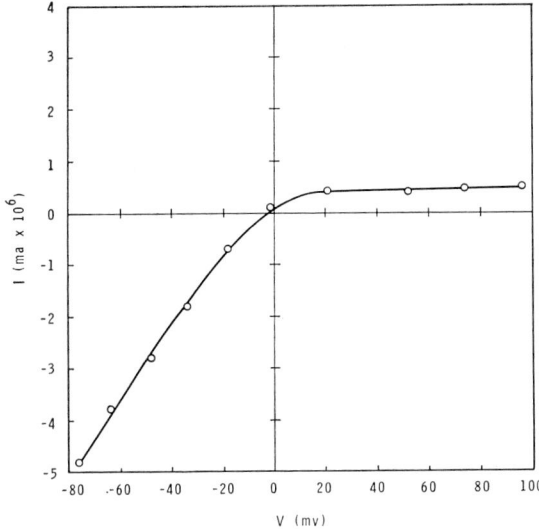

FIGURE 7 Current passing through a Patent Blue-activated sphingomyelin-tocopherol membrane as a function of membrane potential. Both sides buffered at pH 6.9 with 0.005 M histidine—KCl. Current directed away from the dye-rich side is considered positive.

cations and the dyes has yet been detected in aqueous solution so that this phenomenon must arise from an action of the cations on the dye-lipid complex.

At present we feel reasonably sure that the experimental observations we have presented are typical of these membranes, but that much more experimentation is necessary before attempting any comprehensive conceptual understanding of their behavior. Like their counterparts in living systems, these membranes are extremely sensitive to small changes in their environment so that experiments must be repeated several times before the main response pattern can be distinguished among the minor fluctuations. Clearly, however, these relatively simple biochemical membranes exhibit several of the phenomena observed at living-cell boundaries and can serve as model systems for attempting to understand the physical and chemical mechanisms at such interfaces.

Acknowledgments

We would like to acknowledge the technical assistance of Mrs. S. Hughes in carefully preparing the membranes used and also the financial support of Battelle Institute. The EIM used in this work was generously donated by Drs. D. O. Rudin and P. Mueller of the Eastern Pennsyllvania Psychiatric Institute.

References

1. MUELLER, P., and RUDIN, P. O. 1967, 'Action Potential Phenomena in Experimental Bimolecular Lipid Membranes,' *Nature*, **213**, 603–604.
2. SEUFERT, W. D. 1965, 'Induced Permeability Changes in Reconstituted Cell Membrane Structure,' *Nature*, **207**, 174–176.
   MUELLER, P., and RUDIN, D. O. 1967, 'Development of K^+/Na^+ Discrimination in Experimental Bimolecular Lipid Membranes by Macrocyclic Antibiotics', *Biochem. Biophys. Res. Comm.*, **26**, 398–404.
   LAUGER, P., RICHTER, J., LESSLAUER, W. 1967, 'Electrochemistry of Bimolecular Phospholipid Membranes. I. Impedance Measurements in Aqueous Iodine Solutions,' *Ber. Bunsenges. Phys. Chem.*, **71**, 906–910.

PAPER 9

Dispersion forces and stability of lipid bilayers

SHINPEI OHKI

Department of Biophysics, School of Pharmacy and Center for Theoretical Biology, State University of New York at Buffalo Buffalo, New York

I INTRODUCTION

THE INVESTIGATIONS of Gorter and Grendel 1925[1] and Danielli and others (Danielli and Davson, 1934)[2]; Danielli and Harvey, 1934[3] led to the bimolecular lipid leaflet concept of cell biomembrane structure (Figure 1). This model, although constantly being challenged (Korn, 1966[4]), satisfies many of the known electrical and permeability properties of living cell membranes, and has recently received substantial support from electron microscope (Robertson, 1960[5]) and X-ray (Finean, 1962[6]) studies. Attempts by early workers (Danielli, 1936[7]; Dean, Curtis and Cole, 1940[8]; Langmuir and Waugh, 1938[9]) to prepare stable model lipid or proteolipid structures of bimolecular dimensions were unsuccessful. The properties of the basic bimolecular lipid leaflet were therefore unknown and speculations about its possible functional role in biomembranes could not be experimentally tested.

In 1961, Mueller, Rudin, Tien and Wescott[10,11] discovered that solutions of ox brain phospholipid extract, when suspended in an aqueous medium, spontaneously thinned and yielded black films. They also demonstrated that the addition of a proteinaceous material, referred to as E.I.M. (excitability inducing material), imparted on these membranes electrical characteristics similar to those observed in *Valonia* and frog nerve (Mueller, *et al.*, 1962[11], 1963[12]). The properties of these black films were consistent with those anticipated for a bimolecular lipid leaflet and thus, for the first time, a model system had been created that would enable the properties of this unique structure and its interactions with proteins to be examined directly. The demonstration

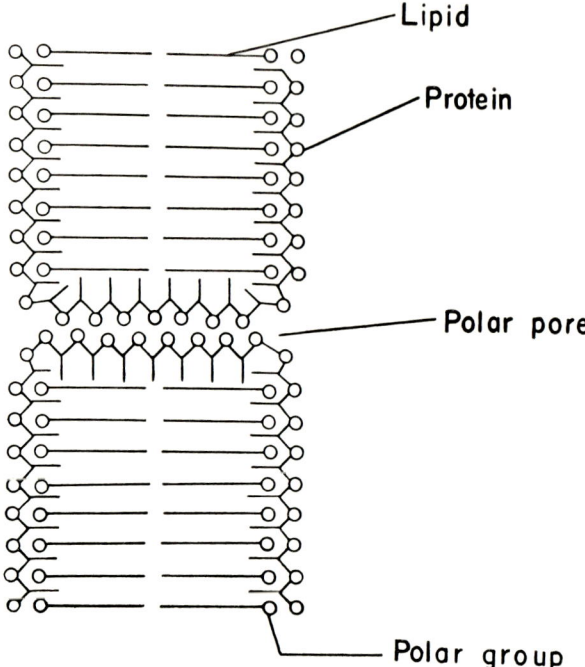

FIGURE 1 Danielli–Davson model of the cell membrane showing the basic bimolecular lipid bilayer-protein sandwich structure and a polar pore.

that the bimolecular lipid leaflet is capable of an independent physical existence has added considerable strength to the bilayer concept, and has opened up a whole new area of research that promises to be far reaching in the quest to understand the structure and function of biological membranes.

Therefore, in order to elucidate the formation and the function of biological membranes, (e.g. selective permeability, excitability and pinocytosis), it is relevant to study the molecular basis of stability and structure of artificial membranes.

When natural lipids or phospholipids, dissolved in organic solvents (chloroform-methanol) are suspended in an aqueous solution, the lipids spread out and give rise to thin membranes. A model of these thin membranes is shown in Figure 2 (a) schematically. It is considered that both outer surfaces are oriented lipid monolayers, and the interior part surrounded by oriented monolayers is a random hydrocarbon liquid. Finally, the thin membrane becomes a bimolecular leaflet which appears to be the most stable state. There

Dispersion forces and stability of lipid bilayers 177

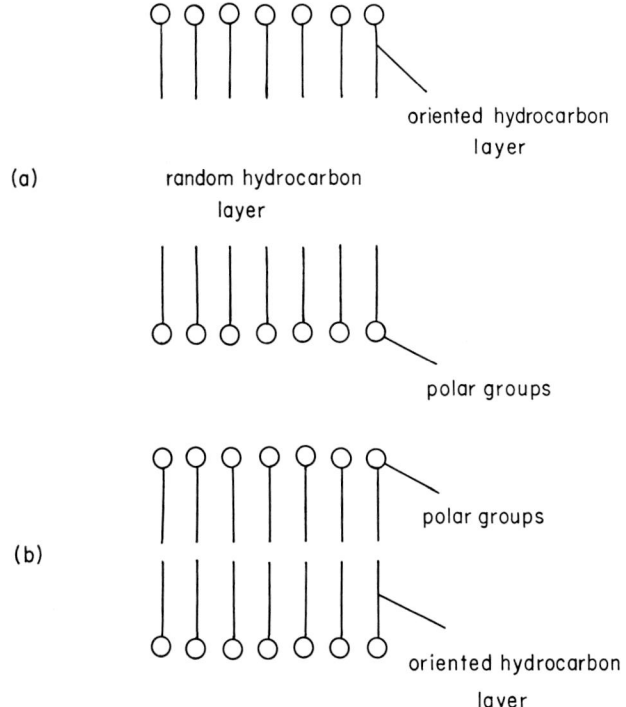

FIGURE 2 (a) A model of thin lipid membrane. (b) Idealized lipid bilayer.

are several indications[13,14,15] that the membranes are bimolecular in thickness (Figure 2 (b)).

In section II, relevant expressions of interlayer interaction for the lipid bilayer system are derived from the second order perturbation theory of intermolecular forces. As one of the factors of the stabilization energies of the lipid bilayers, the interlayer dispersion energy is estimated.

In section III, the outline of the molecular theory for dielectric polarizability, the application of the theory to the lipid bilayer model and the estimation of the thickness of lipid bilayers using the dielectric constant calculated, are given.

In section IV, the possibility of the structural change of the lipid bilayer due to the change of environmental solutions is discussed in terms of the net charge of the polar groups.

In section V, the stability of an asymmetrical membrane is considered in terms of the net charge of polar groups and the degree of chelation of the polar groups with metal ions. An application of the theory to the stability of lipid bilayers is discussed.

II DISPERSION FORCES IN LIPID BILAYERS

The precise nature of the forces driving the thick film to the bimolecular state is not at all well understood. The existence of long range forces has not been established with certainty, although recent studies on thin films by Scheludko[16] are providing increasing evidence for long range electrical repulsive and van der Waals forces in thin films. Tien and Dawidowicz (1966)[14] have suggested that black film formations may be initiated by chance events which bring the opposed hydrocarbon chains sufficiently close together for short range van der Waals interactions to become important. These chance contacts may arise from thermal motion, vibration, local variations in interfacial tension and the presence of impurities. Ohki and Fukuda (1967)[17] have shown that van der Waals forces arising from asymmetric polarization of the hydrocarbon chains can be significant, and that it is reasonable that once the opposing groups are brought into close association they retain this position. The succeeding growth of black film occurs by a process that has been aptly described as a 'zipper-like' action.

According to the second order perturbation theory,[18,19] the dispersion energy between two non-polar molecules is given by

$$W_{\alpha\alpha} = -\sum_{p \neq 0} \sum_{\sigma \neq 0} \frac{|\langle 0(1)0(2)|J(1,2)|\rho(1)\sigma(2)\rangle|^2}{\delta_\rho(1)+\delta_\sigma(2)}$$

$$J(1,2) = -\mathbf{p}(1)\frac{3\mathbf{e}_{12}\mathbf{e}_{12}-1}{(r_{12})^3}\mathbf{p}(2) = -\mathbf{p}(1)\frac{\tilde{T}}{(r_{12})^3}\mathbf{p}(2) \tag{2.1}$$

$$\delta_\rho(1) = E_\rho(1) - E_0(1)$$

$$\delta_\sigma(2) = E_\sigma(2) - E_0(2)$$

where $\mathbf{p}(1)$ is the electric dipole moment of molecule 1; $0(1)$ is the ground state wave function of molecule 1; $\rho(1)$ is the ρth excited state wave function of molecule 1; E_0 is the energy of the ground state; E_ρ is the energy of the ρth excited state; $\mathbf{r}_{12}$ is the distance vector between molecules 1 and 2:

$$\mathbf{r}_{12} = \mathbf{r}_2 - \mathbf{r}_1 = r_{12}\mathbf{e}_{12} \tag{2.2}$$

where r_{12} is the distance between molecules 1 and 2, and $\mathbf{e}_{12}$ is the unit vector of $\mathbf{r}_{12}$.

If the positions of molecules 1 and 2 are fixed, we have

$$\left\langle 0(\tau_1)0(\tau_2) \middle| \mathbf{p}(\tau_1) \frac{\tilde{T}}{(r_{1\,2})^3} \mathbf{p}(\tau_2) \middle| \rho(\tau_1)\sigma(\tau_2) \right\rangle$$

$$= \langle 0(\tau_1) | \mathbf{p}(\tau_1) | \rho(\tau_1) \rangle \frac{\tilde{T}}{(r_{1\,2})^3} \langle 0(\tau_2) | \mathbf{p}(\tau_2) | \sigma(\tau_2) \rangle$$

where τ_1 is the space coordinate of the wave function of molecule 1.

Let us denote $\langle 0(\tau_1) | \mathbf{p}(\tau_1) | \rho(\tau_1) \rangle$ by $\mu_{\rho(1)}$:

$$\langle 0(\tau_1) | \mathbf{p}(\tau_1) | \rho(\tau_1) \rangle \equiv \mu_{\rho(1)} \tag{2.4}$$

Then, Eq. (2.1) is

$$W_{\alpha\alpha} = -\frac{1}{(r_{1\,2})^6} \sum_{\rho \neq 0} \sum_{\sigma \neq 0} \frac{(\mu_{\rho(1)} \tilde{T} \mu_{\sigma(2)})^2}{\delta_{\rho(1)} + \delta_{\sigma(2)}}$$

$$= -\frac{1}{(r_{1\,2})^6} \mathrm{Tr} \left[\sum_{\rho \neq 0} \sum_{\sigma \neq 0} \frac{\tilde{T}(\mu_{\rho(1)} \mu_{\rho(1)}) \tilde{T}(\mu_{\sigma(2)} \mu_{\sigma(2)})}{\delta_{\rho(1)} + \delta_{\sigma(2)}} \right] \tag{2.5}$$

Proof:

$$(\mathbf{X} B \mathbf{Y})^2 = \mathrm{Tr}\,[B(\mathbf{XX})B(\mathbf{YY})]$$

where $\mathbf{X}, \mathbf{Y}$ are vectors and B is a tensor.

Here, we have the following relation:

$$(\mathbf{XX}) \mathbf{e}_i = \mathbf{X}(\mathbf{X} \mathbf{e}_i)$$

$\mathbf{e}_i$: unit vector

Therefore,

$$\mathbf{e}_i B(\mathbf{XX}) = (\mathbf{e}_i B \mathbf{X}) \mathbf{X}$$

$$\mathrm{Tr}\,[B(\mathbf{XX})B(\mathbf{YY})] = \sum_{i=1}^{3} \mathbf{e}_i (B\mathbf{XX}) B(\mathbf{YY}) \mathbf{e}_i$$

$$= \sum_{i=1}^{3} (\mathbf{e}_i B \mathbf{X}) \mathbf{X} B \mathbf{Y}(\mathbf{Y} \mathbf{e}_i)$$

$$= \mathbf{X} B \mathbf{Y} \sum_{i=1}^{3} (\mathbf{e}_i B \mathbf{X})(\mathbf{e}_i \mathbf{Y})$$

With the following relation:

$$\sum_{i=1}^{3} (\mathbf{e}_i B\mathbf{X})(\mathbf{e}_i \mathbf{Y}) = (\mathbf{Y}B\mathbf{X})$$

we have

$$\text{Tr}[B(\mathbf{XX})B(\mathbf{YY})] = (\mathbf{X}B\mathbf{Y})^2$$

If the energy difference δ between the ground state and an excited state can be replaced by the ionization potential of the molecule $\tilde{\delta}$, Eq. (2.5) becomes

$$W_{\alpha\alpha} = -\frac{1}{(r_{1\,2})^6} \frac{1}{\tilde{\delta}_1 + \tilde{\delta}_2} \text{Tr}\left[\sum_{\rho \neq 0}\sum_{\sigma \neq 0} \tilde{T}(\mu_{\rho(1)}, \mu_{\rho(1)})\tilde{T}(\mu_{\sigma(2)}, \mu_{\sigma(2)})\right]$$

$$= -\frac{1}{(r_{1\,2})^6}\frac{\tilde{\delta}_1 \tilde{\delta}_2}{\tilde{\delta}_1 + \tilde{\delta}_2} \text{Tr}\left[\tilde{T}\left(\sum_{\rho \neq 0}\frac{\mu_{\rho(1)}\mu_{\rho(1)}}{\tilde{\delta}_1}\right)\tilde{T}\left(\sum_{\sigma \neq 0}\frac{\mu_{\sigma(2)}\mu_{\sigma(2)}}{\tilde{\delta}_2}\right)\right] \quad (2.6)$$

The polarization vector $\langle \mathbf{p} \rangle$ of a molecule is defined by

$$\langle \mathbf{p} \rangle = \alpha\, \mathbf{E} \quad (2.7)$$

where α is the polarization tensor of a molecule and $\mathbf{E}$ is the electric field acting on the molecule. On the other hand, the quantum-mechanical expression of the polarization is as follows:

$$\mathbf{p}_E = \langle \psi^{(E)} | \mathbf{p} | \psi^{(E)} \rangle \quad (2.8)$$

where ψ^E is the wave function of a molecule with an electric field $\mathbf{E}$. Denote the difference between ψ^E and ψ^0, which is the wave function in the absence of an electric field, by ψ', then Eq. (2.7) is

$$\mathbf{p}_E = \langle \psi^0 | \mathbf{p} | \psi^0 \rangle + \langle \psi^0 | \mathbf{p} | \psi' \rangle + \langle \psi' | \mathbf{p} | \psi^0 \rangle + \langle \psi' | \mathbf{p} | \psi' \rangle \quad (2.9)$$

Since the molecule is nonpolar, $\langle \psi^0 | \mathbf{p} | \psi^0 \rangle = 0$. Therefore, Eq. (2.8) is approximately equal to

$$\mathbf{p}_E = \langle \psi^0 | \mathbf{p} | \psi' \rangle + \langle \psi' | \mathbf{p} | \psi^0 \rangle$$

$$= 2\,\text{Re}\,\langle \psi' | \mathbf{p} | \psi^0 \rangle$$

$$= 2\,\text{Re}\,\langle \psi^0 | \mathbf{p} | \psi' \rangle \quad (2.10)$$

Here, $|\psi'\rangle$ is expressed in terms of the complete set of eigenfunctions of the dipole energy:

$$|\psi'\rangle = \sum_{\rho \neq 0} \frac{\langle \rho | \mathbf{pE} | 0 \rangle}{\delta_\rho} |\rho\rangle \quad (2.11)$$

Dispersion forces and stability of lipid bilayers

For $\mathbf{E} = \text{const}$, Eq. (2.10) becomes

$$|\psi'\rangle = \sum_{\rho \neq 0} \frac{\langle \rho | \mathbf{p} | 0 \rangle \mathbf{E}}{\delta_\rho} |\rho\rangle \qquad (2.12)$$

Insertion of Eq. (2.12) into Eq. (2.10) gives

$$\mathbf{p}_E = 2 \,\text{Re} \left[\sum_{\rho \neq 0} \frac{\mu_\rho \mu_\rho}{\delta_\rho} \mathbf{E} \right] = \langle \mathbf{p} \rangle \qquad (2.13)$$

With Eqs. (2.7) and (2.13), we have the polarizability tensor:

$$\alpha = 2 \sum_{\rho \neq 0} \frac{\mu_\rho \mu_\rho}{\delta_\rho} \qquad (2.14)$$

Therefore, Eq. (2.6) is rewritten as

$$W_{\alpha\alpha} = -\frac{1}{4(r_{12})^6} \frac{\bar{\delta}_1 \bar{\delta}_2}{\bar{\delta}_1 + \bar{\delta}_2} \text{Tr}[\tilde{T}\alpha_1 \, \tilde{T}\alpha_2] \qquad (2.15)$$

When the polarizability tensor is diagonalized, it is expressed by

$$\alpha = \sum_{l=1}^{3} \alpha_l (\underline{\alpha}_l \underline{\alpha}_l) \qquad (2.16)$$

where $\underline{\alpha}_l$ is the unit vector of the lth component of the polarizability tensor. By Eq. (2.16), $\text{Tr}[\tilde{T}\alpha_1 \tilde{T}\alpha_2]$ is rewritten as follows:

$$\text{Tr}[\tilde{T}\alpha_1 \, \tilde{T}\alpha_2] = \text{Tr}\left[\sum_{l=1}^{3} \sum_{m=1}^{3} \alpha_{1l} \alpha_{2m} \tilde{T}(\underline{\alpha}_{1l} \underline{\alpha}_{1l}) \tilde{T}(\underline{\alpha}_{2m} \underline{\alpha}_{2m}) \right]$$

$$= \sum_{l=1}^{3} \sum_{m=1}^{3} \alpha_{1l} \alpha_{2m} \text{Tr}[\tilde{T}(\underline{\alpha}_{1l} \underline{\alpha}_{1l}) \tilde{T}(\underline{\alpha}_{2m} \underline{\alpha}_{2m})]$$

$$= \sum_{l=1}^{3} \sum_{m=1}^{3} \alpha_{1l} \alpha_{2m} [\underline{\alpha}_{1l} \tilde{T} \underline{\alpha}_m]^2 \qquad (2.17)$$

Since $\tilde{T} = 3\mathbf{e}_{12}\mathbf{e}_{12} - 1$, we have

$$\underline{\alpha}_{1l} \tilde{T} \underline{\alpha}_{2m} = 3(\underline{\alpha}_{1l} \mathbf{e}_{12})(\underline{\alpha}_{2m} \mathbf{e}_{12}) - \underline{\alpha}_{1l} \underline{\alpha}_{2m} \qquad (2.18)$$

Therefore, Eq. (2.17) is

$$\text{Tr}[\tilde{T}\alpha_1 \, \tilde{T}\alpha_2] = \sum_{l=1}^{3} \sum_{m=1}^{3} \alpha_{1l} \alpha_{2m} [3(\underline{\alpha}_{1l} \mathbf{e}_{12})(\underline{\alpha}_{2m} \mathbf{e}_{12}) - \underline{\alpha}_{1l} \underline{\alpha}_{2m}]^2 \qquad (2.19)$$

or Eq. (2.15) is

$$W_{\alpha\alpha} = -\frac{1}{8(r_{12})^6} \frac{\bar{\delta}_1 \bar{\delta}_2}{\bar{\delta}_1 + \bar{\delta}_2} \left\{ \sum_{l=1}^{3} \sum_{m=1}^{3} \alpha_{1l} \alpha_{2m} [3(\underline{\alpha}_{1l} \mathbf{e}_{12})(\underline{\alpha}_{2m} \mathbf{e}_{12}) - \underline{\alpha}_{1l} \underline{\alpha}_{2m}]^2 \right\} \qquad (2.20)$$

Let us consider an application of the calculation of asymmetrical molecular interaction to a lipid bilayer model. It is supposed that the hydrocarbon chain in the bilayer system is oriented along the normal to the surface of the membrane (Figure 2 (b)), and that each oriented hydrocarbon is an assembly of oriented C_2H_4 molecular units of which the polarizability tensor is axially symmetrical with respect to the normal to the layer. Ohki and Fukuda[17] have calculated the polarizability of such a C_2H_4 molecular unit by using the bond polarizability approximation.

If C_2H_4 molecular units are arranged as shown in Figure 3, the polarizability tensor is diagonal in this cartesian coordinate system (X, Y, Z):

$$\alpha_{C_2H_4} = \begin{pmatrix} \alpha_{xx} & 0 & 0 \\ 0 & \alpha_{yy} & 0 \\ 0 & 0 & \alpha_{zz} \end{pmatrix}$$

$$= \begin{pmatrix} \alpha_\perp & 0 & 0 \\ 0 & \alpha_\perp & 0 \\ 0 & 0 & \alpha_\| \end{pmatrix}$$

$$= \begin{pmatrix} 2.96 & 0 & 0 \\ 0 & 2.96 & 0 \\ 0 & 0 & 5.48 \end{pmatrix} \quad (2.21)$$

In the bilayer system, $\alpha_1 = \alpha_2$ because the bilayer is composed of the same hydrocarbon chains:

$$\left.\begin{array}{l} \alpha_{11} = \alpha_{xx} = \alpha_{21} = \alpha_\perp \\ \alpha_{12} = \alpha_{yy} = \alpha_{22} = \alpha_\perp \\ \alpha_{13} = \alpha_{zz} = \alpha_{23} = \alpha_\| \end{array}\right\} \quad (2.22)$$

Therefore, the interaction energy due to dispersion forces between two C_2H_4 molecular units arranged as shown in Figure 4 is

$$W_{\alpha\alpha} = -\frac{1}{8}\frac{I}{(r_{1\,2})^6} \sum_{m=1}^{3} \left\{ \alpha_m^2 [1 - 6(\mathbf{e}_{1\,2}\underline{\alpha}_m)^2] + \left[\sum_{m=1}^{3} 3\alpha_m(\mathbf{e}_{1\,2}\underline{\alpha}_m)^2\right]^2 \right\} \quad (2.23)$$

where the ionization potential δ is denoted by I.

The dispersion energy $W_{\alpha\alpha}(R . d)$ per unit area of the bilayer between two oriented hydrocarbon layers of infinite extent, as shown in Figure 4, can be calculated by integrating Eq. (2.23) as follows:

$$W_{\alpha\alpha}(R, d) \, dS_2 = \int_{-\infty}^{\infty} dx_1 \int_{-\infty}^{\infty} dy_1 \int_{R}^{R+d} dz_1$$

$$\int_{R}^{R+d} dz_2 \, W_{\alpha\alpha}(x_1, y_1, z_1, x_2, y_2, z_2) \rho_1 \rho_2 \, dS_2$$

$$= -\pi I \left(\frac{\alpha_L + \alpha_T}{8}\right)^2 \rho_{C_2H_4}^2 \left[\frac{1}{R^2} - \frac{2}{(R+d)^2} + \frac{1}{(R+2d)^2}\right] \quad (2.24)$$

where ρ_1 and ρ_2 are the densities of C_2H_4 molecular units in each layer. In the bilayer, $\rho_1 = \rho_2 = \rho_{C_2H_4}$. R is the separation between two oriented hydrocarbon layers and d is the thickness of the hydrocarbon layer (as shown in Figure 4). $\rho_{C_2H_4}$ is a function of b and the cross-sectional area per hydrocarbon

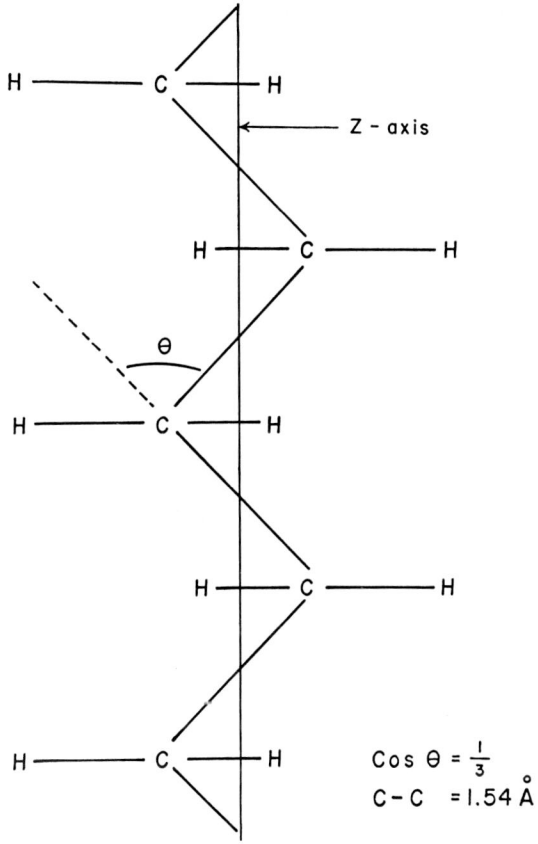

FIGURE 3 Geometry of tetrahedral hydrocarbon chain.

molecule A. The numerical values of the density $\rho_{C_2H_4}$ with respect to various values of b and A (area per molecule) are shown in Table II. In this calculation, the longitudinal length of a C_2H_4 molecular unit is estimated to be 2.514 Å from a knowledge of the length and the angle of the C—C bond (θ_{C-C}), which are shown in Table I and Figure 3.

Similarly, we can calculate the dispersion energy between the oriented hydrocarbon and the random hydrocarbon phases. The interaction energy

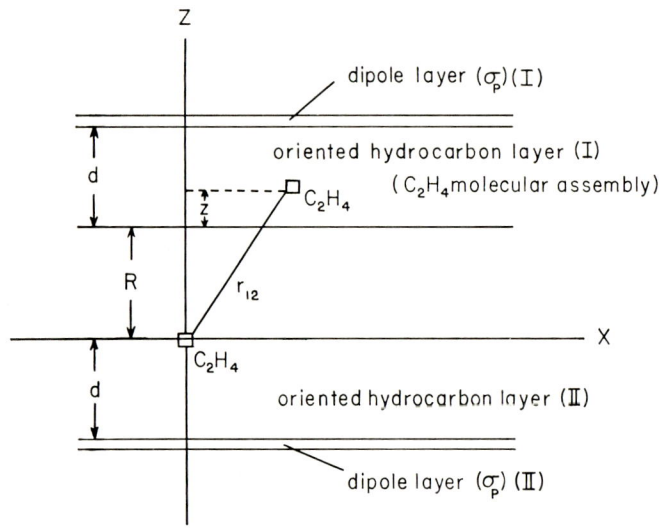

FIGURE 4 Coordinate and dimension of the lipid bilayer model.

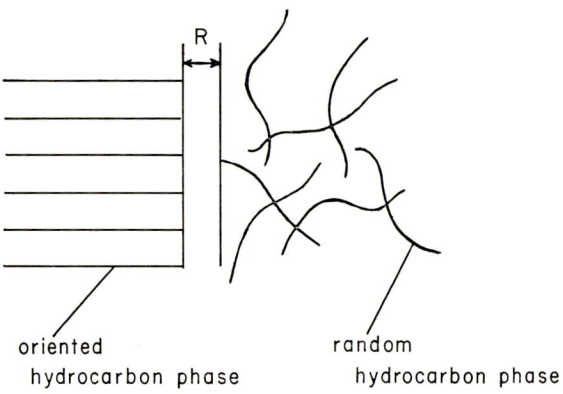

FIGURE 5 Sketch of the system of oriented/random hydrocarbon phases.

between two half infinite oriented and random hydrocarbon phases separated by the distance R (Figure 5) is

$$W_{\alpha\alpha OR} = -\frac{\pi}{32}\rho_O \rho_R I \alpha_R (\alpha_L + \alpha_T)\frac{1}{R^2} \quad (2.25)$$

where ρ_O and ρ_R are the densities of C_2H_4 hydrocarbon units in the oriented hydrocarbon and the random hydrocarbon phases, respectively. α_R is the average polarizability of the random hydrocarbon phase which is

$$\alpha_R = \tfrac{1}{3}(\alpha_\| + 2\alpha_\perp) \quad (2.26)$$

TABLE I The numerical values of various physical quantities.

$I = 13$ ev (Ionization potential for $C_2H_4^{(40)}$)
$\alpha_L(C_2H_4) = 5.72\,\text{Å}^3$
$\alpha_T(C_2H_4) = 2.96\,\text{Å}^3$ (reference 17)
$b = 2.514\,\text{Å}$ (reference 17)
$\rho_{Ran}(C_2H_4) = 1.54 \times 10^{22}/\text{cm}^3$ (calculated from the data of n-decane in reference 47)
C–C $= 1.54\,\text{Å}$
$\theta_{C-C} = \cos^{-1}\tfrac{1}{3}$

The ratio g of the interaction energy between two oriented hydrocarbon phases to that between an oriented and a random hydrocarbon phase at the same separation is

$$g = \frac{W_{\alpha\alpha O\cdot O}}{W_{\alpha\alpha O\cdot R}} = \frac{1}{2}\frac{\rho_O}{\rho_R}\frac{\alpha_L + \alpha_T}{\alpha_R} \quad (2.27)$$

With the numerical values in Table I we can obtain the value of g with respect to the area per molecule (Table II).

From Eqs. (2.24) and (2.27) and Tables I and II we can calculate the dispersion energy between two oriented hydrocarbon layers and the energy difference ($\Delta W_{\alpha\alpha} = W_{\alpha\alpha OO} - W_{\alpha\alpha OR}$) between (i) two oriented hydrocarbon layers and (ii) oriented-random hydrocarbon phases with various separation R and area per molecule A, respectively (Tables III and IV).

The interaction energy between a permanent dipole $\mathbf{p}$ of the polar group and an induced dipole of a C_2H_4 molecular unit is obtained by the same procedure as in the calculation of dispersion energy:

$$W_{\alpha p} = -\frac{1}{2(r_{1\,2})^6}\left\{\sum_{m=1}^{3}\alpha_m[3(\mathbf{p}\mathbf{e}_{1\,2})(\underline{\alpha}_m\mathbf{e}_{1\,2}) - \mathbf{p}\underline{\alpha}_m]^2\right\}$$

$$= -\frac{p}{2(r_{1\,2})^6}\left\{\sum_{m=1}^{3}\alpha_m[3(\mathbf{p}\mathbf{e}_{1\,2})(\underline{\alpha}_m\mathbf{e}_{1\,2}) - \delta_{1m}]^2\right\} \quad (2.29)$$

G

TABLE II A: the area per hydrocarbon chain; $\sigma_{C_2H_4}$: the surface density of hydrocarbon chain; $\rho_{C_2H_4}$: the density of C_2H_4 hydrocarbon unit; $g = \dfrac{W_{\alpha\alpha OO}}{W_{\alpha\alpha OR}}$

A (Å²)	$\sigma_{C_2H_4} = \dfrac{1}{A}(\times 10^{14}/\text{cm}^2)$	$\rho_{C_2H_4} = \dfrac{1}{bA}(\times 10^{22}/\text{cm}^3)$	$g = \left(\dfrac{1}{2}\dfrac{\rho_O}{\rho_R}\left\|\dfrac{\alpha_L + \alpha_T}{\alpha_R}\right\|\right)$
20	5.000	1.988	1.44
21	4.761	1.894	1.38
22	4.545	1.808	1.31
23	4.349	1.729	1.25
24	4.166	1.657	1.20
25	4.000	1.591	1.15
26	3.846	1.529	1.11

TABLE III The absolute value of the dispersion energy $|W_{\alpha\alpha}(R.A)|$ between oriented/oriented hydrocarbon layers with area per hydrocarbon chain A and various separation R.

| A (Å²) | R (Å) | | | | | |
	1.8	2.0	2.2	2.4	2.6	2.8
20	91.85	74.40	61.49	51.67	44.02	37.96
21	83.28	67.46	55.75	46.84	39.91	34.41
22	75.90	61.47	50.80	42.69	36.37	31.36
23	69.43	56.23	46.47	39.05	33.29	28.69
24	63.78	51.66	42.70	35.88	30.57	26.36
25	58.78	47.61	39.35	33.06	28.17	24.29
26	54.35	44.38	36.35	30.57	26.05	22.46

ergs/cm²

TABLE IV The absolute value of the dispersion energy difference $|\Delta W_{\alpha\alpha}|$ between oriented/oriented hydrocarbon layers and oriented/random hydrocarbon phases with respect to A and R.

| A (Å²) | R (Å) | | | | | |
	1.6	1.8	2.0	2.2	2.4	2.6
20	33.52	28.07	22.74	18.79	15.79	13.48
21	29.03	22.93	19.58	15.35	12.90	11.00
22	22.74	17.97	14.55	12.03	10.11	8.61
23	17.58	13.89	11.25	9.30	7.81	6.66
24	13.46	10.63	8.61	7.12	6.00	5.10
25	10.10	7.67	6.22	5.14	4.12	3.68

ergs/cm²

where p and $\mathbf{p}$ are the magnitude and the unit vector of the dipole moment $\mathbf{p}$ respectively, and δ_{1m} is the Kronecker δ:

$$\delta_{1m} = \begin{cases} 1 & m = 1 \\ 0 & m \neq 1 \end{cases}$$

The polarization energy between a dipole layer I per unit area and the other hydrocarbon monolayer II (Figure 4) is calculated by the following integration:

$$\overline{W}_{\alpha p} dS_2 = \int_{-\infty}^{\infty} dx_1 \int_{-\infty}^{\infty} dy_1 \int_{r+d}^{r+2d} dz_1 W_{\alpha p} \rho_{C_2H_4} \sigma_p dS_2$$

$$= -\frac{\pi}{8} p^2 (\alpha_L + \alpha_T) \rho_{C_2H_4} \sigma_p \left\{ \frac{1}{(R+d)^3} - \frac{1}{(R+2d)^3} \right\} dS_2 \quad (2.30)$$

where σ_p is the surface density of the dipole layer. For the bilayer system the total polarization energy $\overline{W}_{\alpha p}^b$ is twice $\overline{W}_{\alpha p}$:

$$\overline{W}_{\alpha p}^b = 2\overline{W}_{\alpha p} = -\frac{\pi}{4} p^2 (\alpha_L + \alpha_T) \rho_{C_2H_4} \sigma_p \left\{ \frac{1}{(R+d)^3} - \frac{1}{(R+2d)^3} \right\} \quad (2.31)$$

With the value in Table I for a permanent dipole moment of a polar group, the numerical values of the polarization energy with respect to various phases at a separation R and area per molecule A are shown in Table V.

The interaction energy between two dipoles is

$$W_{p_1 p_2} = \frac{1}{(r_{1\,2})^3} [\mathbf{p}_1 \mathbf{p}_2 - 3(\mathbf{p}_1 \mathbf{e}_{1\,2})(\mathbf{p}_2 \mathbf{e}_{1\,2})] \quad (2.32)$$

where p_1 and p_2 are the dipole moments. Let us suppose that the dipoles rotate freely around the z axis, then the average dipole moment may be taken to be the z component of each dipole. In the bilayer membrane, if we assume that a dipole in one dipole layer (I) is of the same magnitude, opposed and parallel to a dipole in the other dipole layer (II) ($\theta_1 = \theta_2 = \theta$), the dipole interaction energy is

$$\left. \begin{aligned} W_{pp} &= \frac{p^2}{(r_{1\,2})^3} (3\cos^2 \theta - 1) \\ &= \frac{p^2}{(z^2 + x^2)^{3/2}} \left(3\frac{z^2}{z^2 + x^2} - 1 \right) \\ \cos \theta &= \frac{z}{\sqrt{(z^2 + x^2)}}, \quad p_1 = p_2 \equiv p \end{aligned} \right\} \quad (2.33)$$

TABLE V The absolute value of the polarization energy $|W_{\alpha p}|$ between the dipole layer of polar groups of unit area and the oriented hydrocarbon layer with respect to A and R.

A (Å²)	R (Å²)						
	1.6	1.8	2.0	2.2	2.4	2.6	2.8
26	0.0212	0.0207	0.0203	0.0198	0.0194	0.0189	0.0185
25	0.0229	0.0224	0.0219	0.0214	0.0209	0.0205	0.0200
24	0.0248	0.0243	0.0237	0.0232	0.0227	0.0222	0.0217
23	0.0270	0.0265	0.0259	0.0253	0.0247	0.0242	0.0236
22	0.0295	0.0289	0.0282	0.0276	0.0270	0.0264	0.0258
21	0.0324	0.0318	0.0310	0.0304	0.0290	0.0290	0.0283
20	0.0358	0.0351	0.0342	0.0335	0.0327	0.0320	0.0313

ergs/cm²

where z is the distance between two dipole layers and x is the projection of the distance between two dipoles on the dipole layers. The energy for a dipole layer per unit area, interacting with a two-dimensional dipole layer at a distance z, is

$$\overline{W}_{pp} = 2\pi \int_0^{x_0} x W_{pp} \sigma_p^2 \, dx \quad \text{(for a finite layer)}$$
$$= 0 \quad \text{(for an infinite layer)} \quad (2.34)$$

The numerical values of $\overline{W}_{pp}$ is shown in Figure 6. The energy due to polarization interaction or dipole–dipole interaction is not significant for the bilayer system, because the distance between interacting molecules is much greater than in dispersion interaction.

Let us consider a membrane made up of a thick layer of lipid; that is, at the interface there will be a layer of lipid molecules oriented with their polar groups in contact with water, and with the non-polar chain in contact with bulk lipids. The bulk lipid molecules will be oriented more or less at random (Figure 7). However, the polar groups of the lipids are in an unfavorable state in the bulk lipids, because the polar groups are hydrophilic and will more likely associate with water than in the hydrocarbon phase. There must be some diffusion of the polar groups from bulk lipid phase out to the interface with water. Moreover, if the membrane is supported with hydrophilic materials, a buoyant force and an interfacial tension at the border of the wall will be the cause of the thinning of the thick lipid layer. If the thickness of the membrane is steadily decreased by the above causes, as well as by thermal motion and vibration at the region at which the two oriented molecular layers come into contact, then since the dispersion force between oriented/oriented hydrocarbon layers is more attractive than that between oriented/random hydrocarbon phases, it follows that once oppositely oriented hydrocarbon layers make contact with each other, they retain this position.

If we use 48 Å^2 as the area per lipid molecule according to the discussion of the following section (III), and 1.6 Å $\sim$ 2.0 Å for a separation between two phases (the distance between the ends of CH_3 and CH_3 tails), the dispersion

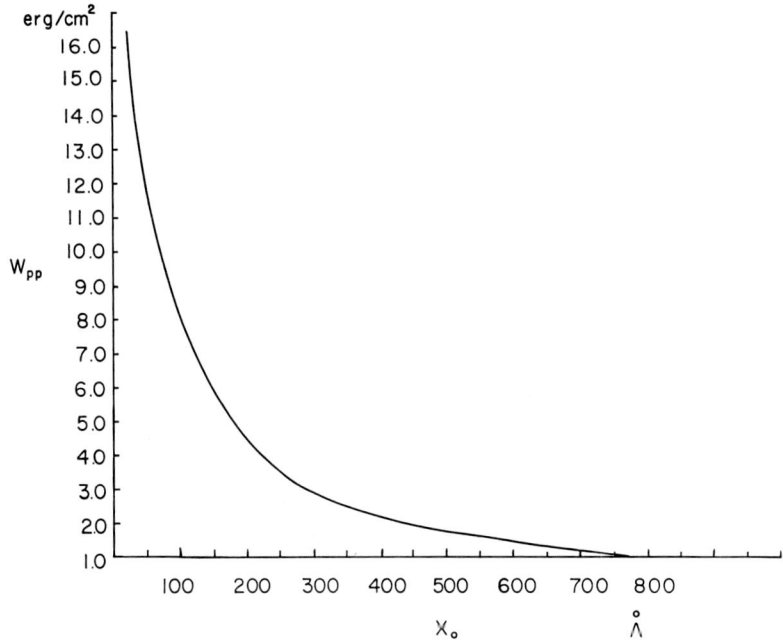

FIGURE 6 Dipole–dipole interaction energy between a dipole layer of unit area and a dipole layer of finite size with a separation $Z = 50$ Å; x_0 is the diameter of the dipole layer of a finite disc.

energy difference $\Delta W_{\alpha\alpha}$ will be in the range 8.6 $\sim$ 13.5 erg/cm.^{2} Therefore, unless a greater energy than $\Delta W_{\alpha\alpha}$ is applied to the molecule, the bilayer state is more stable than the thick lipid membrane. If further decreases in thickness are made, a rapid rise in surface free energy will occur. This arises from the fact that when the membrane is bimolecular, all the polar groups are at the water surface, so that, assuming constant density, further decreases in thickness will create a new surface consisting of a hydrocarbon-water interface. This new interface has a greater interfacial tension, about 50 erg/cm^2, while in the bilayer state, the interfacial tension is about 1 erg/cm^2 for lecithin bilayers. It is energetically unfavorable for the lipid bilayer to be thinner than the bilayer thickness.[20] There is also strong experimental evidence to support this bimolecular layer through X-ray diffraction measurements.[15]

In order to maintain the bimolecular state, the dispersion energy between oriented/oriented hydrocarbon layers is significant; but the interlayer polarization energy (interlayer dipole-induced dipole interaction) and the interlayer

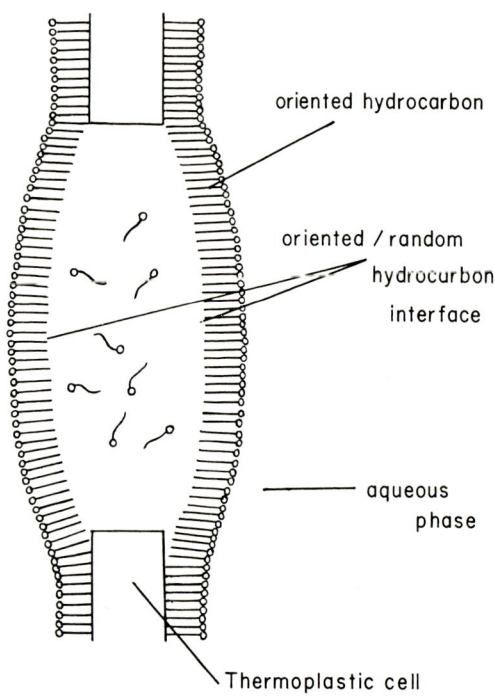

FIGURE 7 Diagrammatic representation of the formation of a bimolecular lipid membrane.

dipole–dipole interaction energy are negligibly small compared with the interlayer dispersion energy. However, if we consider the case of a bilayer of a finite size with the permanent dipoles on the surface of one side of the membrane being oriented opposite to the dipoles on the surface of the other side of the membrane, the interaction energy between two dipole layers is not negligible but plays a part in the stability of the bilayer. Figure 6 shows the interaction energy between a dipole layer of one unit area and a dipole layer of a finite size (its diameter = x_0) with a separation of 50 Å. One sees that the repulsive forces due to dipole–dipole interaction become greater as the size of the bilayer becomes smaller. The repulsive forces may be competitive with the attractive forces due to the dispersion interaction so that the bilayer state becomes unstable.

III DIELECTRIC CONSTANT AND REFRACTIVE INDEX OF LIPID BILAYERS

Several reports concerning the thickness of lipid bilayers, which is one of the important features of the structure of these molecular arrangements, have been presented by several authors.[13,14,21,22,23,24] Determination of the thickness of the bilayer has been made by capacitance or optical measurements, and the thickness has been estimated by using the bulk values for the dielectric constant or the refractive index determined by the Brewster angle measurements.

Since the bilayer may be considered as an oriented hydrocarbon molecular assembly, the dielectric constant and the refractive index must be anisotropic. From Kirkwood's theory[25] of dielectric polarization, the dielectric constant and the refractive index for a model of a lipid bilayer are calculated. Using the calculated values of these dielectric constants and refractive indices and two independent experimental results (capacitance and optical measurements) simultaneously, the degree of the anisotropy of hydrocarbon chains as well as the thickness of the lecithin bilayer are deduced.

Consider a dielectric consisting of nonpolar molecules with asymmetrical polarizability. The polarization $\mathbf{P}$ is equal to the average induced dipole moment per unit volume. If the dielectric consists of N molecules of total volume V, the polarization $\mathbf{P}$ is expressed by

$$\mathbf{P} = \frac{N\mathbf{p}}{v} = \rho\mathbf{p} \tag{3.1}$$

where $\mathbf{p}$ is the average dipole moment per molecule and ρ is the number density of the molecules per unit volume. Neglecting the electrical saturation effect and the inhomogeneity in the local field, the polarization for a molecule is given by

$$\mathbf{p}_i = \alpha \mathbf{E}_i \tag{3.2}$$

where α, the polarizability of a molecule, is a tensor, and E_i is the electric field acting at the center of molecule i. The electric field $\mathbf{E}_i$ can be expressed by the sum of the external field $\mathbf{D}_i$ and the local field $\mathbf{E}_i^{loc}$:

$$\mathbf{E}_i = \mathbf{D} + \mathbf{E}_i^{loc} \tag{3.3}$$

The local field is produced by electric dipole moments of the surrounding molecules,

$$\mathbf{E}_i^{loc} = -\sum_{\substack{k=1 \\ \neq i}}^{N} T_{ik} \mathbf{p}_k \tag{3.4}$$

$$T_{ik} = \nabla_i \nabla_k \left(\frac{1}{r_{ik}}\right) = \frac{1}{(r_{ik})^3} [1 - 3\mathbf{e}_{ik} \mathbf{e}_{ik}] \tag{3.5}$$

where $\mathbf{r}_{ik}$ is the displacement vector between molecules i and k. The unit vector of $\mathbf{r}_{ik}$ is defined as $\mathbf{e}_{ik}$; r_{ik} is the distance between molecules i and k; 1 is the unit tensor. Insertion of Eq. (3.5) into Eq. (3.4) gives

$$\mathbf{E}_i = \mathbf{D} - \sum_{\substack{k=1 \\ \neq i}}^{N} T_{ik} \mathbf{p}_k \tag{3.6}$$

From Eqs. (3.2) and (3.6), we obtain the following relation

$$\mathbf{p}_i + \alpha \sum_{\substack{k=1 \\ \neq i}}^{N} T_{ik} \mathbf{p}_{ik} = \alpha \mathbf{D} \tag{3.7}$$

The formal solution of Eq. (3.7) is given by

$$\mathbf{p}_i = (1 + \alpha T)^{-1} \alpha \mathbf{D} \tag{3.8}$$

where T may be regarded as the matrix of a linear transformation T in the $3N$ dimensional configuration space of the entire system, the electric moments $\mathbf{p}_i$ being the projections of a $3N$ dimensional vector on the subconfiguration space of molecule i.

On the other hand, according to electrostatics, we have the following relations:

$$\rho \mathbf{p} = \mathbf{P} = \frac{\varepsilon - 1}{4\pi} \mathbf{E} \tag{3.9}$$

$$\mathbf{D} = \varepsilon \mathbf{E} \tag{3.10}$$

where ε is the average dielectric constant which is a tensor. From Eqs. (3.8) and (3.10), we have:

$$\left(1 - \frac{1}{\varepsilon}\right) = \frac{4\pi \rho \mathbf{p}}{\mathbf{D}} = \frac{4\pi \rho \alpha}{1 + \alpha T} \tag{3.11}$$

If the polarization tensor is diagonalized with respect to the x, y, and z axes (in cartesian coordinates), we have

$$\varepsilon = \begin{pmatrix} \varepsilon_{xx} & 0 & 0 \\ 0 & \varepsilon_{yy} & 0 \\ 0 & 0 & \varepsilon_{zz} \end{pmatrix}$$

$$\equiv \begin{pmatrix} \varepsilon_x & 0 & 0 \\ 0 & \varepsilon_y & 0 \\ 0 & 0 & \varepsilon_z \end{pmatrix} \quad (3.12)$$

In this case, when an external electric field is applied along the x axis, the dielectric constant is given by

$$\varepsilon_x = \frac{1}{1-(4\pi\rho\alpha_{xx}/1+\alpha_{xx}T_{xx})} \quad (3.13)$$

and similarly for the other axes.

The electric field is

$$\mathbf{E}_i = [1+\alpha T]^{-1}\mathbf{D} = \left[1 - \alpha \sum_{\substack{k=1 \\ \neq i}} T_{ik} + \alpha^2 \sum_{\substack{k,l=1 \\ \neq i}} T_{ik}T_{kl} - \ldots\right]\mathbf{D} \quad (3.14)$$

If the dielectric is a bulk phase of an infinite size we have approximately the following relations:

$$\sum_{kl} T_{ik}T_{kl} = \left(\sum_{\substack{k=1 \\ \neq i}} T_{ik}\right)^2$$

$$\sum_k \sum_l \sum_n T_{ik}T_{kl}T_{ln} = \left(\sum_{\substack{k=1 \\ \neq i}} T_{ik}\right)^3 \quad (3.15)$$

$$\vdots$$

where

$$T_{ik} = \sum_{k \neq i}\left(\frac{1}{(r_{ik})^3} - \frac{3e_{ik}e_{ik}}{(r_{ik})^3}\right)$$

For a bilayer model system (Figure 8), T has been calculated on the basis of the above considerations.[26] The numerical value of the component parallel to the hydrocarbon chain ($T_\parallel$) is a function of the area per molecule and the thickness of the bilayer, while the component perpendicular to the hydrocarbon chain $T_\perp$ is a function of the area per molecule only, provided that the chain of a molecule is sufficiently large compared with a C_2H_4 molecular unit.

In the capacitance measurements, since an electric field is applied perpendicular to the bilayer, only the parallel (to the chain) component $\varepsilon_\parallel$ of the dielectric constant of the bilayer pertains to the capacitance measured. On the other hand, in the optical measurements, the intensity of reflected light is

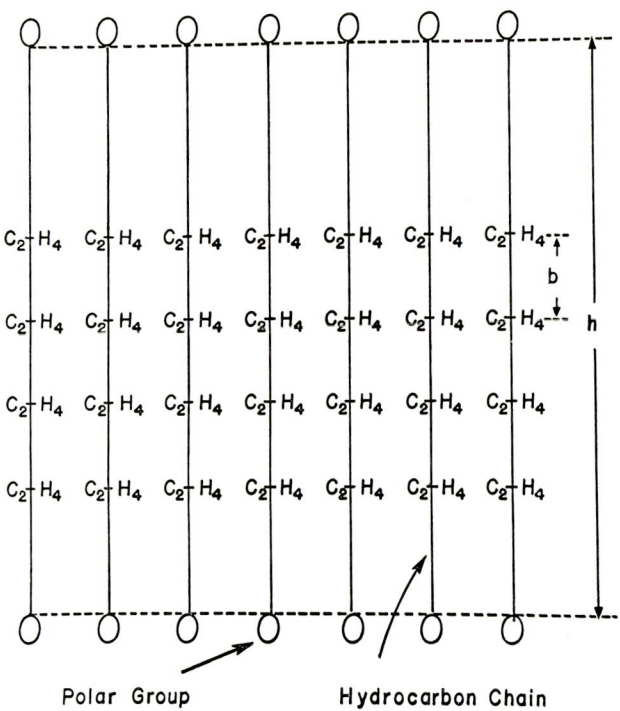

FIGURE 8   The simplest model of a lipid bilayer.

related to both refractive indices (i.e. components parallel and perpendicular $(n_\parallel, n_\perp)$ to the hydrocarbon chain). In the special case when the incident ray is normal to the surface of the membrane surface, only the perpendicular component $n_\perp$ of refractive index pertains to the reflectance. Usually, all measurements of reflectivity of the bilayer membrane have been made at a small angle of incidence. Thus, these two experiments (capacitance and optical reflectance measurements) are independent of each other.

If we use the formula of a parallel plate capacitor for the capacitance of a bilayer, we have

$$C = \frac{\varepsilon_\parallel}{4\pi h} \qquad (3.16)$$

where C is the capacitance, $\varepsilon_\parallel$ is the parallel component of the dielectric constant, and h is the thickness of the hydrocarbon portion of a bilayer membrane. $\varepsilon_\parallel$ is given by

$$\varepsilon_\parallel = \frac{1}{1-(4\pi\rho\alpha_\parallel/1+T_\parallel \alpha_\parallel)}$$

$$= \frac{1}{1-\dfrac{4\pi\rho\alpha_\parallel}{1+[A'-(B'/h)]\alpha_\parallel}} \qquad (3.17)$$

where A' and B' are numerical constants which depend on the area per hydrocarbon chain of a bilayer. The numerical values of A' and B' with respect to the area per hydrocarbon chain are shown in Table VI. From Eqs. (3.16) and (3.17), we obtain

$$h = \frac{1+A'\alpha_\parallel+4\pi CB'\alpha_\parallel+\sqrt{[(1+A'\alpha_\parallel+4\pi CB'\alpha_\parallel)^2-16\pi C(1+\alpha_\parallel A'-4\pi\rho\alpha_\parallel)]}}{8\pi C(1+A'\alpha_\parallel-4\pi\rho\alpha_\parallel)}$$

$$(3.18)$$

The relation between the thickness and the area per hydrocarbon chain is shown in Figure 9 and Table VII for the value $0.38\ \mu F/cm^2$ of a lecithin bilayer (Hanai, Haydon and Taylor)[14] and for the value $0.45\ \mu F/cm^2$ of a lecithin bilayer (Tien 1967)[23], respectively. Different measured capacitances produce a family of similar curves.

The optical properties of a lipid film in an aqueous solution are fundamentally the same as those of thin transparent solid film. The interference phenomonon produced when monochromatic light is reflected from the two faces of a thin dielectric plate may be described as follows (Heavens 1955[27]; Landau & Lifshitz 1960[28]). The geometrical figure of the system is shown in Figure 9, where h is the thickness, n_0 is the refractive index of the medium (aqueous solution), $n_\parallel$ and $n_\perp$ are the parallel (perpendicular to the layer) and perpendicular (parallel to the layer) components of the dielectric constant of the bilayer, θ_0 is the angle of the incident ray of the layer, and θ_1 is the angle of the reflected ray in the layer. The two reflected rays, A and B, have undergone a relative phase shift which results in the observed interference phenomenon. The path difference of the two rays is $2n_b h \cos\theta_1 + \lambda/2$. Thus the

TABLE VI Numerical values of dielectric constant for lipid bilayer model with respect to area per hydrocarbon chain and thickness.

$b = 2.514 \text{ Å}$

$\alpha_\| = 5.72 \text{ Å}^3$ (for C_2H_4 unit, Ohki and Fukuda (1967))

$$\varepsilon_\| = \cfrac{1}{1 - \cfrac{4\pi\rho\alpha_\|}{1 + \alpha_\| T_\|}}$$

$$T_\| = A' - \frac{B'}{h}$$

$a^2 (\text{Å}^2)$	$\rho(=\rho_{C_2H_4})$	$A' (\text{Å}^{-3})$	$B' (\text{Å}^{-2})$
20		0.2454	0.8954
21		0.2353	0.8689
22		0.2253	0.8426
23		0.2152	0.8162
24		0.2055	0.7900
25		0.1951	0.7639
26		0.1869	0.7310

$a^2 (\text{Å}^2)$	$\varepsilon_\perp^{(17)}$	$\varepsilon_\| (h = 60 \text{ Å})$	$\varepsilon_\| (h = 50 \text{ Å})$
20	4.0277	2.6079	2.6394
21	3.3976	2.5092	2.5374
22	2.9776	2.4297	2.3889
23	2.6765	2.3651	2.3889
24	2.4512	2.3128	2.3350
25	2.2761	2.2708	2.2915
26	2.1316		

TABLE VII

$$h = \frac{1 + A'\alpha_\| + 4\pi CB'\alpha_\| + \sqrt{[(1 + A'\alpha_\| + 4\pi CB'\alpha_\|)^2 - 16\pi C(1 + A'\alpha_\| - 4\pi\rho\alpha_\|)B'\alpha_\|]}}{8\pi C(1 + A'\alpha_\| - 4\pi\rho\alpha_\|)}$$

$a^2 (\text{Å}^2)$	$C = 0.38 \, \mu F$ (Hanai et al., 1964) h (Å)	$C = 0.45 \, \mu F$ (Tien 1968) h (Å)
20	60.09	51.82
21	58.17	49.93
22	56.34	48.39
23	54.97	47.19
24	53.93	46.15
25	52.97	45.33
26	51.94	44.10

phase shift 2δ is $4\pi n_b \cos\theta_1/\lambda + \pi$ where λ is the wave length of the light used and n_b is the refractive index of the bilayer.

Substitution of δ into the Fresnel formula for the amplitude E of the reflected ray gives

$$E = \frac{\gamma_1 + \gamma_2 \exp(-2i\delta)}{1 + \gamma_1\gamma_2 \exp(-2i\delta)} E_0 \qquad (3.19)$$

where E_0 is the amplitude of the incident ray, γ_1 and γ_2 are the reflective coefficients at the front surface (F) and the back surface (B); $r_1 = n_0 - n_1/n_0 + n_1$, $r_2 = n_1 - n_0/n_0 + n_1$.

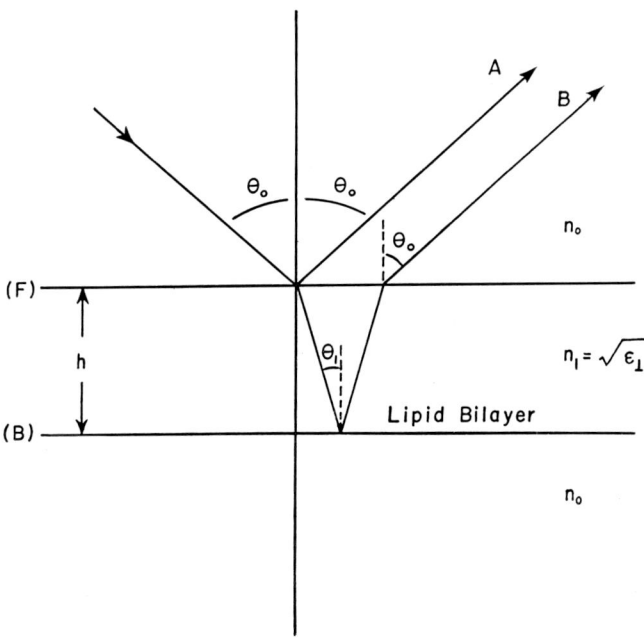

FIGURE 9 Sketch of geometrical relation between rays transmitted and reflected by a thin dielectric plate of thickness h. θ_0 is the incident angle and θ_1 is the internal angle of reflection. n_0 and n_1 are the refractive indices of the medium and the plate, respectively.

In our lipid bilayer model, the optical axis coincides with the molecular axis. We shall take it as the z axis in cartesian coordinates, denoting the corresponding principal values of the dielectric constant ε by $\varepsilon_\parallel$. The directions of the other two principal axes, in a plane perpendicular to the optical axis, are arbitrary, and the corresponding principal values, which we denote by $\varepsilon_\perp$, are equal.

If we insert $\varepsilon_x = \varepsilon_y = \varepsilon_\perp$, $\varepsilon_z = \varepsilon_\parallel$ into Fresnel's equation:

$$n^2(\varepsilon_x n_x^2 + \varepsilon_y n_y^2 + \varepsilon_z n_z^2) - [n_x^2 \varepsilon_x(\varepsilon_y + \varepsilon_z) + n_y^2 \varepsilon_y(\varepsilon_x + \varepsilon_z) + n_z^2 \varepsilon_z(\varepsilon_x + \varepsilon_y)]$$
$$+ \varepsilon_x \varepsilon_y \varepsilon_z = 0 \qquad (3.20)$$

We obtain the quadratic equations:

$$\begin{cases} n^{o2} = \varepsilon_\perp \\ \dfrac{n_z^{ex2}}{\varepsilon_\perp} + \dfrac{n_x^{ex2} + n_y^{ex2}}{\varepsilon_\parallel} = 1 \end{cases} \qquad (3.21)$$

Thus, there are two types of waves propagated in a uniaxial crystal. One is called the ordinary wave with respect to which the crystal behaves like an isotropic body of refractive index $n^o = \sqrt{\varepsilon_\perp}$. The other is called the extraordinary wave. The magnitude of this wave vector depends on the angle it makes with the optical axis:

$$\frac{1}{n^{ex}} = \frac{\sin^2 \theta_1}{\varepsilon_\parallel} + \frac{\cos^2 \theta_1}{\varepsilon_\perp} \qquad (3.22)$$

From Eq. (3.22) it is easily seen that when the angle of incidence is small the extraordinary ray behaves the same as the ordinary ray.

The reflectivity R, defined as the ratio of the reflected energy $I(=E^2 = E_s^2 + E_p^2)$ to the incidence energy $I_0 = E_0^2$, is:

$$\left.\begin{aligned} R &= \frac{I}{I_0} = \frac{E_s^2 + E_p^2}{E_0^2} \\ \frac{E_s^2}{E_{so}^2} &= \frac{\gamma_{1s}^2 + 2\gamma_{1s}\gamma_{2s}\cos 2\delta + \gamma_{2s}^2}{1 + 2\gamma_{1s}\gamma_{2s}\cos 2\delta + \gamma_{1s}^2\gamma_{2s}^2} \\ \frac{E_p^2}{E_{po}^2} &= \frac{\gamma_{1p}^2 + 2\gamma_{1p}\gamma_{2p}\cos 2\delta + \gamma_{2p}^2}{1 + 2\gamma_{1p}\gamma_{2p}\cos 2\delta + \gamma_{1p}^2\gamma_{2p}^2} \end{aligned}\right\} \qquad (3.23)$$

where the suffixes s and p refer to the perpendicular and parallel components to the plane of incidence, and the γ's are the reflective coefficients. For the lipid bilayer in the solution, $|\gamma_1| = |\gamma_2| \equiv \gamma$. γ is composed of the components of $\gamma_\parallel$ and $\gamma_\perp$.

According to Blodgett and Langmuir (1937)[29], the reflection coefficients γ_s and γ_p of light polarized in directions perpendicular and parallel to the plane of incidence are given by

$$\left.\begin{aligned} \gamma_s &= \frac{n_0 \cos \theta_0 - (n_\perp^2 - n_0^2 \sin \theta_0)^{\frac{1}{2}}}{n \cos \theta_0 + (n_\perp^2 - n_0^2 \sin \theta_0)^{\frac{1}{2}}} \\ \gamma_p &= \frac{n_0(n_\parallel^2 - n_0^2 \sin^2 \theta_0)^{\frac{1}{2}} - n_\perp n_\parallel \cos \theta_0}{n_0(n_\parallel^2 - n_0^2 \sin^2 \theta_0)^{\frac{1}{2}} + n_\perp n_\parallel \cos \theta_0} \end{aligned}\right\} \qquad (3.24)$$

If the angle of incidence is small (less than 25°), γ_s and γ_p are reduced to

$$\gamma_s = \gamma_p = \frac{n_\perp - n_0}{n_\perp + n_0} \equiv \gamma_\perp \tag{3.25}$$

In this case, the reflectivity R is

$$R = \frac{I}{I_0} = \frac{E_\perp^2 + E_\parallel^2}{E_0^2} = \frac{\gamma_{11}^2 + 2\gamma_{11}\gamma_{21}\cos 2\delta + \gamma_{21}^2}{1 + 2\gamma_{11}\gamma_{21}\cos 2\delta + \gamma_{11}^2\gamma_{21}^2} = \frac{4\gamma_\perp^2 \sin^2 \delta}{1 - 2\gamma_\perp^2 + 4\gamma_\perp^2 \sin^2 \delta + \gamma_\perp^4} \tag{3.26}$$

For lipid films in an aqueous solution (0.1 N NaCl) the Fresnel coefficient $\gamma_\perp$ is quite small. Therefore, Eq. (3.26) reduces to

$$R = 4\gamma_\perp^2 \sin^2 \delta = 4\gamma_\perp^2 \sin^2\left[\frac{2\pi n_\perp h \cos \theta_1}{\lambda}\right] \tag{3.27}$$

Recently, Cherry and Chapman (1968)[24] have measured the absolute reflective intensity at an incident angle 10°. They obtained the reflectivity of a lecithin bilayer in solutions of various refractive indices. In a medium of 0.1 N NaCl the reflectivity R was 5.90×10^{-5} and the refractive index n_0 of the solution was 1.334. Using the reflectivity of the bilayer, the refractive index of the solution and our refractive index, calculated for a lipid bilayer model, we can obtain the thickness with respect to area per hydrocarbon chain. The relation between the thickness and the area per molecule which satisfies Cherry's experimental result, is shown in Figure 10.

For thin layers showing interference fringes, the phase difference should satisfy the following relation:

$$\delta = \frac{2\pi n_b h \cos \theta}{\lambda} = k\frac{\pi}{2} \tag{3.28}$$

where k is an integer. When $k = 1$ the film gives the maximum reflection before the transition to the black state. Therefore, Eq. (3.29) reduces to

$$R_{silv} = 4\gamma_{silv}^2 \sin^2 \delta_{silv} = 4\gamma_{silv}^2 \tag{3.29}$$

for the film of the maximum reflection which is observed in the silvery film, where

$$\gamma_{silv} = \frac{n_{silv} - n_0}{n_{silv} + n_0}$$

Since the silvery film is composed almost entirely of liquid hydrocarbon, we may use the refractive index 2.09 (Hanai, Haydon and Taylor[21] 1964) for a liquid hydrocarbon mixture as the refractive index of the silvery film.

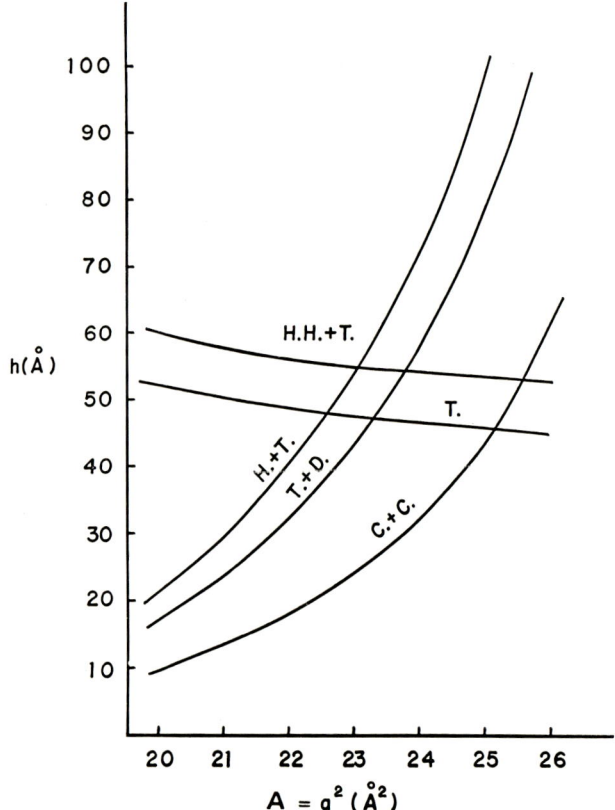

FIGURE 10 The relation between the thickness of lecithin bilayer and the average area per hydrocarbon chain. The values (h, a) on the solid line are used to fit the experimental data.

H.H.T. Hanai, et al. (1964)[13]   T-D Tien, et al. (1966)[14]
T Tien (1968)[23]   C-C Cherry, et al. (1968)[16]
H-T Huang (1965)[21]

For the bilayer, Eq. (3.27) is

$$R_b = 4\gamma_b^2 \sin^2 \delta_b = 4\gamma_\perp^2 \sin^2 \delta_b \tag{3.30}$$

where the suffixes silv and b refer to the silvery state and the bilayer state. The ratio R_b/R_{silv} is expressed by

$$\frac{R_b}{R_{silv}} = \frac{\gamma_b^2}{\gamma_{silv}^2} \sin^2\left[\frac{2\pi n_b h \cos \theta_1}{\lambda}\right] = \frac{\gamma_\perp^2}{\gamma_{silv}^2} \sin^2\left[\frac{2\pi n_\perp h \cos \theta_1}{\lambda}\right] \tag{3.31}$$

for the case of a small incident angle.

TABLE VIII Experimental data for capacitance and reflectivity measurements.

Lipid	Aqueous phase	Capacitance	Reflectivity	
Lecithin (egg)	Various	0.38 µF/cm		Hanai, et al. (1964)
Lecithin (egg)	0.1 N NaCl	0.45 µF/cm		Tien (1968)
Lecithin (egg)	0.1 N NaCl		$\frac{I_b}{I_s} = 0.0618$	Huang, et al. (1965)
Glycerol distearate	0.1 N NaCl		$\frac{I_b}{I_s} = 0.0424$	Tien, et al. (1966)
Lecithin (egg)	0.1 N NaCl		$R = \frac{I_b}{I_0} = 5.90 \times 10^{-5}$	Cherry, et al. (1968)

With the experimental results of relative intensity measurements (Table VIII), we can obtain the thickness as a function of area per hydrocarbon chain for each case (Huang and Thompson, 1965,[21] 1966;[22] Tien, et al., 1966[14]). The results are shown in Figure 10 and Table IX. Relying upon two independent experimental data in Table VIII, we can deduce the bilayer thickness and the area per hydrocarbon chain from the intersection of two curves in Figure 10, and Table X.

Thus, the dimension of the lecithin bilayer is in the following range:

thickness 54 ∼ 48 Å and area per molecule 51 ∼ 47 Å²

The corresponding dielectric constant and refractive index are

$$\varepsilon_\parallel = 2.32 \sim 2.29$$
$$\varepsilon_\perp = 2.59 \sim 2.25$$
$$n_\parallel = 1.52 \sim 1.51$$
$$n_\perp = 1.60 \sim 1.50$$

The dielectric constant and the refractive index are not so anisotropic for the bilayer. That is, the hydrocarbon chain is more or less oriented perpendicular to the layer. But as a whole the anisotropy of the dielectric constant and the refractive index is small.

IV BILAYER—MICELLE TRANSFORMATION

Structural change of the phospholipid bilayers may be closely related to the function of the biological membranes, such as the excitability of the membrane. Since the lipids have several ionizable head groups, changes of the

TABLE IX

$$R = \frac{I}{I_0} = 4r_b^2 \sin^2\left(\frac{2\pi n_b h \cos\theta_1}{\lambda}\right)$$

$$\frac{I_b}{I_s} = \left(\frac{r_b}{r_s}\right)^2 \sin^2\left(\frac{2\pi n_b h \cos\theta_1}{\lambda}\right)$$

$n_b = 1.334$ (for 0.1 N NaCl)

$n_b = \sqrt{\varepsilon_\perp}$

$n_s = \sqrt{\varepsilon^{\text{mixture}}} = 2.07 = 1.4387$ (Hanai, et al., 1964)

From Cherry and Chapman (1968)

a^2 (Å^2)	$n_b = \sqrt{\varepsilon_\perp}$	$r_b = \dfrac{n_b - n_0}{n_b + n_0}$	$\cos\theta_1$	h
20	2.0068	0.2014	0.9933	9.63
21	1.8432	0.1602	0.9920	13.20
22	1.7254	0.1279	0.9909	17.68
23	1.6360	0.1016	0.9899	23.50
24	1.5656	0.0798	0.9890	31.30
25	1.5084	0.0614	0.9881	42.26
26	1.4600	0.0450	0.9871	59.65

a^2 (Å^2)	n_b	$r_b = \dfrac{n_b - n_0}{n_b + n_0}$	$r_s = \dfrac{n_s - n_0}{n_s + n_0}$	$\theta_0 = 20°$ $\cos\theta_1$	h (from Huang-Thompson 1965)	h (from Tien 1966)
20	2.0069	0.2014	0.0377	0.9743	21.38	17.12
21	1.8432	0.1602	0.0377	0.9695	29.40	23.56
22	1.7254	0.1279	0.0377	0.9650	39.54	31.67
23	1.6360	0.1016	0.0377	0.9611	52.60	42.22
24	1.5656	0.0798	0.0377	0.9574	70.11	56.39
25	1.5084	0.0614	0.0377	0.9540	94.57	76.32

TABLE X Thickness, area per hydrocarbon chain, dielectric constant and refractive index of the lipid bilayer.

	HHT and HT	HHT and TD	HHT and CC	T and HT	T and TD	T and CC
h	54.5	54.0	52.0	47.5	46.5	45.0
a^2	23.1	23.8	25.6	22.6	23.3	25.2
$\varepsilon_\parallel$	2.34	2.32	2.23	2.41	2.36	2.29
$\varepsilon_\perp$	2.65	2.59	2.20	2.79	2.60	2.25
$n_\parallel$	1.53	1.52	1.49	1.55	1.53	1.51
$n_\perp$	1.62	1.60	1.48	1.67	1.61	1.50

environmental solution (such as pH and salt content) may cause structural alteration of the membrane as a result of the dissociation of polar groups. There are a few experiments to support this interpretation in the monolayer[30] and on the phospholipid vesicle.[31] For example, phosphatidyl choline has zwitterionizable groups. Therefore, in solutions of extremely low pH, the phosphatidyl choline bilayer would be charged positively. On the other hand, in a solution of high pH, the bilayer would be charged negatively. Consequently, the large electrostatic repulsive forces due to the dissociation of the polar groups would make the bilayer membrane unstable in the solution.

According to previous experiments on the phospholipid bilayers, bilayer formation greatly depends upon the condition of the environmental solution (e.g. pH and the concentration of various ions). It was shown[32] that in 0.1 N NaCl, a phosphatidyl choline bilayer is stable within the range of pH 1–9. However, in solutions of pH lower than 1, or higher than 9, the bilayer becomes unstable and tends to break up.

When the phospholipids are dispersed in the aqueous solution, they could have several possible forms:

1. spherical micelle
2. cylindrical micelle
3. bilayer
4. lamellar, the structure of which is essentially the same as that of the bilayer
5. bubble membrane (or vesicles), the structure of which is also essentially the same as that of the bilayer.

There have been several experimental studies on the transformations among these structural forms in the liquid–crystal phases.[33]

Some investigators[34,35] propose that spherical micelles transform to a cylindrical form in a soap solution when the number of ions becomes larger than can be accommodated in a sphere. This phase transformation is due to the concentration of soap. Recently, Parsegian[36] proposed a theory of phase transition of a plane membrane to a cylindrical micelle on a liquid–crystal phase of lipids in terms of their volume fractions in an aqueous solution.

Here, a thermodynamical analysis of transformation among bilayer, cylindrical micelle and spherical micelle is described in terms of the charge of the polar groups, which is related to the pH values of the solution.

Let us suppose that there are three possible states of lipids in the aqueous solutions, that is, the bilayer, the cylindrical micelle, and the spherical micelle (Figure 11). For each case, we may define the total free energy G as follows:

$$G = G^{bulk} + G^{surface} = U - ST + PV + G^{surface} \qquad (4.1)$$

where G^{bulk} is the free energy of the hydrocarbon phase which is not in contact

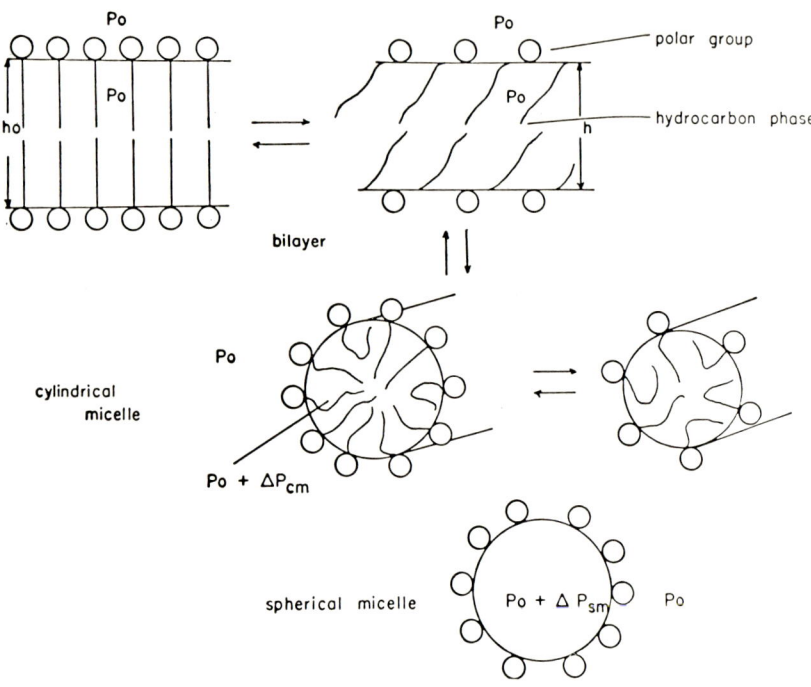

FIGURE 11 A schematic sketch of bilayer, cylindrical micelle and spherical micelles.

with the aqueous phase, and U, S, P and V are the internal energy, the entropy, the pressure and the volume of the bulk lipids, and G^{surface} is the free energy of the interface between the hydrocarbon and the aqueous phases.

It should be noted that the free energy of the aqueous solution does not change in spite of the existence of an electric field. Therefore, the change of the pressure of the bulk lipids is compensated by the electrostatic energy. Thus, the pressure is given by

$$P = P_0 + \frac{\varepsilon_0 \kappa^2}{8\pi} \psi^2 \qquad (4.2)$$

where P_0 is the pressure at infinite distance from the lipid, ψ is the electrical potential, and $-(\varepsilon_0 \kappa^2/4\pi)\psi$ is the charge density.

(a) Bilayer

For the bilayer state, the specific free energy is

$$G_b = G_b^{\text{bulk}} + G_b^{\text{surface}} \tag{4.3}$$

where G_b^{bulk} is the free energy per lipid molecule for the bulk hydrocarbon in the bilayer and G_b^{surface} is the free energy per lipid molecule at the interface of the bilayer. When the polar groups acquire a net charge, the bilayer apparently spreads out in the plane of the bilayer and becomes thinner.

If we assume that the total volume does not change and that the lipid molecules have restricted freedom of movement such as in a liquid crystal, the free energy G_b^{bulk} of the bulk lipid may be considered to be independent of its thickness. In order to express the free energy change of an extending bilayer in terms of its net charge, it is sufficient to know the relative free energy G_b^{rel} of the bilayer. This is defined as follows:

$$G_b^{\text{rel}}(A, q) = G_b - G_b^{\text{bulk}} = E_b(A, q) + \sigma_{wp} A_0 + \sigma_{wh} A_1 \tag{4.4}$$

Here E_b is the energy per molecule due to the electrostatic interaction of the polar groups, being a function of the net charge; q is the degree of the dissociation of a polar group; A is the surface area per molecule of the bilayer; σ_{wp} and A_0 are the surface tension and the initial surface area per molecule of the bilayer with the polar group of zero net charge, respectively; and σ_{wh} and A_1 are the surface tension and the area per molecule of the hydrocarbon-water interface created by the expansion of the film. Since the interfacial tension σ_{wp} of the bilayer is much smaller ($\sim 1 \text{ erg/cm}^2$ for a lecithin bilayer[23] in $0.1 n$ NaCl ($\sim$pH 6)) than σ_{wh} ($\sim 52 \text{ erg/cm}^2$ for n-Tetradecane[37]), we may ignore the term $\sigma_{wp} A_0$. Then we have

$$G_b^{\text{rel}}(A, q) = E_b(A, q) + \sigma_{wh} A_1 \tag{4.5}$$

When the polar groups have net charges, the bilayer extends and, since the volume of lipids is assumed not to be changed, the bilayer becomes thinner. The effective area of water-hydrocarbon interface A_1 created by the expansion of the surface is

$$A_1 = A - A_0 \tag{4.6}$$

And the energy due to the interfacial tension of the hydrocarbon water interface is

$$\sigma_{wh} A_1 = \sigma_{wh}(A - A_0) \tag{4.7}$$

The energy E_b of the electrostatic interaction due to the polar groups is obtained as follows. It may be assumed here that the charge of the polar group is uniformly distributed on both sides of the surfaces of the membrane. Then, the potential at a point p at a distance x from the surface is expressed by Eq. (A.11). Therefore, from Eq. (A.11) ψ_b at a point on the surface is

$$\psi_b = \frac{4\pi\sigma}{\varepsilon_0 \kappa} = \frac{4\pi e q}{\varepsilon_0 \kappa A} \tag{4.8}$$

where ε_0 is the dielectric constant of the aqueous solution, and κ is the Debye–Hückel constant of the solution. The electrostatic energy per polar group due to all the remaining polar groups is one half of $eq\,\psi_b$:

$$E_b = \tfrac{1}{2}\psi_b\, eq = \frac{2\pi e^2 q^2}{\varepsilon_0 \kappa A} \tag{4.9}$$

Thus, the specific relative free energy for the bilayer state is

$$G_b^{\mathrm{rel}} = \sigma_{wh}(A - A_0) + \frac{2\pi e^2 q^2}{\varepsilon_0 \kappa A} \tag{4.10}$$

With the following numerical values

$$\left.\begin{array}{l} e = 4.8 \times 10^{-10}\ \text{e.s.u.} \\ \varepsilon_0 = 80 \\ \sigma_{wh} = 50\ \text{dyn/cm} \\ A_0 = 48\ \text{Å}^2 \\ \kappa = 0.10\ \text{Å}^{-1} \quad \text{(for 0.1 N NaCl solution at 300°K)} \end{array}\right\} \tag{4.11}$$

the relative free energy per lipid molecule of the bilayer state is expressed in terms of the degree q of dissociation of a polar group and the average surface area per molecule A.

The dimension of the stable state of the bilayer for a given net charge per polar group is obtained by taking the minimum value of Eq. (4.10) with respect to A. The value A obtained

$$A = \left\{\begin{array}{ll} q\sqrt{\dfrac{2\pi e^2}{\varepsilon_0 \kappa \sigma_{wh}}} & \text{for}\quad q > A\sqrt{\dfrac{\varepsilon_0 \kappa \sigma_{wh}}{2\pi e^2}} \equiv q_c \\ A_0 & \text{for}\quad q < q_c \end{array}\right. \tag{4.12}$$

is the average surface area per lipid molecule at the stable state for the bilayer configuration. The relation between A and q is shown in Figure 12.

Dispersion forces and stability of lipid bilayers

The relative free energy of the stable state is obtained by substituting Eq. (4.12) into Eq. (4.10):

$$G_b^{rel} = \begin{cases} \dfrac{2\pi e^2 q^2}{\varepsilon_0 \kappa A_0} & \text{for} \quad q < q_c \\ q\sqrt{\left(\dfrac{8\pi e^2 \sigma_{wh}}{\varepsilon_0 \kappa}\right)} - \sigma_{wh} A_0 & \text{for} \quad q > q_c \end{cases} \quad (4.13)$$

The relation between G_b^{rel} and q is illustrated in Figure 13.

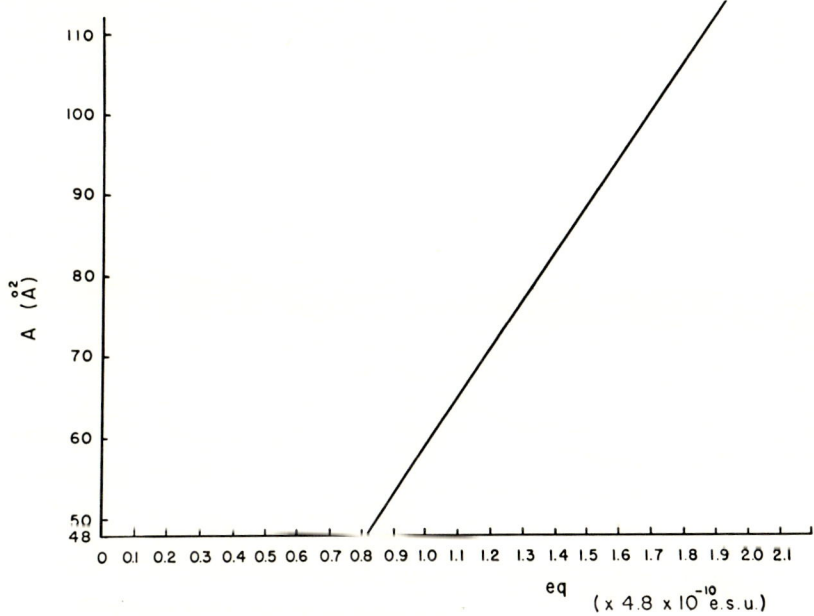

FIGURE 12 The numerical relation between net charge eq and the average area A per molecule, at which the relative free energy of a bilayer state has a minimum value.

(b) *Cylindrical micelle*

For the cylindrical micelle state, the specific free energy G_{cm} is

$$G_{cm} = G_{cm}^{bulk} + G_{sm}^{surface} \quad (4.14)$$

where G_{cm}^{bulk} is the free energy of the hydrocarbon phase in the cylindrical micelle and $G_{cm}^{surface}$ is the free energy of the interface between the hydrocarbon and the aqueous phase of the cylindrical micelle.

Since the interface of the cylindrical micelle is curved, there is a pressure difference Δp_{cm} between the two phases divided by the interface. The bulk free energy G_{cm}^{bulk} is no longer the same as that of the bilayer state. The difference between the two bulk free energies of the cylindrical micelle and the bilayer states is

$$G_{cm}^{bulk} - G_b^{bulk} = \Delta P_{cm} \bar{v} \qquad (4.15)$$

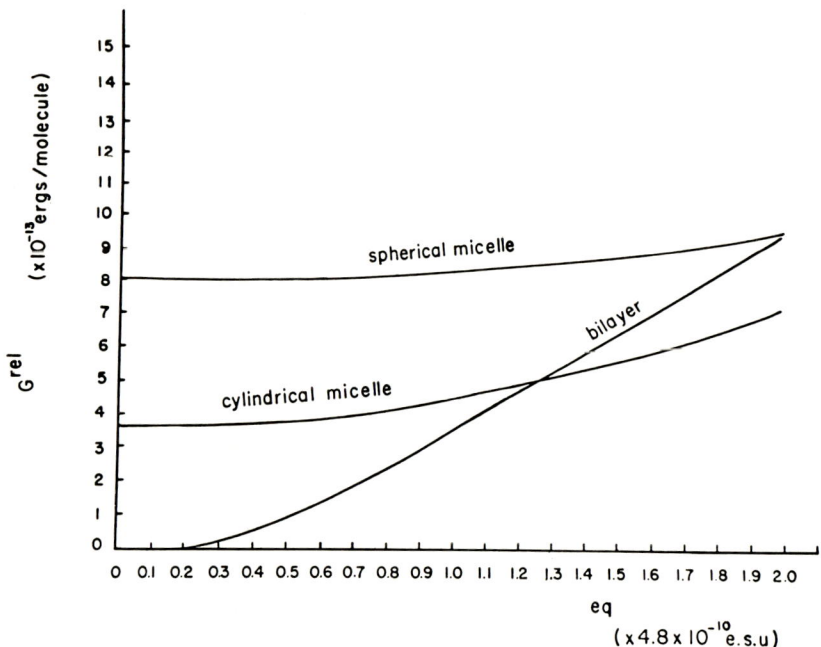

FIGURE 13 Specific relative free energies in the bilayer, cylindrical micelle and spherical micelle states with respect to the net charge eq.

where $\bar{v}$ is the volume per lipid molecule. Therefore, the relative free energy per lipid molecule of the cylindrical micelle defined by

$$G_{cm}^{rel} \equiv G_{cm} - G_b^{bulk} \qquad (4.16)$$

is written as

$$G_{cm}^{rel} = G_{cm}^{surface} + \Delta P_{cm} \bar{v} \qquad (4.17)$$

The surface free energy $G_{cm}^{surface}$ is

$$G_{cm}^{surface} = E_{cm}(R, q) + \sigma_{wp} A_0 + \sigma_{wh} A_2 \cong E_{cm}(R, q) + \sigma_{wh} A_2 \qquad (4.18)$$

Here E_{cm} is the energy per lipid molecule of electrostatic interaction; R is the radius of the cylinder; A_0 is the initial surface area per molecule in the bilayer state with the polar groups of zero net charge; the term $\sigma_{wp} A_0$ can be neglected as before; and A_2 is the area per molecule of hydrocarbon-water interface which is estimated as follows.

Suppose the maximum radius of the cylindrical micelle is the fully extended length of a lipid molecule ($l_0 \sim 25$ Å)[38], that is:

$$R \leq l_0 \tag{4.19}$$

The volume per unit length of the cylindrical micelle is

$$V = \pi R^2 \tag{4.20}$$

Since the volume of a lipid molecule $\bar{v}$ may be $l_0 A_0$, the number N_{cm} of phospholipids (or polar groups) in the specific volume of the cylindrical micelle is

$$N_{cm} = \frac{V}{\bar{v}} = \frac{\pi R^2}{l_0 A_0} \tag{4.21}$$

Then, the surface area per molecule is

$$A = \frac{2\pi R}{N_{cm}} = \frac{2 l_0 A_0}{R} \tag{4.22}$$

Therefore, the interfacial area A of the hydrocarbon-water interface is

$$A_2 = A - A_0 = A_0 \left(\frac{2 l_0}{R} - 1 \right) \tag{4.23}$$

Thus, we have

$$\sigma_{wh} A_2 = \sigma_{wh} A_0 \left(\frac{2 l_0}{R} - 1 \right) \tag{4.24}$$

The potential energy at a point on the surface is obtained from Eq. (B.8) in Appendix II:

$$\psi_{cm} = \frac{4\pi \sigma K_0(\kappa R)}{\varepsilon_0 \kappa K_1(\kappa R)} \tag{4.25}$$

where σ is the density of the surface charge which is

$$\sigma = \frac{eqR}{2 l_0 A_0} \tag{4.26}$$

and ε_0 is the dielectric constant of the aqueous phase. Therefore, the electrostatic energy per lipid molecule of a cylindrical micelle at the surface is

$$E_{cm} = \tfrac{1}{2}eq\psi_{cm} = \frac{\pi e^2 q^2 K_0(\kappa R) R}{\varepsilon_0 \kappa K_1(\kappa R) l_0 A_0} \quad (4.27)$$

The pressure difference ΔP_{cm} is expressed by

$$\Delta P_{cm} = P - \frac{\varepsilon_0 (\nabla \psi)^2}{8\pi} + \frac{\sigma_{cm}}{R} - P_0 \quad (4.28)$$

Here the first two terms are the pressure and Maxwell's tension at the surface of the cylinder, respectively; σ_{cm} is the average surface tension. They are given by

$$P - P_0 = \frac{\varepsilon_0 \kappa^2}{8\pi}\psi^2 = \frac{(4\pi\sigma)^2}{8\pi\varepsilon_0}\left[\frac{K_0(\kappa R)}{K_1(\kappa R)}\right]^2 \quad (4.29)$$

$$\frac{\varepsilon_0 (\nabla \psi)^2}{8\pi} = \frac{(4\pi\sigma)^2}{8\pi\varepsilon_0} \quad (4.30)$$

$$\sigma_{cm} \cong \frac{\sigma_{wh} A_2}{A} = \sigma_{wh} R\left(\frac{1}{R} - \frac{1}{2l_0}\right) \quad (4.31)$$

Insertion of Eqs. (4.18) and (4.28) into Eq. (4.17) gives

$$G_{cm}^{rel} = \frac{\pi e^2 q^2 R^2}{2\varepsilon_0 l_0 A_0}\left(\frac{2K_0(\kappa R)}{\kappa R K_1(\kappa R)} + \frac{K_0^2}{K_1^2} - 1\right) + 3\sigma_{wh} A_0 \left(\frac{l_0}{R} - \frac{1}{2}\right) \quad (4.32)$$

As in the bilayer case, the free energy of the cylindrical micelle of the stable state for a given net charge per molecule is obtained by taking the minimum value of Eq. (4.32) with respect to R. The value R which gives the minimum is the radius of the cylinder of the stable state. Since Eq. (4.32) decreases as R increases from 0 to l_0, the radius of the cylinder is l_0. Hence the relative specific free energy of the stable cylindrical micelle is

$$G_{cm}^{rel}(q) = \frac{\pi e^2 q^2 l_0}{2\varepsilon_0 A_0}\left(\frac{2K_0(\kappa l_0)}{\kappa l_0 K_1(\kappa l_0)} + \frac{K_0(\kappa l_0)}{K_1(\kappa l_0)} - 1\right) + \frac{3}{2}\sigma_{wh} A_0 \quad (4.33)$$

The numerical relation between the relative free energy G_{cm}^{rel} and the net charge eq per molecule is shown in Figure 13.

(c) Spherical micelle†

For a spherical micelle, the relative free energy per molecule is derived in a manner similar to the case of the cylindrical micelle

$$G_{sm}^{rel}(q) = \frac{2\pi e^2 q^2 l_0}{9\varepsilon_0 A_0} \frac{\kappa l_0 + 2}{(\kappa l_0 + 1)^2} + \frac{10}{3}\sigma_{wh} A_0 \qquad (4.34)$$

The numerical relation between $G_{sm}^{rel}(q)$ and q is shown in Figure 13.

It is seen from Figure 13 that the lipid in the aqueous solution has the lowest specific free energy in the bilayer state among the three different states (bilayer, cylindrical micelle and spherical micelle) within the range of the net charge of a polar group $0 \sim 1.25|e|$. For the value of the net charge over $1.25|e|$, however, the specific free energy in the cylindrical micelle has the lowest value among these three states. The free energy for the spherical micelle state is always the highest in the range of the net charge of a polar group $0 \sim 2|e|$.

Therefore, it is clear that the formation of the spherical micelle, in the aqueous solution of pH range corresponding to the net charge per molecule $0 \sim 2|e|$, is not very likely to occur. It is deduced that the bilayer state is the most stable state for the lipid molecule in the aqueous solution over the range of the net charge per molecule $0 \sim 1.25|e|$, and, in the range of the net charge per molecule $1.25 \sim 2.0|e|$, the lipids would form the cylindrical micelle.

Since the energy per molecule due to thermal agitation is roughly estimated to be several kT ($kT = 4.2 \times 10^{-14}$ erg at $T = 300°K$), the bilayer would become unstable in a solution of pH corresponding to a net charge of the order of magnitude of $|e|$ per molecule, where the difference of the energies between the bilayer and the cylindrical micelle states is comparable to the energy of thermal agitation. At this point, the bilayer can transform to the cylindrical micelle. However, since the difference between the energies of the bilayer and the cylindrical states in the range of the net charge ($1.0 \sim 1.5|e|$) is of the same order of magnitude as the energy due to thermal agitation, both the bilayer and the cylindrical micelle states may coexist in the solution. It may occur that the very unstable bilayer state exists even in a solution of high or low pH which corresponds to a net charge $-e$ or $+e$, of the polar group dissociated. However, in almost all cases, in these pH ranges, the bilayer will break into the solution and form pieces of the bilayer form or the cylindrical micelle form.

Experimental results[32] show (Figure 14) that the specific resistance of the lipid bilayer has a maximum value in a solution with pH $4 \sim 5$ for the phosphatidyl choline bilayer and pH $3 \sim 4$ for the phosphatidyl serine bilayer.

† A detailed calculation will be published in *J. Colloid and Interface Sci.*

These pH's may correspond to the isoelectric points for these phospholipids. In the solution of pH range away from the isoelectric point, the specific resistance of the bilayer gradually decreases because of the thinning of the membrane due to the large electrical repulsive forces between the polar groups

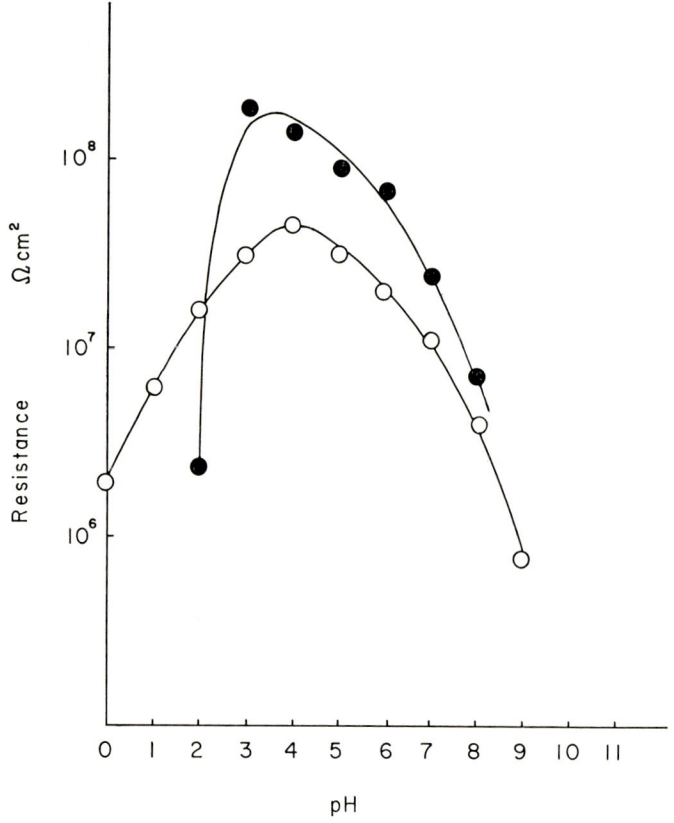

FIGURE 14 Specific resistance of lipid bilayer membranes as a function of pH. ○ phosphatidyl choline bilayer; ● phosphatidyl serine bilayer.

dissociated. We can also deduce from capacitance measurements[39] that the bilayer extends due to the dissociation of the polar groups with respect to pH of the solution, and becomes thinner. The phosphatidyl choline bilayer in 0.1 N sodium chloride was stable within the range of pH 1 ~ 9. However, in a solution of pH lower than 1 or higher than 9, the bilayer became unstable and tended to break. Since a phosphatidyl choline has a zwitterionic group, in a solution of extremely high or low pH the polar group would have at most one negative charge (an electronic charge) or one positive charge, respectively. For the phosphatidyl choline bilayer, even if the polar group is dissociated fully the net charge per molecule would be one electronic charge. In this case, according to our theoretical results, both the bilayer and the cylindrical micelle states could coexist in the aqueous solution. However, in almost all cases of complete dissociation, the bilayers would not keep their large form, and break into the solution. From Figures (13) and (14) it is suggested that the dissociation of a charge of $-e$ per molecule might correspond to pH $9 \sim 10$ and $+e$ to pH $0 \sim 1$.

For the phosphatidyl serine bilayer, the membrane was stable within the pH range 2–8 in 0.1 N sodium chloride solution. However, in the solution of lower pH than 2 or greater than 8.5, the membrane did not form. Since a phosphatidyl serine molecule possesses three dissociable ionic groups (i.e. one positively dissociable group and two negatively dissociable groups), the electrostatic repulsive forces due to charged groups would be greater than those of the phosphatidyl choline bilayer, and make the membrane more unstable. From our experiments, it was found that the stability of the phosphatidyl serine bilayer depended more strongly upon pH. Presumably, the dissociation of a positive charge would take place around pH $1 \sim 2$, and the isoelectric point may be around pH 3 where the membrane resistance is largest. In the range of pH $4 \sim 8$, the molecule would be charged negatively ($-e < eq < 0$) and around pH 8.5, the two polar groups of a molecule would begin to dissociate negatively ($eq = -2e$). Consequently, at pH greater than 8.5, the phosphatidyl serine bilayer cannot keep its form and breaks into the solution. Therefore, the dissociation of the net charge of e per phosphatidyl serine molecule might correspond to pH 2 and the dissociation smaller than -3 to pH greater than 8.

It is suspected that as the pH value is varied away from the isoelectric point, the bilayer extends and becomes thinner. If the net charge per polar group becomes greater than $|e|$, the bilayer begins to form cylindrical micelles or develops holes across the membrane which have the same structure as the cylindrical micelle. A gradually increasing number of holes in the membrane or cylindrical micelles would be produced until the bilayer form disappears. In the course of this process, it is likely from our results that the size of a hole or a cylindrical micelle remains constant, while the number of holes increases as the resistance of the bilayer decreases.

From Figure 12 we may say that even if the bilayer is broken in the range of the net charge $0 \sim |e|$ (electronic charge) per polar group, the lipid might not form micelles, but rather might form small spherical bubbles which have essentially the bilayer structure, or alternatively, might form the lamellar structure[40] in the aqueous system.

It is concluded here that the change of the net charge of lipid polar groups due to the change of the environmental solutions gives rise to the structural change of the lipid bilayers. This in turn may be related to the drastic change of the function of the biological membranes, such as membrane excitation.[41,42]

V STABILITY OF ASYMMETRICAL MEMBRANE

Asymmetrical phospholipid membranes can be considered in many ways. For example, by breaking a certain polar group from a lipid molecule with the addition of some enzymes (or proteins), we may transform a symmetrical phospholipid bilayer into an asymmetrical bilayer. By changing the salt content or concentration (or pH) in the aqueous solution on one side of the membrane, a membrane asymmetrical with respect to the surface of the membrane would be prepared. The asymmetry of these membranes arises from changing only the polar head groups or the degree of dissociation of the polar groups on the surface of the membranes. Another type of asymmetrical membrane might also be produced by putting two different phospholipid monolayers or vesicles together. There have been a few attempts to form these membranes.[4,344] The importance of the asymmetrical membrane would be in the understanding of the function of the biological membrane. Here, we shall restrict the asymmetry of the membrane to the change of the polar head group due to the change of the environmental solution.

Let us suppose the following model system for an asymmetric membrane. The membrane is composed of a monomolecular layer, each molecule of which is a long chain of equal length and has two dissociable polar head groups at both ends of a molecule. The molecules lie with their axes normal to the plane of the membrane such as in a unimolecular crystal, while the head groups coordinate in two outer surfaces facing the aqueous phases (Figure 9). The head group is able to chelate with a metal ion, depending on both the concentration of metal ions and the degree of dissociation of the polar group. It is assumed that if a polar head group chelates with a metal ion, the polar group loses its net charge, or the polar group does not receive any electrostatic interaction from the other polar groups charged.

Let us designate the position of the head groups on the surface of the membrane as 'sites'. We may denote the number of total sites on one surface of the membrane by N. When N_1 sites out of N are chelated with metal ions, the number of the sites which are not occupied by the metal ions is

$$N - N_1 = N_2 \tag{5.1}$$

We may define the average probability with which a polar group is chelated with the metal ion as follows:

$$x = \frac{N_1}{N} \qquad (5.2)$$

From Eqs. (5.1) and (5.2), we have

$$\left.\begin{array}{l} N_1 = Nx \\ N_2 = N(1-x) \end{array}\right\} \qquad (5.3)$$

As extreme cases we see that if all the sites are chelated with the metal ions, $x = 1$, and that if all the sites are not chelated, $x = 0$.

Let us suppose the following case: On one surface of the membrane there is a probability of chelation per site x_0, and on the other surface of the membrane, there are no chelated sites. The total free energy for this system is given by

$$F(x_0) = F^{\text{bulk}} + F^{\text{surface}}(x_0) \qquad (5.4)$$

where F^{bulk} is the free energy of the inside phase of the membrane, F^{surface} is the free energy of both surfaces of the membrane which is expressed by

$$F^{\text{surface}}(x_0) = [E^{\text{sites}}_{(x_0)} - (ST)^{\text{sites}} + F^{\text{surface}}_0] \qquad (5.5)$$

where E^{sites} is the electrostatic interaction energy due to the dissociated polar groups on the surface; $(ST)^{\text{sites}}$ is the entropy term due to the mixing of the sites; and F^{surface}_0 is the free energy of the surface excluding the contribution from the polar groups. If we assume that F^{surface}_0 is not affected no matter whether the polar groups chelate with the metal ions or not, F^{surface} varies only with the variation of probability x_0.

Suppose that Nx out of Nx_0 chelated molecules turn over from one side to the other side, or Nx out of Nx_0 chelated sites exchange with Nx unchelated sites from one side to the other side of the membrane. The energy of electrostatic interaction due to charged polar group is

$$E^{\text{sites}} = E_1 + E_2 + E_{12} \qquad (5.6)$$

where E^{sites}_1 and E^{sites}_2 are the energies of electrostatic interaction among the polar groups on the surfaces of one side and the other of the membrane, respectively. E^{sites}_1, E^{sites}_2 may be expressed by

$$\begin{array}{l} E^{\text{sites}}_1 = \tfrac{1}{2} Neq\, \psi_1 (1-x) \\ E^{\text{sites}}_2 = \tfrac{1}{2} Neq\, \psi_2 (1-x_0+x) \end{array} \qquad (5.7)$$

where ψ_1 and ψ_2 are the electrostatic potentials at a site on side 1 with surface charge density σ and on side 2 with surface charge density σ', respectively. ψ_1 and ψ_2 are given as follows (24)

$$\left. \begin{aligned} \psi_1 &= \frac{4\pi}{\varepsilon_0 \kappa} \frac{\sigma + \sigma' + \sigma'(\varepsilon_0 \kappa h/\varepsilon_1)}{2 + (\varepsilon_0 \kappa h/\varepsilon_1)} \\ \psi_2 &= \frac{4\pi}{\varepsilon_0 \kappa} \frac{\sigma + \sigma' + \sigma(\varepsilon_0 \kappa h/\varepsilon_1)}{2 + (\varepsilon_0 \kappa h/\varepsilon_1)} \end{aligned} \right\} \quad (5.8)$$

where ε_0 and ε_1 are the dielectric constants of aqueous and lipid phases, respectively, and σ is the average charge density due to dissociated polar groups. If the area per molecule is A and the net charge per molecule is eq, we have

$$\sigma = \frac{eq}{A}(1-x)$$

$$\sigma' = \frac{eq'}{A}(1-x_0+x) \quad (5.9)$$

Then, the total energy due to the electrostatic interaction of the polar groups is

$$\begin{aligned} E^{\text{sites}}(x_0) &= E_1^{\text{sites}}(x_0) + E_2^{\text{sites}}(x_0) \\ &= \frac{2\pi e^2 q^2}{\varepsilon_0 \kappa A} \frac{1}{2+(\varepsilon_0 \kappa h/\varepsilon_1)} \Big\{ (1-x_0+x)^2 + (1-x)(1-x_0+x) \\ &\quad + \frac{\varepsilon_0 \kappa h}{\varepsilon_1}(1-x_0+x)^2 + (1-x)^2 + (1-x)(1-x_0+x) \\ &\quad + \frac{\varepsilon_0 \kappa h}{\varepsilon_1}(1-x)^2 \Big\} \end{aligned} \quad (5.10)$$

The partition function of the total system is

$$\begin{aligned} Z &= e^{-F/kT} \\ &= {}_{Nx_0}C_{Nx} \, {}_NC_{Nx_0} \exp\left(-\frac{E^{\text{ele}}}{kT}\right) \exp\left(-\frac{F_0^{\text{surface}}}{kT}\right) \exp\left(-\frac{F^{\text{bulk}}}{kT}\right) \\ &\equiv {}_{Nx_0}C_{Nx} \, {}_NC_{Nx_0} \exp\left(-\frac{E^{\text{ele}}}{kT}\right) \exp\left(-\frac{F^0}{kT}\right) \end{aligned} \quad (5.11)$$

where ${}_{Nx_0}C_{Nx} \, {}_NC_{Nx_0}$ is the number of various distributions of the sites, and

$$F_0 \equiv F^{\text{bulk}} + F_0^{\text{surface}} \quad \text{which does not depend on } x_0 \quad (5.12)$$

The free energy of the membrane is

$$F = -kT \log z = E^{\text{sites}}_{(x_0)} - kT \log {}_{Nx_0}C_{Nx} \, {}_NC_{Nx_0} + F_0 \quad (5.13)$$

Dispersion forces and stability of lipid bilayers

Using the approximation $N! = N \log N$, we have

$$F = E^{\text{sites}}_{(x_0)} - kTN \left\{ \log \frac{1}{1-x_0} - x_0 \log \frac{x_0 - x}{1 - x_0} - x \log \frac{x}{x_0 - x} \right\} + F_0 \quad (5.1)$$

Since we assume that F_0 is not affected by the change of the polar groups, it is sufficient to know the relative free energy F^{rel} per molecule in order to express the free energy change of the membrane due to the change of the polar group. The relative specific free energy is defined as follows

$$\frac{F - F_0}{N} \equiv F^{\text{rel}} = \varepsilon - kT \left\{ \log \frac{1}{1-x_0} - x_0 \log \frac{x_0 - x}{1 - x_0} - x \log \frac{x}{x_0 - x} \right\} \quad (5.15)$$

where $\varepsilon = E^{\text{sites}}/N$.

When the total number of sites chelated with metals is a constant x_0, the stationary state of the system can be determined by

$$\frac{\partial F^{\text{rel}}}{\partial x} = 0 \quad (5.16)$$

With the value of x_1 which satisfies Eq. (5.15), the relative free energy of a molecule is a minimum. We call this state $F(x_1)$ a stationary state of the membrane. The difference ΔF between the specific free energies at $x = x_0$ and $x = x_1$ is expressed in terms of x_0 and q by Eqs. (5.9), (5.14) and (5.15):

$$\Delta F = F(x_0) - F(x_1) \quad (5.17)$$

The numerical values of ΔF with respect to the probability of chelation of site x_0 and the degree of dissociation of the polar groups q can be calculated. The numerical value of ΔF is shown with various values of q for $x_0 = 1$ in Figure 15.

According to Salem,[45,46] the London van der Waals dispersion energy per molecule in lipid monolayer films is given by

$$W_{\text{dis}} = \frac{n}{2} \frac{A}{4l^2 D^4} \rho \left(3 \tan^{-1} + \frac{\rho}{1+\rho^2} \right), \quad \rho = \frac{L}{D} \quad (5.18)$$

where n is the number of the nearest neighbour hydrocarbon chains, A is the coefficient of the dispersion interaction between two basic molecular units (CH_2 hydrocarbon units), l is the length of a CH_2 unit, D is the mutual distance of two hydrocarbon chains, L is the length of a hydrocarbon chain and N_c is the number of hydrocarbon molecules.

Since the phospholipid is composed of two aliphatic hydrocarbon chains, if we use 70 Å^2 for the area per lipid molecule, the mutual distance of two nearest hydrocarbon chains D is estimated to be 5.92 Å for a square lattice packing. If we apply the formula of Eq. (5.19) to the bilayer membrane with the following numerical values:

A: 1340 Å^6 Kcal/mol for CH_2—CH_2[46]

l: 1.26 Å

N_c: 36 (= 18 × 2)

D: 5.92 Å (for square packing with area per molecule 70 Å^2).

Thus, the cohesive energy of a hydrocarbon chain in the lipid molecular layer due to the dispersion forces of neighboring hydrocarbon chains is

$$W_{disp} = 4.04 \times 10^{-13} \text{ erg} \tag{5.19}$$

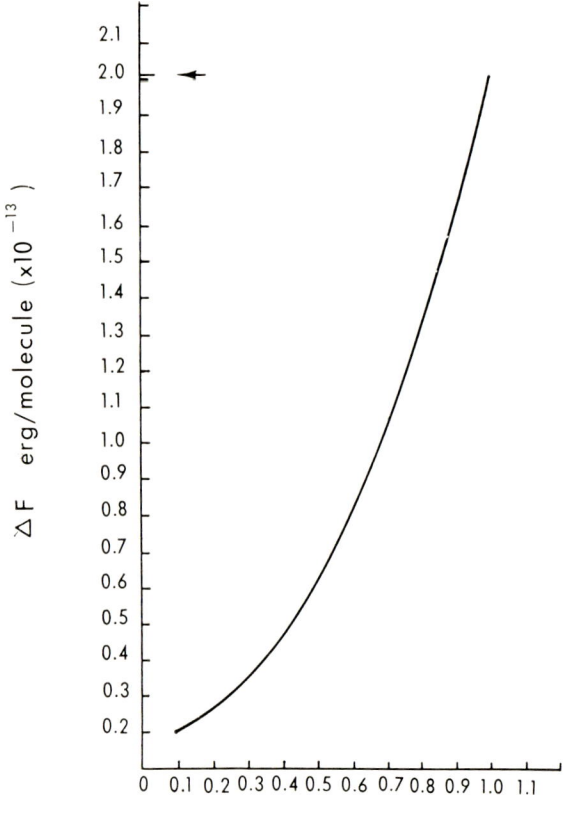

FIGURE 15   The numerical value of ΔF with respect to q for the case $x_0 = 1.0$.

In the square lattice molecular arrangements, if the energy, greater than $\frac{1}{2} W_{dis}$, which corresponds to the energy to break two nearest neighboring chains is given to a lipid molecule, the membrane can break. That is, if ΔF is greater than $\frac{1}{2} W_{dis}$, the membrane may break. As shown in Figure 16, for a given concentration of chelation there is a critical degree of dissociation q_c of the polar group at which the membrane breaks:

$$q_c = 0.95 \qquad x_0 = 1 \qquad (5.20)$$

Admittedly the analysis given here is over-simplified and contains a number of assumptions (such as the non-spherical character of CH_2 groups, orienta-

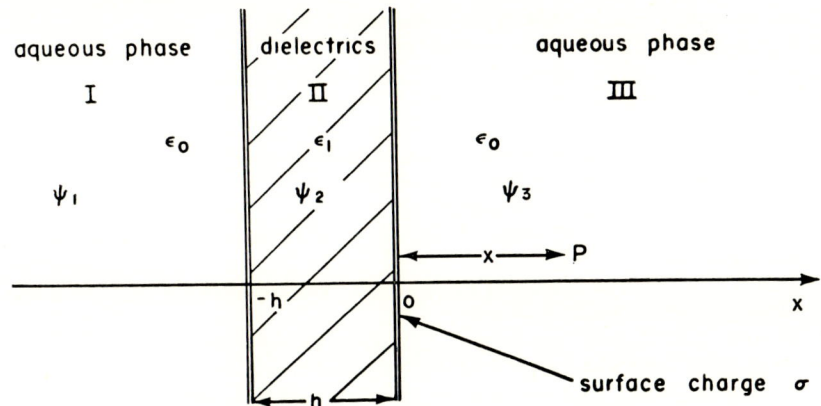

FIGURE 16 A scheme of a dielectric layer in an aqueous solution.

tion effect, and the neglect of the microscopic dielectric constant) which will be valid only in extreme cases. However, the treatment is qualitatively adequate to indicate the significance of the instability of the asymmetrical membrane. The instability of the asymmetrical membrane with divalent metals (e.g. Ca^{++}) might be chiefly related to the excitation phenomena of the biological membranes.

References

1. GORTER, E., and GRENDEL, F. 1925, *J. Exptl. Med.*, **41**, 439.
2. DANIELLI, J. F., and DAVSON, H. A. 1935, *J. Cell. Comp. Physiol.*, **5**, 495.
3. DANIELLI, J. F., and HARVEY, E. N. 1934, *J. Cell. Comp. Physiol.*, **5**, 483.
4. KORN, E. D. 1966, *Science*, **153**, 1491.
5. ROBERTSON, J. D. 1960, *Pro: in Biophysics and Biophysical Chem.*, **10**, 343.
6. FINEAN, J. B. 1962, *Circulation*, **26**, 1151.
7. DANIELLI, J. F. 1936, *J. Cell. Comp. Physiol.*, **7**, 393.
8. DEAN, R. B., CURTIS, H. J., and COLE, K. S. 1940, *Science*, **91**, 50.

9. LANGMUIR, I., and WAUGH, D. F. 1938, *J. Gen. Physiol.*, **21**, 745.
10. MUELLER, P., RUDIN, D. O., TIEN, H. T., and WESCOTT, W. C. 1962, *Circulation*, **26**, 1167.
11. MUELLER, P., RUDIN, D. O., TIEN, H. T., and WESCOTT, W. C. 1962, *Nature*, **194**, 979.
12. MUELLER, P., and RUDIN, D. O. 1963, *J. Theoret. Biol.*, **4**, 268.
13. HANAI, T., HAYDON, D. A., and TAYLOR, J. 1964, *Proc. Roy. Soc.*, **281A**, 377.
14. TIEN, H. T., and DAWIDOWICZ, E. A. 1966, *J. Colloid and Interface Sci.*, **22**, 438.
15. LEVINE, Y. K., BAILEY, A. I., and WILKINS, M. H. F. 1968, *Nature*, **220**, 577.
16. SHELUDKO, A. 1967, *Advances Colloid and Inter. Sci.*, **1**, 391.
17. OHKI, S., and FUKUDA, N. 1967, *J. Theoret. Biol.*, **15**, 362.
18. MIDZUNO Y., and KIHARA, T. 1956, *J. Phys. Soc., Japan*, **11**, 1045.
19. HIRSCHFELDER, J. O., CURTISS, C. F., and BIRD, R. B. 1964, 'Molecular Theory of Gases and Liquids', John Wiley & Sons, New York.
20. DANIELLI, J. F. 1966, *J. Theoret. Biol.*, **12**, 439.
21. HUANG, C., and THOMPSON, T. E. 1965, *J. Molec. Biol.*, **13**, 183.
22. THOMPSON, T. E., and HUANG, C. 1966, *J. Molec. Biol.*, **16**, 576.
23. TIEN., H. T. 1967, *J. Phys. Chem.*, **71**, 3395.
24. CHERRY, R. J., and CHAPMAN, D., private communication.
25. KIRKWOOD, J. G. 1936, *J. Chem. Phys.*, **4**, 592.
26. OHKI, S. 1968, *J. Theoret. Biol.*, **19**, 97.
27. HEAVENS, O. S. 1955, 'Optical Properties of Thin Solid Films'. Butterworth, London.
28. LANDAU, L. D., and LIFSHITZ, E. M. 1960, 'Electrodynamics of Continuous Media', London; Addison-Wesley.
29. BLODGETT, K. B., and LANGMUIR, I. 1937, *Phys. Rev.*, **51**, 964.
30. SHAH, D. O., and SCHULMAN, J. H. 1967, *J. Lipid Res.*, **8**, 215.
31. PAPAHADJOPOULOS, D. 1968, *Biochem. Biophys. Acta*, **163**, 240.
32. OHKI, S., and GOLDUP, A. 1968, *Nature*, **217**, 459.
33. KAVANAU, J. L. 1965, Structure and Function in Biological Membrane Vol. 1 (Holden-Day, Inc.).
34. LUZZATI, V., and MUSTACCHI, H. 1958, *Disc. Faraday Soc.*, **25**, 43.
35. GÖTZ, K. G., and HECKMANN, K. 1958, *J. Colloid. Sci.*, **13**, 266.
36. PARSEGIAN, A. 1966, *Trans. Faraday Soc.*, **62**, 848.
37. FOWKES, F. 'Molecular Forces at Interfaces' presented at Symposium on Properties of Surfaces and Surface Coatings Related to Paper and Wood, State University College of Forestry at Syracuse University, October 25–26, 1965.
38. ANSELL, G. B., and HAWTHORNE, J. N. 'Phospholipids', Elsevier Publishing Co., New York, 1964.
39. OHKI, S. 1969, *Biophysical J.*, **9**, 1195.
40. LUZZATI, V., and FUSSON, F. 1962, *J. Cell. Biol.*, **12**, 207.
41. HODGKIN, A. L., and HUXLEY, A. F. 1952, *J. Physiol.*, **117**, 500.
42. TASAKI I., and SINGER, I. 1966, *Ann. N. Y. Acad. Sci.*, **137**, 792.
43. TAKAGI, M., AZUMA, K., and KISHIMOTO, V. 1965, *Ann. Rept. Biol.*, **13**, 107.
44. PAPAHADJOPOULOS, D. P., and OHKI, S. 1969, *Science*, **164**, 1075.
45. SALEM, L. 1960, *Molec. Phys.*, **3**, 441.
46. SALEM, L. 1962, *J. Chem. Phys.*, **37**, 2100.
47. WEAST, R. C. 'Handbook of Chemistry and Physics', 49th edition, The Chemical Rubber Co., Ohio, 1968.

APPENDIX I

We consider the following system: a dielectric layer of infinite extent is surrounded by aqueous phases (Figure 16). Let us denote these phases by I, II and III, and their dielectric constants by ε_0, ε_1 and ε_0, respectively, and the thickness of the layer by h. If there are continuous charge distributions σ on the surface of the dielectric layer, the electrostatic potentials for these three phases can be obtained by solving the following Poisson's equations

$$(\nabla^2 - \kappa^2)\psi_1 = 0 \quad \text{for phase I} \tag{A.1}$$

$$\nabla^2 \psi_2 = 0 \quad \text{for phase II} \tag{A.2}$$

$$(\nabla^2 - \kappa^2)\psi_3 = 0 \quad \text{for phase III} \tag{A.3}$$

where $\nabla^2 \equiv \partial^2/\partial x^2$. ψ_1, ψ_2 and ψ_3 are

$$\psi_1 = a_1 e^{\kappa x} \tag{A.4}$$

$$\psi_2 = a_2 x + b_2 \tag{A.5}$$

$$\psi_3 = a_3 e^{-\kappa x} \tag{A.6}$$

respectively, where κ is the Debye Hückel constant, and a_1, a_2, a_3 and b_2 are constants to be determined. The boundary conditions of the electrostatic potentials at the two surfaces are as follows:

$$\psi_1(-h) = \psi_2(-h)$$

$$\varepsilon_0 \left.\frac{\partial \psi^2}{\partial x}\right|_{x=-h} = \varepsilon_1 \left.\frac{\partial \psi^2}{\partial x}\right|_{x=-h} + 4\pi\sigma \tag{A.7}$$

and

$$\psi_2(0) = \psi_3(0)$$

$$\varepsilon_1 \left.\frac{\partial \psi_2}{\partial x}\right|_{x=0} = \varepsilon_0 \left.\frac{\partial \psi_3}{\partial x}\right|_{x=0} + 4\pi\sigma \tag{A.8}$$

where σ is the surface charge density. With the above boundary conditions and Eqs. (A.4), (A.5), and (A.6), we have

$$\psi_1 = \frac{4\pi\sigma}{\varepsilon_0 \kappa} e^{\kappa(x+h)} \tag{A.9}$$

$$\psi_2 = \frac{4\pi\sigma}{\varepsilon_0 \kappa} \tag{A.10}$$

$$\psi_3 = \frac{4\pi\sigma}{\varepsilon_0 \kappa} e^{-\kappa x} \tag{A.11}$$

If there is a continuous charge distribution σ at the interface between the phases II and III, the electrostatic potentials in the three phases are obtained by a procedure similar to the above calculation:

$$\psi_1^a = \frac{4\pi\sigma \, e^{\kappa(x+h)}}{\varepsilon_0 \kappa [2+(\varepsilon_0 \kappa h/\varepsilon_1)]} \tag{A.12}$$

$$\psi_2^a = \frac{4\pi\sigma}{\varepsilon_1 \kappa [2+(\varepsilon_0 \kappa h/\varepsilon_1)]} X + \frac{4\pi\sigma [1+(\varepsilon_0 \kappa h/\varepsilon_1)]}{\varepsilon_0 \kappa [2+(\varepsilon_0 \kappa h)/\varepsilon_1]} \tag{A.13}$$

$$\psi_3^a = \frac{4\pi\sigma [1+(\varepsilon_0 \kappa h/\varepsilon_1)]}{\varepsilon_0 \kappa [2+(\varepsilon_0 \kappa h/\varepsilon_1)]} e^{-x} \tag{A.14}$$

APPENDIX II

Let us suppose that a cylindrical dielectric is in an aqueous solution and that there is a continuous charge distribution σ on the surface (Figure 17). The electrostatic potentials of the inside and the outside of the cylinder are obtained by solving the following Poisson's equations:

$$\nabla^2 \psi_1 = 0 \quad \text{inside} \tag{B.1}$$

$$(\nabla^2 - \kappa^2)\psi_2 = 0 \quad \text{outside} \tag{B.2}$$

where

$$\nabla^2 \equiv \frac{\partial^2}{\partial r^2} + \frac{1}{r}\frac{\partial}{\partial r}$$

The solutions for ψ_1 and ψ_2 are

$$\psi_1 = a_1 \tag{B.3}$$

$$\psi_2 = a_2 K_0(\kappa r) \tag{B.4}$$

where a_1 and a_2 are constants to be determined by the boundary conditions. K_0 is the modified Bessel function of the zeroth order.

With the following boundary conditions for ψ_1 and ψ_2

$$\psi_1(R) = \psi_2(R)$$

$$\varepsilon_1 \frac{\partial \psi_1}{\partial r}(R) = \varepsilon_0 \frac{\partial \psi_2}{\partial r}(R) + 4\pi\sigma \tag{B.5}$$

Dispersion forces and stability of lipid bilayers

where R is the radius of the cylindrical dielectric, we have the electrostatic potentials of the inside and the outside of the cylinder as follows:

$$\psi_1 = \frac{4\pi\sigma K_0(\kappa R)}{\varepsilon_0 \kappa K_1(\kappa R)} \tag{B.6}$$

$$\psi_2 = \frac{4\pi\sigma K_0(\kappa r)}{\varepsilon_0 \kappa K_1(\kappa R)} \tag{B.7}$$

where r is the distance of the point p from the center of the cylinder. Therefore, the electrical potential at the surface of the cylinder is

$$\psi_2 = \frac{4\pi\sigma K_0(\kappa R)}{\varepsilon_0 \kappa K_1(\kappa R)} \tag{B.8}$$

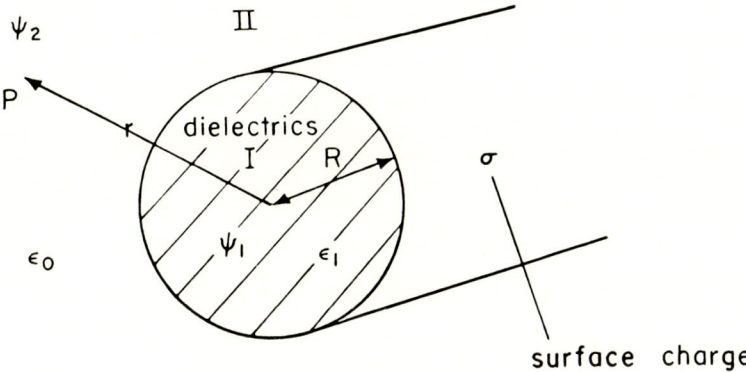

FIGURE 17 A scheme of a cylindrical dielectric in an aqueous solution.

APPENDIX III

In the case of a spherical dielectric with surface charge σ immersed in an aqueous phase, (Figure 18) the potentials of the inside and the outside of the spherical dielectric are obtained by solving the following Poisson's equations

$$\nabla^2 \psi_1 = 0 \quad \text{inside} \tag{C.1}$$

$$\nabla^2 \psi_2 - \kappa^2 \psi_2 = 0 \quad \text{outside} \tag{C.2}$$

where

$$\nabla^2 \equiv \frac{\partial^2}{\partial r^2} + \frac{2}{r}\frac{\partial}{\partial r}$$

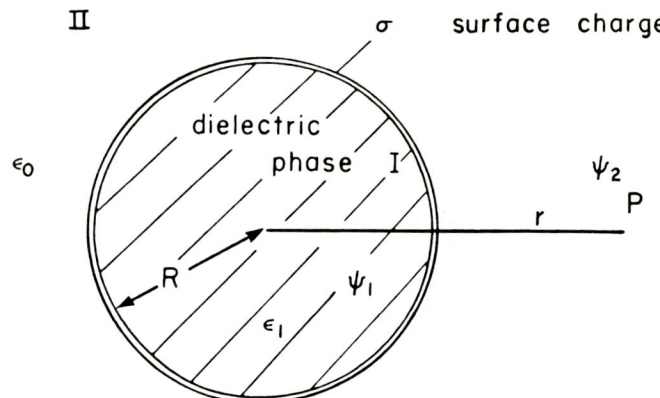

FIGURE 18 A scheme of a spherical dielectric in an aqueous solution.

The solutions ψ_1 and ψ_2 for Eqs. (C.1) and (C.2) are

$$\psi_1 = a_1 \tag{C.3}$$

$$\psi_2 = a_2 \frac{e^{-\kappa r}}{r} \tag{C.4}$$

With the boundary conditions of the continuity of ψ and

$$\varepsilon_1 \left.\frac{\partial \psi}{\partial r}\right|_{r=R} = \varepsilon_0 \left.\frac{\partial \psi}{\partial r}\right|_{r=R} + 4\pi\sigma$$

we have:

$$\psi_1 = \frac{4\pi\sigma R}{\varepsilon_0(1+\kappa R)} \tag{C.5}$$

$$\psi_2 = \frac{4\pi\sigma R^2 e^{-\kappa(r-R)}}{\varepsilon_0(1+\kappa R)r} \tag{C.6}$$

where σ is the surface charge density, ε_0 and ε_1 are the dielectric constants of the aqueous phase and the spherical dielectric, respectively. R is the radius of the spherical dielectric, and r is the distance of a point p from the center of the spherical micelle (Figure 18). Therefore, the electrostatic potential at the surface of the spherical dielectric is expressed by

$$\psi_2 = \frac{4\pi\sigma R}{\varepsilon_0(1+\kappa R)} \tag{C.7}$$

DISCUSSION

HILL   You had one slide that showed the relative free energies of three different kinds of micelles. Was that for a particular salt concentration? And does the appearance of a slide change with the salt concentration?

OHKI   Yes, salt concentration can be changed. Since we have been doing the experiment using 0.1 molar sodium chloride solution, I have calculated those values with 0.1 molar; but there is a concentration dependence.

ADAM   May I amplify this question. It will also depend on the concentration of phospholipids in the suspension because Luzzatti has shown that you have different configurations of the system according to the relative amount of lipids and of water.

OHKI   Only the area of a molecule is taken into account for molecular arrangement and so I don't take into account any structure, for example whether it is an external or square lattice. I calculated those free energies with respect to the area per molecule and net charge.

I'm sure you are thinking about a hexagonal lattice or square lattice or some other structure.

ADAM   No, I understand you compare different configurations of the system, either the bilayer or let's say the cylindrical micelles. Luzzatti has shown that nature verifies one of these—either the bilayer or the micelle model only for certain relative amounts of water and phospholipids—and if you want to compare your calculations with experiments then maybe you should take into account this proportion of lipid to water. You have always very much of water and very little of phospholipids in your theoretical model?

SCHLÖGL   Have you tried to make a guess how many pores you would have as an average per square centimeter?

OHKI   Yes, I did, but I can't tell you right now how many pores we have in a unit area of the membrane. But the important thing is the size of a pore. The radius of a pore is almost constant and small. So, even if a small pore is made, capacitance of the membrane will not change. In fact, membrane capacitance doesn't change due to environmental condition changes, while the membrane conductance changes greatly.

SCHLÖGL   Let me ask the following question: Would it not be possible to conceive of a pore as sort of a negative micelle, so as to obtain an estimate on the order of magnitude of free energy?

OHKI   I am sorry, I can't give you the answer.

PAPER 10

Ion transport across lipid bilayer membranes

P. LÄUGER
Fachbereich Biologie, Universität, Konstanz

ARTIFICIAL BIMOLECULAR MEMBRANES are now studied in many laboratories as models for biological membranes. Compared with the highly complex functions of the cell membranes, the model appears, of course, rather primitive. However, even in this early stage, lipid bilayer membranes have proved to be very suitable to isolate some of the many biological membrane phenomena and to study them separately. Besides this, these membranes are very interesting objects from a purely physicochemical point of view.

In the following I want to discuss some approaches to a theory of ion transport across lipid bilayer membranes. However, I do not intend to speak here about excitability inducing molecules or the generation of specific ion channels by cyclic antibiotics. Rather, I wish to restrict myself to the discussion of a much more basic problem, namely the passage of simple ions across the unmodified lipid film, and I hope to give you an indication of how much remains to be done even at this level.

The most easily observable property related to ion transport is the electrical conductance of the membrane. From the experimental results which are now available, two distinct features emerge:

(a) the membrane conductivity is in general very low, but strongly dependent on the nature of the ion;
(b) the current-voltage characteristic of the membrane is non-linear.

Point (a) is illustrated by Table I in which the membrane conductivity, λ_0, in the limit of vanishing voltage is compared for different anions.

The conductivity is extremely low in solutions of small ions like Cl^- (for comparison, the equivalent bulk conductivity σ_0 is also given which is obtained by multiplying λ_0 with the membrane thickness $d \simeq 70$ Å). Thus the membrane may be regarded as a rather perfect dielectric. However, the membrane conductivity is increased by many orders of magnitude in the presence of certain anions, some of which are listed in Table I. A common property of these ions is the delocalisation of the electric charge, i.e. the charge is spread over a large part of the molecule. The charge delocalisation may be considered to be the ultimate reason for the enhanced membrane conductivity. This is

TABLE I. Membrane conductivity, λ_0, for different anions. c_- = concentration of the anion in the aqueous phase (pH $\gg pK_a$ in all cases). $\sigma_0 \equiv \lambda_0 \cdot d$ is the equivalent bulk conductivity (d = membrane thickness)

ion	c_- (M)	λ_0 (ohm^{-1} cm^{-2})	σ_0 (ohm^{-1} cm^{-1})	reference
Cl$^-$	10^{-2}	$5 \cdot 10^{-10}$	$4 \cdot 10^{-16}$	Hanai, Haydon and Taylor (15)
O_2N-C$_6$H$_2$(NO$_2$)$_2$-O$^{\ominus}$	10^{-4}	$1.4 \cdot 10^{-7}$	$1 \cdot 10^{-13}$	Libermann, and Topaly (16)
F$_3$CO-C$_6$H$_4$-N-N=C(CN)$_2^{\ominus}$	$3 \cdot 10^{-5}$	$\sim 10^{-6}$	$\sim 10^{-12}$	"
Cl$_4$C$_6$-N-N-CF$_3^{\ominus}$	10^{-4}	$4 \cdot 10^{-6}$	$3 \cdot 10^{-12}$	"

easily understood by the introduction of a simple theoretical model for the membrane (Figure 1). The mechanical behavior of artificial bilayer membranes suggests that the hydrocarbon chains forming the interior of the film are in a more or less disordered liquid-like state. As a first approximation we may therefore consider the membrane as a homogeneous liquid hydrocarbon film of dielectric constant ε_m interposed between two aqueous phases. This model has already been proved to be useful in the interpretation of water permeability (1) and carrier mediated ion transport (2).

The membrane conductivity is proportional to the concentration, c_m^-, of anions in the membrane, if we assume the contribution of the cations to the current to be negligible:

$$\lambda_0 = A c_m^- \qquad (1)$$

c_m^- is given by the partition coefficient of the anion, which in turn is determined by the energy ΔE required to transfer the ion of radius r and charge $-e_0$ from the aqueous phase to the lipid phase. The major term in ΔE is the electrostatic energy which can be calculated by means of the Born process:

$$\left. \begin{array}{l} \lambda_0 = Ac_- \exp\left(-\dfrac{\Delta E}{kT}\right) \\[2mm] = Ac_- \exp\left[-\dfrac{q_0}{r}\left(\dfrac{1}{\varepsilon_m}-\dfrac{1}{\varepsilon}\right)\right] \\[2mm] q_0 \equiv \dfrac{e_0^2}{2kT} = 282\ \text{Å}\,(25^\circ\text{C}) \end{array} \right\} \quad (2)$$

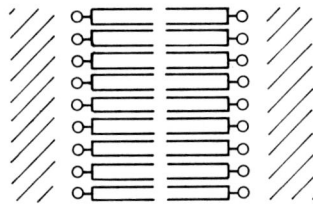

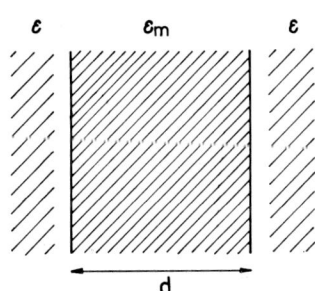

FIGURE 1 Hydrocarbon film of dielectric constant ε_m as a model for the membrane.

(c_- = anion concentration in the external solutions). It is the large value of the parameter q_0, combined with the low dielectric constant, ε_m, of the membrane, which makes the membrane conductivity so extremely small. On the other hand, an increase in r, the equivalent ion radius, has a tremendous effect on the partition coefficient and hence on the conductivity (3). Ions in which the charge is delocalized have a large equivalent radius and are therefore expected to increase the membrane conductivity. This is indeed what has been observed.

In order to get a little more information on the electrochemistry of the bimolecular membrane we may study its current-voltage characteristic. In most experiments a strongly non-linear current-voltage curve is observed. The ohmic region extends to about 30 mV; above this voltage the current increases much more rapidly than the voltage (Figure 2).

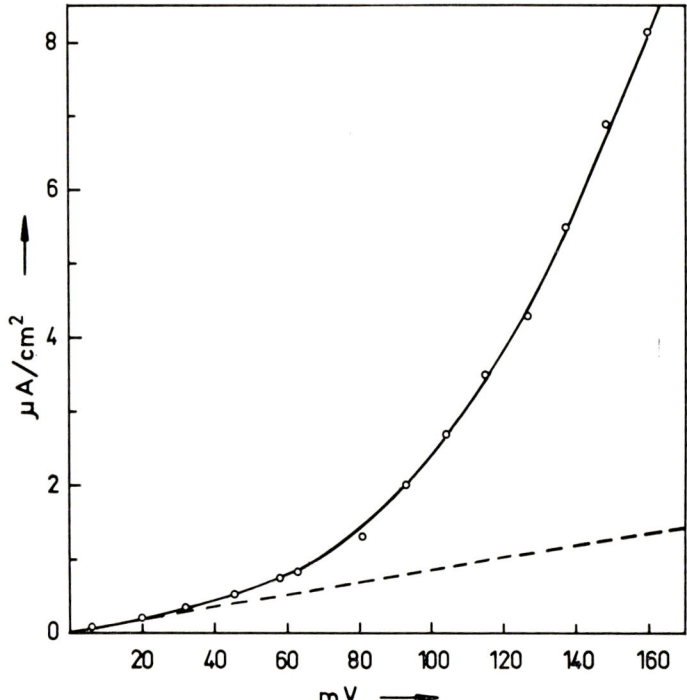

FIGURE 2 Current-voltage characteristic of a dioleoyllecithin membrane in 10^{-2} M 2, 4-dinitrophenol (pH = 4.42, t = 25 °C).

Non-linear response to the applied voltage is a well known phenomenon with nerve membranes, and it seems therefore to be worthwhile to study the non-linear behavior of the model system in some detail. A closer inspection shows that there are several different mechanisms, each of which may lead to a non-linear current-voltage characteristic. Here I want to discuss only two of them, namely

(a) the injection of ions into the membrane;
(b) the dissociation field (or Wien) effect.

Injection of charge carriers is a common phenomenon in solid state physics. With a lipid membrane in aqueous phase the situation is somewhat different, because the charge carriers are (as far as we know) always ions and not electrons. Let me consider first a limiting case: a lipid membrane in which the ions of the aqueous electrolyte are completely insoluble. Such a membrane is an ideal dielectric. If we apply an external voltage, the potential varies linearly with distance across the membrane (see Figure 3). However, the voltage does not drop entirely across the membrane itself. A part of the potential drop occurs in the two diffuse space charge layers at the membrane surfaces. If we now go back to the real membrane in which the ions have a finite but very small solubility, the picture remains essentially correct. At

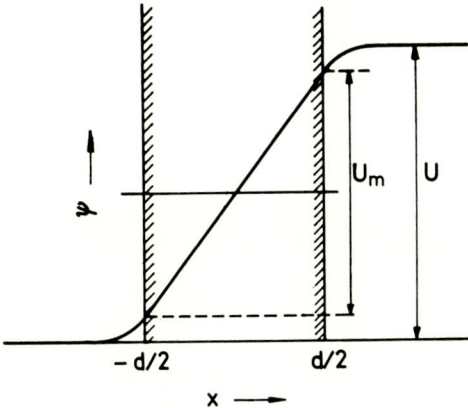

FIGURE 3 Dielectric film of thickness between two electrolyte solutions. ψ = electric potential.

every point in the membrane the space charge density is so low that the potential profile is still linear. This is the so-called constant field approximation originally introduced by Goldman (4). A second simplification becomes possible by virtue of the fact that the current density in the system is extremely small. This means that the external phases which have a high conductivity remain practically near equilibrium. We can therefore calculate the ion concentrations at the outer surface of the membrane from the Boltzmann equation. According to the potential difference between the interphase and the solution, the concentration of positive ions is raised at one interphase and the concentration of negative ions at the other. Correspondingly, positive and negative ions are injected into the membrane from both sides. This is shown schematically in Figure 4.

As a consequence of the ion injection, the membrane conductivity becomes a function of the voltage. This function can be calculated in closed form by solving the Nernst–Planck equations for the membrane interior (5). The ion injection effect has been treated in a very general form by Bruner (6). The calculation shows that the shape of the conductance function depends strongly on the total ionic strength of the aqueous solution. The non-linearity is pronounced only at concentrations below 10^{-3} M, whereas at higher concentrations the membrane should approach an ohmic behavior. This prediction can

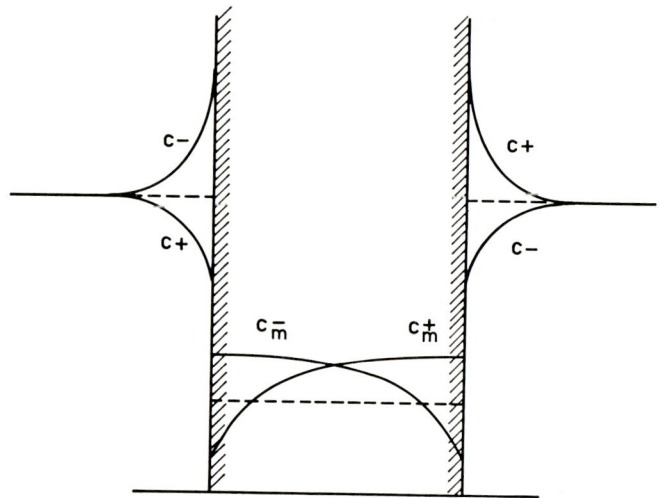

FIGURE 4 Injection of ions into the membrane. Cations and anions are assumed to be identical except for the charge sign.
– – – – concentration profiles at zero voltage ($U = 0$)
———— concentration profiles at $U > 0$.

be tested easily by the experiment. However, in all systems so far studied, the expected linear current-voltage relation at high ionic strengths is not observed. In solutions of 2,4-dinitrophenol the voltage dependence of the conductivity is even more pronounced at high ionic strength (5). Although there is little doubt about the existence of the ion injection mechanism, the current-voltage characteristic must be dominated, at least at higher concentrations, by a second effect.

A mechanism which should be seriously considered as a possible source of non-linear behavior of the bilayer membrane is the Wien effect (7). As a consequence of the low dielectric constant of the membrane, an appreciable part of the ions is present in the form of neutral ion pairs which cannot contribute to the electric current. However, with a membrane thickness of ~ 100 Å and voltages of ~ 100 mV, the electric field strength in the membrane is in the order of 10^5 V/cm. At these high fields, the dissociation constant of ion pairs in the membrane is markedly increased. The number of charge carriers and hence the conductivity become thus a function of the field strength. The change in conductivity of a macroscopic phase due to the field dissociation can be calculated from the theory of Onsager (8). As is shown in Figure 5, the

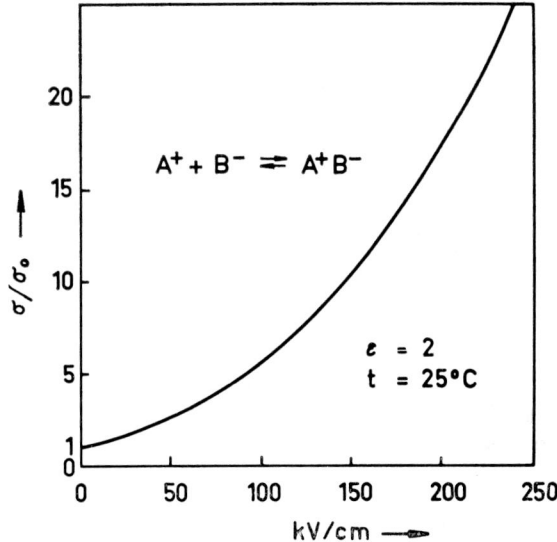

FIGURE 5 Change in conductivity of a macroscopic phase due to the Wien effect, as calculated after the theory of Onsager (8). The ions A^+, B^- are assumed to be identical except for the charge sign. σ_0 = conductivity in the limit of low field strength.

effect is very great for a phase of low dielectric constant and would be amply sufficient to account for the non-linear behavior of the membrane.

However, in the case of a thin film interposed between two aqueous solutions, the situation is somewhat complicated by the existence of boundary conditions at the interphases. The most simple boundary condition is obtained with an assumption that the exchange of ions across the water/lipid interface is very rapid. Then the ion concentration at the lipid side of the boundary remains for all field strengths at a fixed value and depends upon the partition coefficient of the ion. The concentration profile inside the membrane is determined by the combined effects of ion transport down the gradient of the electrochemical potential and the field-induced dissociation of ion pairs.

The degree to which the membrane conductivity is changed by the Wien effect depends essentially on the ratio of two relaxation times. We may first define a diffusional relaxation time τ_{diff} which is the time required to re-establish the equilibrium between membrane and the aqueous phases after a sudden change of the ion concentration in the membrane:

$$\tau_{\text{diff}} = \frac{(d/2)^2}{2D} \tag{3}$$

(d = membrane thickness, D = diffusion coefficient of the ions in the membrane; for simplicity, it is assumed in the following that cations and anions have identical properties except for the charge sign). As pointed out by Onsager (8), the Wien effect itself is governed by another relaxation time τ_{chem}. This is the time required for the approach to stationary ion concentrations in a macroscopic phase if an electric field is suddenly applied. As the concentration, c, of free ions can be assumed to be small compared with the concentration of ion pairs, the chemical relaxation time is given by the approximation:

$$\tau_{\text{chem}} \approx \frac{1}{2ck_r} \tag{4}$$

k_r is the rate constant for the recombination of ions which is in the order of 10^{13} cm^3 mole^{-1} s^{-1}. The order of magnitude of the ion concentration c can be obtained from the ohmic conductivity of the membrane with an estimated diffusion coefficient $D = 10^{-6}$ cm^2 s^{-1}.

The ratio

$$\alpha \equiv \frac{\tau_{\text{diff}}}{\tau_{\text{chem}}}$$

of the two relaxation times then assumes the value $\alpha \simeq 10^{-7}$ s/10^{-1} s = 10^{-6}. This would mean that the influence of the Wien effect on the membrane conductivity is quite negligible, because the ions produced by field dissociation rapidly diffuse out of the membrane.

This result is a consequence of the rather unrealistic assumption that the ion transport across the membrane/water interface is completely unhindered. When an ion moves from the aqueous phase into the membrane it has to cross the zone in which the polar head groups of the lipid molecules are located. These head groups rather strongly interact with each other by dipole–dipole forces and hydrogen bonds. The passage of an ion across the polar layer therefore involves a certain activation energy. We may thus replace the simple square well potential energy curve of the ion by a better approximation in which an energy barrier of height E_s is located at each membrane surface; correspondingly, barriers of height E_i are associated with the migration of the ion in the interior of the membrane (Figure 6). The ohmic resistance, $1/\lambda_0$, of the membrane is then the sum of two terms, the first of which is related to the membrane interior and the second to the surface barriers:

$$\frac{1}{\lambda_0} = \frac{1}{\lambda_i} + \frac{2}{\lambda_s} \tag{5}$$

On the basis of Eyrings theory of transport (9), the following approximative relation may be derived:

$$\frac{\lambda_s}{\lambda_i} \simeq \frac{d}{l} \exp\left(-\frac{E_s - E_i}{kT}\right) \tag{6}$$

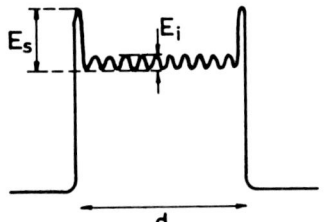

FIGURE 6 Potential energy of an ion in the membrane.

in which l is the mean jump length of the ion in the membrane. From measured activation energies for the diffusion in polar and non-polar media, the difference $E_s - E_i$ may be estimated to be several Kcals/mole. With $d/l \sim 10$, the ratio λ_s/λ_i becomes of the order of 10^{-2}. This means that the rate-limiting step in the ion migration across the membrane is the passage over the surface barrier.

We may now return to the Wien effect. The barrier at the interface has the effect that the ions produced by field dissociation are more or less trapped within the membrane. The concentration of free charge carriers in the membrane is therefore substantially raised by the electric field. This expectation is supported by the mathematical evaluation of the model (5). The membrane conductivity can be obtained as a function of voltage by a perturbation treatment of the transport equation. An example is shown in Figure 7 in which the conductivity is plotted against the reduced voltage. With an assumed value of 10^{-2} for the ratio λ_s/λ_i the theoretical curve is in qualitative agreement with the experimental conductance function. This result gives indeed some evidence for the idea that the non-linear behavior of the bilayer membrane is caused by the Wien effect. However, we should bear in mind that this interpretation rests entirely on the assumption of a high energy barrier at the lipid/water interphase. An independent test of the hypothesis should be possible by the measurement of the electrical relaxation time of the membrane. Experiments of this type are now in preparation.

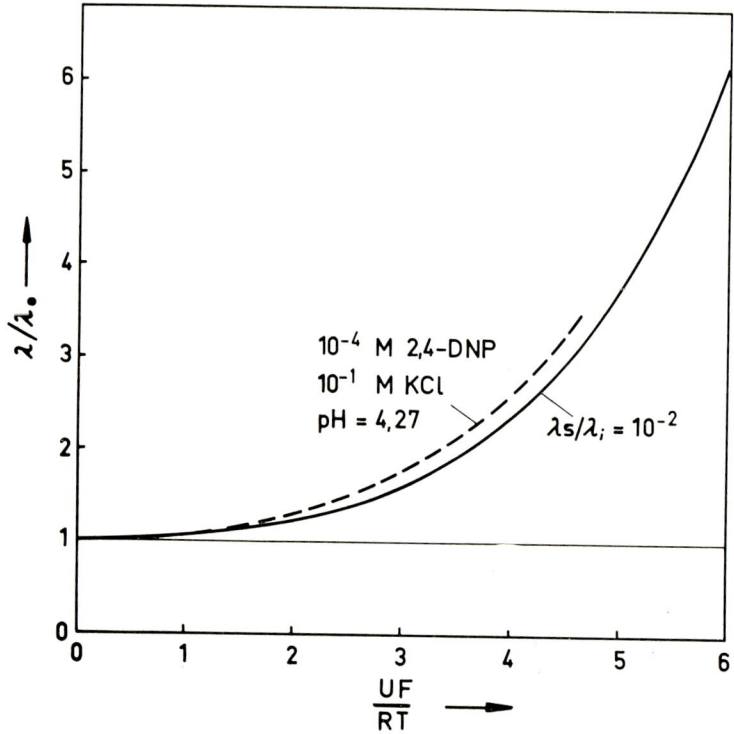

FIGURE 7 Influence of the Wien effect on the membrane conductivity. λ = integral membrane conductivity, λ_0 = membrane conductivity in the limit of low field strength. UF/RT = reduced voltage.

Finally, I wish to make a few remarks on the influence of certain neutral molecules on the electrical conductivity of the bilayer membrane. As I mentioned before, the membrane conductivity is very low in solutions of simple electrolytes like KCl or KJ. If, however, J_2 is added to the external solutions, the conductivity increases by many orders of magnitude (10). A similar observation is made in the experiments with uncoupling agents like 2,4-dinitrophenol (see Table I). It is well-known that certain lipid-soluble weak acids uncouple phosphorylation from the oxidation of substrates in mitochondria. Two years ago, Bielawski, Thompson, and Lehninger (11) and later Libermann and co-workers (12) were able to show that these substances have a large effect on the conductivity of artificial lipid membranes. Surprisingly, it turned out that the membrane conductivity depends not only on the concentration of the lipid soluble anion A^- in the aqueous phase, but also on the concentration of the undissociated acid HA. This is manifested by the observation that at a given total concentration of the uncoupler in the external solution, the membrane conductivity λ_0 is a function of the pH $\times$ λ_0 has a maximum at a pH approximately equal to the pK of the acid, where the dissociated and the undissociated form of the molecule are present in equal concentrations.

How can we explain the very strong influence of a neutral molecule like J_2 on the membrane conductivity? The solution of this problem seems to be given again by the relation between the conductivity and the potential energy of an ion in the membrane:

$$\lambda_0 \sim \exp\left[-\frac{q_0}{r}\left(\frac{1}{\varepsilon_m}-\frac{1}{\varepsilon}\right)\right] \tag{7}$$

As ε is about 80 and ε_m of the order of 2, λ_0 depends strongly on the dielectric constant, ε_m, of the membrane. It is well-known that J_2 is soluble in a lipid phase; the equilibrium concentration in hexane being about forty times the concentration in water. Besides this, J_2 (and also 2,4-dinitrophenol) forms a rather stable molecular complex with lipids like lecithin (13, 14). When the membrane takes up J_2 which is a strongly polarisable molecule, the dielectric constant is shifted from ε_m to a higher value $\varepsilon_m + \Delta\varepsilon_{cm}$. Accordingly, the equilibrium ion concentration in the membrane is increased, and the conductivity assumes a new value λ'_0:

$$\lambda'_0 = \lambda_0 \exp\left[-\frac{q_0}{r}\left(\frac{1}{\varepsilon_m+\Delta\varepsilon_m}-\frac{1}{\varepsilon_m}\right)\right] \simeq \lambda_0 \exp\left(\frac{q_0}{r}\frac{\Delta\varepsilon_m}{\varepsilon_m^2}\right) \tag{8}$$

with $r = 2$ Å and $\varepsilon_m = 2$:

$$\lambda'_0 \simeq \lambda_0 \exp(35\Delta\varepsilon_m) \tag{8a}$$

Therefore, a small change of ε_m has a drastic effect on the membrane conductivity (14).

In experiments with J_2 in the aqueous phase, the electrical resistance and capacitance of a lecithin membrane were measured simultaneously (10). By the addition of J_2 to the exterial solutions the membrane capacitance increased from 0.33 to 0.51 $\mu F/cm^2$. This corresponds to a change in dielectric constant from 2.0 to 3.1. If an ion radius of $r = 3.8$ Å is assumed, the observed conductivity increase from $\lambda_0 \simeq 10^{-7}$ mhos/cm^2 to $\lambda_0 \simeq 10^{-1}$ mhos/cm^2 is correctly described by Eq. (8).

This consideration provides also an explanation for the pH dependence of the membrane resistance in the presence of uncoupler substances. Here the acid HA in the aqueous phase has a twofold function: in its neutral form it is taken up by the membrane thereby increasing the dielectric constant; besides this, it supplies the lipid soluble charge carrier A^- by dissociation. At high pH values, the concentration of the neutral molecule HA is very small; correspondingly the dielectric constant and the conductivity of the membrane are low. At the other end of the pH scale, the membrane contains a relatively large amount of the undissociated acid and has an increased dielectric constant, but now the concentration of the charge carrier is small. The conductivity goes therefore through a maximum at a mean pH, where HA and A^- have comparable concentrations.

Finally, I wish to thank my collaborators E. Bamberg, B. Ketterer, Dr. B. Neumcke, Dr. G. Stark, and Dr. D. Walz for their contributions to this work.

References

1. HANAI, T., and HAYDON, D. A. 1966, *J. Theor. Biol.*, **11**, 370.
2. EISENMAN, G., CIANI, S., and SZABO, G. (in press).
3. FINKELSTEIN, A., and CASS, A. 1968, *J. Gen. Physiol.*, **52**, 145s.
4. GOLDMAN, D. E. 1943, *J. Gen. Physiol.*, **27**, 37.
5. WALZ, D., BAMBERG, E., and LÄUGER, P. 1969, *Biophys. J.*, **9**, 1150; NEUMCKE, B., WALZ, D., and LÄUGER, P., *Biophys. J.* (in press).
6. BRUNER, L. J. 1965, *Biophys. J.*, **5**, 867, 887; 1967, **7**, 947.
7. MIYAMOTO, V. K., and THOMPSON, T. E. 1967, *J. Coll. Interf. Sci*, **25**, 16.
8. ONSAGER, L. 1934, *J. Chem. Phys.*, **2**, 599.
9. JOHNSON, F. H., EYRING, H., POLISSAR, u. M. J. The Kinetic Basis of Molecular Biology, Chapter 14, Wiley 1954.
10. LÄUGER, P., RICHTER, J., and LESSLAUER, W. 1967, *Ber. Bunsenges. physik. Chem.*, **71**, 906.
11. BIELAWSKI, J., THOMPSON, T. E., and LEHNINGER, A. L. 1966, *Biochem. Biophys. Res. Com.*, **24**, 948.
12. SKULACHEV, V. P., SHARAF, A. A., and LIBERMAN, E. A. 1967, *Nature*, **216**, 718.
13. ROSENBERG, B., and JENDRASIAK, G. L. 1968, *Chem. Phys. Lipids*, **2**, 47.
14. ROSENBERG, B., and BHOWMIK, B. B., in press.
15. HANAI, T., HAYDON, D. A., and TAYLOR, J. 1965, *J. Gen. Physiol.*, **48**, No. 5, part 2, p. 59.
16. LIBERMAN, E. A., and TOPALY, V. P. 1968, *Biochim. Biophys. Acta*, **163**, 125.

DISCUSSION

KALKWARF Could I hear once more what you felt were the charged carriers in this membrane which I believe was a lecithin membrane with both the dinitrophenol and potassium chloride present.

LAUGER Yes, it was a membrane of synthetic dioleoyl lecithine. In the presence of potassium chloride alone, the membrane conductance is very low, but if you add dinitrophenol the current-determining ion is the dinitrophenolate ion. This is in your sense a charged carrier across the membrane.

KALKWARF Do you measure transport of dinitrophenolate across the barrier then to determine this? Or is it the difference in potential tha tone is looking at here?

LÄUGER Yes, besides the conductivity, you can measure the membrane potential if you have different concentrations of the dinitrophenolate ion on both sides of the membrane and this again gives the indication that the charge carrying ion is the dinitrophenolate ion.

KALKWARF I see. The reason I was probing here is that many compounds have now been found which can, in my terminology, activate a bilayer membrane, but they need not be the charge carrier. They seem to make the membrane susceptible for other ions to pass through. But I gather here the dinitrophenol is actually going through, perhaps in addition to the potassium or some other charged carrier, or do you feel it carries the entire current?

LÄUGER The question may be conclusively answered by tracer experiments which are now in preparation.

SCHLÖGL What was the idea that you need the Wien effect to understand the non-linearity? I thought you already get a non-linearity as a consequence of the redistribution of ions.

LÄUGER If the ion redistribution mechanism would be operative, then we would expect that at high ionic strengths the non-linearity vanishes. But this is not the case. We made the experiment with dinitrophenol at a low total ionic strength and at a high total ionic strength and the non-linearity was about the same in both cases. This speaks clearly against the redistribution or ion injection effect, at least at high ionic strengths. There must be another mechanism present.

SNELL It seems to me that your explanation could be in terms of the fact that the dinitrophenol is indeed a charged carrier for potassium. You have not ruled out that it increases the conductivity for the opposite charged ion and this would be in line with your observation that at higher ionic strengths the non-linearity indeed increases, without introducing the Wien effect at all.

LÄUGER But how would you then explain the experiment with the same concentration of potassium chloride on both sides of the membrane but different concentrations of dinitrophenolate concentration. In this case you observe almost a Nernst potential with the minus sign on the dilute side. I think this speaks against this carrier function of the nitrophenalate ion. Would you suggest a neutral complex of dinitrophenol with potassium?

SNELL That's what I was suggesting, but I don't understand your experiment—what the observations are here.

LÄUGER This experiment indicates that the main charge carrier is the dinitrophenolate ion.

SNELL That speaks for a distribution of the dinitrophenol cating through the membrane. Then, when you pass current what does this show, when you change the potential? How about the conductivity?

LÄUGER At low voltages, the conductivity is determined mainly by the dinitrophenolate ion. At higher fields, ion pairs ($DNP^-\ K^+$ and/or $DNP^-\ H^+$) may dissociate as a consequence of the Wien effect, so that the conductivity becomes in part cationic.

EISENMAN Provided the partition coefficients for the organic phase sufficiently favor the dinitrophenolate over the potassium ion or the chloride, for a membrane of this thickness and this dielectric constant, one would expect an excess of the species in the membrane phase as has been deduced by Sergio Ciani in developing the theory that I will be discussing in my lecture.

LÄUGER Yes, but the charge excess is quite low. You may calculate the approximate concentration of the dinitrophenolate in the membrane from the conductivity and this gives a very low charge.

MYSELS In that K_r in your second relaxation time, is that a diffusion-limited reaction or did you put some energy barrier into it?

LÄUGER This K_r can be calculated with the assumption of a diffusion controlled ion recombination.

MYSELS And then for concentration, what did you put in? On what basis?

LÄUGER The concentration is obtained from the ohmic conductivity, λ_0, of the membrane with an estimated value of the diffusion coefficient D of the ion in the membrane. With $\lambda_0 = 10^{-6}\,\text{ohm}^{-1}\,\text{cm}^{-2}$ and $D \simeq 10^{-6}\,\text{cm}^2\,\text{sec}^{-1}$

the ion concentration is calculated to be about 10^{-13} mole/cm^3. You have a relation between the ohmic conductivity on the one hand and the concentration and the mobility of the ion on the other hand. One can only estimate the diffusion coefficient but this diffusion coefficient should be not very different from the value for a small molecule in a higher hydrocarbon like hexadecane. The relation between the membrane conductivity, λ_0, and the ion concentration, c, is given by the equation $\lambda_0 \simeq cDF^2/dRT$ (d = membrane thickness).

EISENMAN In relation to this I'd like to emphasize a point I will be making on Friday, which is that we have evidence that the diffusion of an antibiotic molecule within the membrane is rapid compared to its rate of crossing the membrane-solution interfaces. This finding seems to be consistent with the kind of picture that you have proposed. Indeed, our picture of the essence of the membrane, like yours, is that it can be represented by a liquid hydrocarbon phase, of some 50 Å thickness.

PAPER 11

Role of water structure in various membrane systems

W. DROST-HANSEN

Division of Physical and Chemical Oceanography, Institute of Marine Sciences, University of Miami, Miami, Florida

I INTRODUCTION

THE STUDY of membranes is presently a very popular intellectual occupation The reason for this is not difficult to understand; membranes play an enormous role in fields ranging from crass technology to the most subtle, delicate aspects of biological functioning. No wonder then, that a phenomenal mass of information about membrane properties is presently in existence. Unfortunately, however, our conceptual understanding is lagging far behind. Many theories have been proposed for the functioning of membranes, particularly biological ones. The theoretical descriptions of membrane phenomena have included contributions from such illustrious authors as Eyring and co-workers (concerned with processes on the molecular level) to the systematic, gross description in terms of phenomenological parameters by Katchalsky, Kedem and co-workers. However, in spite of all these attempts, our understanding, on the molecular level, of the functioning of membranes remains fragmentary, incomplete and possibly in part incorrect. It is to be hoped that this Membrane Conference will further theoretical treatments of membrane processes; but such progress, of necessity, must be based on models of membranes. In this connection one may recall the statement by Linderström-Lang: 'One well established and generally accepted method of treating systems which are complicated beyond comprehension is to construct simple models and see whether they fit the systems in question. If they do, you will immediately become suspicious, and so will your colleagues most certainly, with the result that a blooming literature springs up (or breaks out) dealing with the problem of how you have managed to make all your errors cancel one another. If they do not fit, the beauty of the models themselves may shine for years untainted by the squalid awkwardness of reality.' To this, we may add the recent statement by Kenneth S. Cole about the construction of membrane models. 'If we fail to look carefully and worry effectively about the exceptions, we may too long postpone the appearance of some radical, undreamed of, unifying concept.' The purpose of the present Note is to call attention to some squalid aspects of reality, in the form of exceptions, which can no longer be neglected.

For several years, the present author and his co-workers[†] have studied the problem of the structure of water and more recently, the problem of the structure of water near interfaces. While the structure of bulk water remains, to a large extent, the puzzle it has been for years, it is possible that some advances may be made within a reasonable span of time regarding the structure of water near interfaces. This is due to the likely existence of stabilized water structures which in turn may owe their stability to the presence of an interface. The problem of the possible stability and likely extent of changed water structures near interfaces has been debated for years. Thanks primarily to the recent work in Russia by Boris Deryagin and co-workers, the question of the existence of water in states different than those classically recognized, has now become generally accepted and, thus, it becomes appropriate for us to apply the notion of changed water structures near interfaces in our construction of conceptual models for membrane phenomena.

In the present Note, some evidence is presented on which the conclusion is based that water near interfaces is different from that of bulk water and that the interface produces, at least in some cases, a stabilization of structured units of water, capable of undergoing higher-order phase transitions. Eventually it will be necessary for any model and any theoretical treatment of membrane processes to allow for this phenomenon and, indeed, incorporate it into the detailed, quantitative models.

As to the structure of water near interfaces, we present here merely some of the evidence for the existence of structured elements of one kind or another and stress, in particular, the effects of temperature on such structures. It is interesting to note that in spite of the very large amount of experimental data available on membrane systems, often specifically designed to test one theoretical model or another for membrane behavior, relatively few studies have been carried out as a function of temperature, although this would seem one of the most obvious ways of testing at least one simple aspect of the pro-

† See Drost-Hansen, W., 1965 through 1968; Drost-Hansen and Oppenheimer, 1960, and Drost-Hansen and Thorhaug, 1967.

posed models. Even fewer studies have been made at closely-spaced temperature intervals. As a result, the existence of higher-order phase transitions in the structure of the water associated with many types of membranes and the obvious influence of such on membrane properties has gone unnoticed.

II THERMAL ANOMALIES IN VICINAL WATER

Figure 1 through 5 show some examples taken from recent publications suggesting thermal anomalies in surface properties. On the basis of these, and

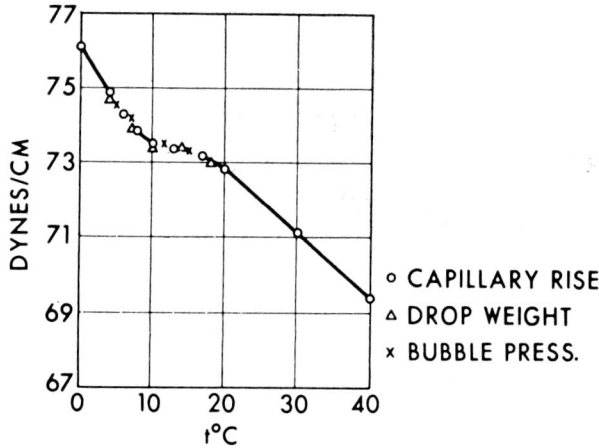

FIGURE 1 Surface tension of water. Data by Bodson and Timmermans. Similar results obtained with heavy water (D_2O).

a large number of other studies (discussed elsewhere), it is suggested that evidence exists for more or less abrupt changes at different, discrete temperatures. Previously, the present author had suggested that such anomalies were a general occurrence in all aqueous solutions. It now appears that the anomalies are primarily found in the properties of water near interfaces—where the concept of an interface should be taken in the broadest possible sense; i.e. the air/water interface, water/immiscible organic liquid interface, water/solid interface or the interface between water and some large macromolecular solute molecules. For a discussion of the various examples shown, the reader is referred to the articles by the present author.

The fact that abrupt thermal changes are observed in various properties is interpreted here as evidence for higher-order phase transitions and the existence of such imply the existence of (relatively large) structured entities, capable of undergoing order-disorder transformations.

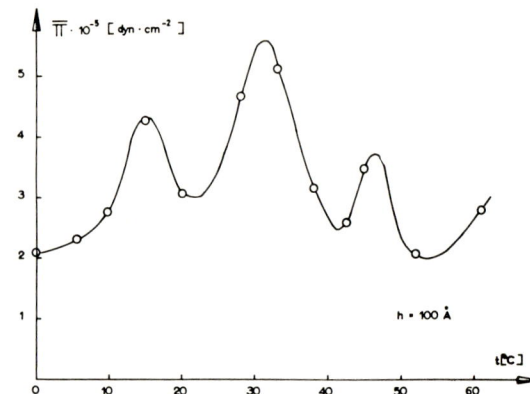

FIGURE 2 Disjoining pressure of water between highly polished quartz surfaces, separated 100 Å. Data by Peschel and Aldfinger (personal communication), 1968.

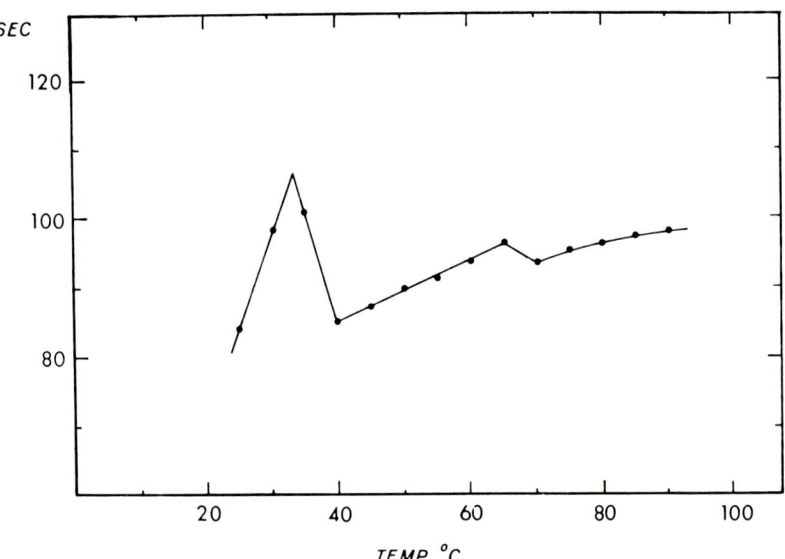

FIGURE 3 Viscous damping of water in vibrating 'hairpin' quartz capillary viscometer. Data by Forslind, 1966.

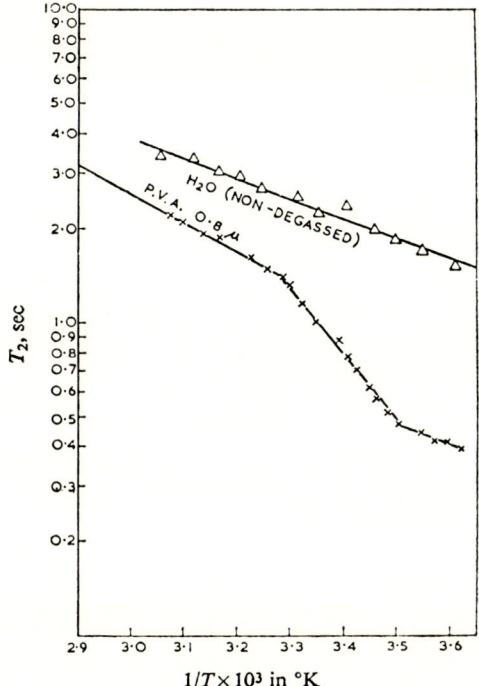

FIGURE 4 Temperature dependence of spin–spin relaxation time, T_2, for polyvinyl acetate particles in aqueous suspension. Johnson, et al., 1966.

III THERMAL ANOMALIES IN MEMBRANE SYSTEMS

Relatively few membrane systems have been studied at closely-spaced temperature intervals. However, in those cases where such measurements have been made, evidence is also available which suggests that membrane properties may undergo abrupt changes at or near the same temperatures where the anomalies were observed, in general, for aqueous interfacial systems. Figures 6 through 10 show some examples.

IV BIOLOGICAL IMPLICATIONS

Thermal anomalies in the properties of water near interfaces and in membrane properties have been observed both for aqueous systems consisting primarily of an interface in contact with pure water as well as dilute aqueous solutions and rather concentrated solutions of both electrolytes and non-electrolytes.

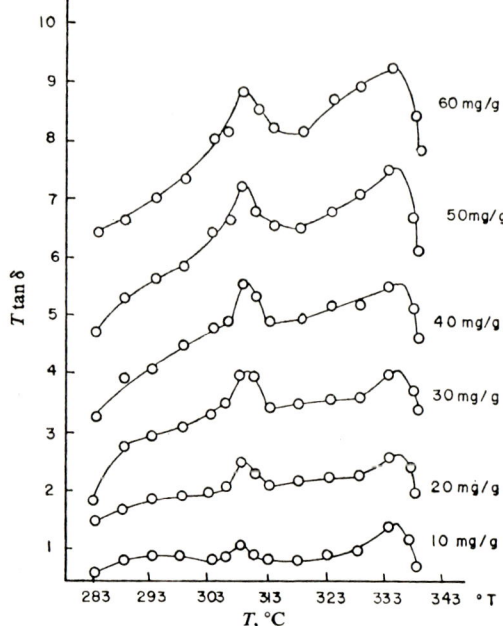

FIGURE 5 Ttanδ versus temperature for water adsorbed on chondroitin-4-sulphate. Lubesky, *et al.*, 1967.

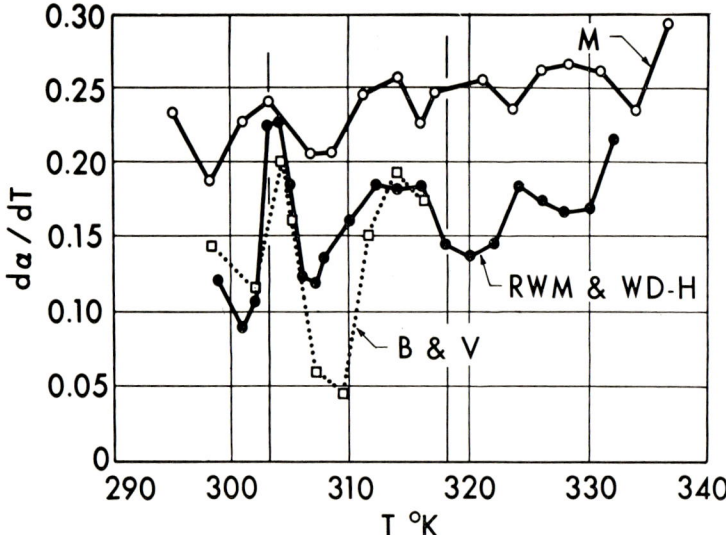

FIGURE 6 Entropy of surface formation for water. Reproduced from Drost-Hansen, 1965B.

Role of water structure in various membrane systems

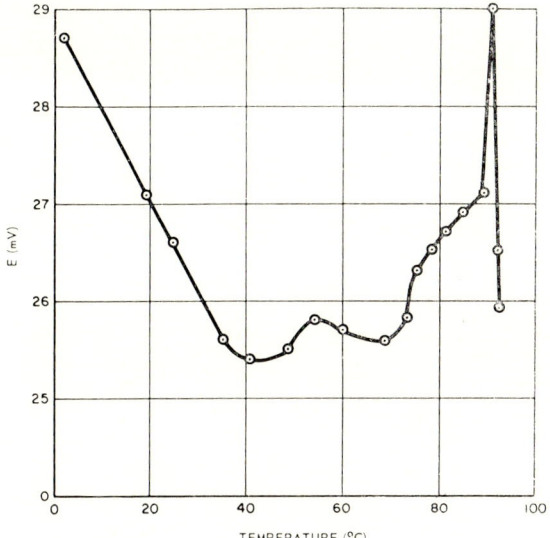

FIGURE 7 Effect of temperature on membrane bi-ionic potential; collodion-potassium oleate membrane. Nelson and Blei, 1966.

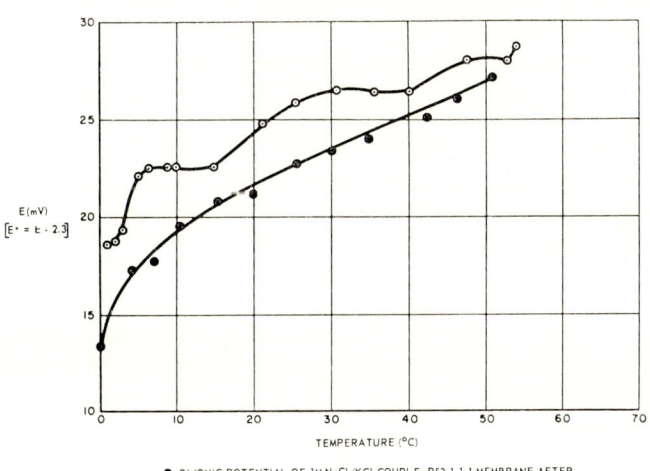

● BI-IONIC POTENTIAL OF 1M NaCl/KCl COUPLE, PS3-1-1-1 MEMBRANE AFTER ADDING 5 DROPS OF CH$_3$Cl TO EACH HALF CELL AND EQUILIBRATING FOR 24 HOURS

○ BI-IONIC POTENTIAL AFTER WASHING MEMBRANE WITH NaCl/KCl (1M) REPEATEDLY

FIGURE 8 Effect of temperature on bi-ionic potential in polysoap membrane. Solid circles—with chloroform present; open circles—after removing the chloroform. Nelson and Blei, 1966.

Hence, it is reasonable to assume that the anomalies might be manifested also in the properties of biologic systems. Indeed, for biologic systems, the surface-to-volume ratio is high. While other causes for the existence of large temperature effects on biologic systems exist (such as transformational changes of proteins or phase transitions in lipid systems) a vast number of biological systems exhibit anomalies far more abrupt than can readily be

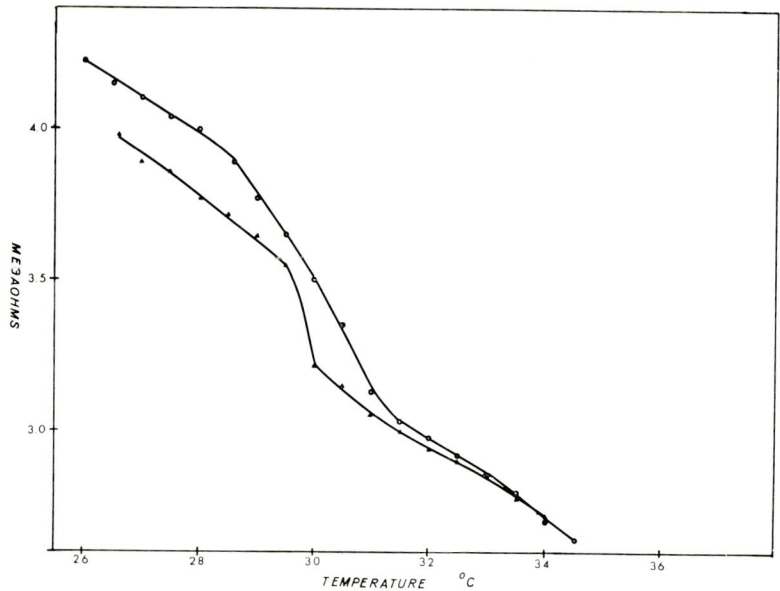

FIGURE 9 Resistance of Millipore filter, impregnated with toluene, in 0.1 molar KCl. Drost-Hansen, *et al.*, 1968, unpublished.

explained in terms of existing, kinetic theories of molecular biology. Furthermore, many, though not all, of the anomalies occur at or near the temperatures where the anomalies are observed in the properties of water near interfaces and water in membranes. Hence, it is natural to suggest that in these biologic systems the anomalies are also manifestations of changes in vicinal water structures.

Figures 11 through 16 show some examples of this type. The notion of such phase transitions playing a dominant role in determining temperature responses of biologic systems has been elaborated upon by the present author in a number of papers, and include the prediction of minimum, optimum, and maximum (lethal) temperatures for growth and other processes.

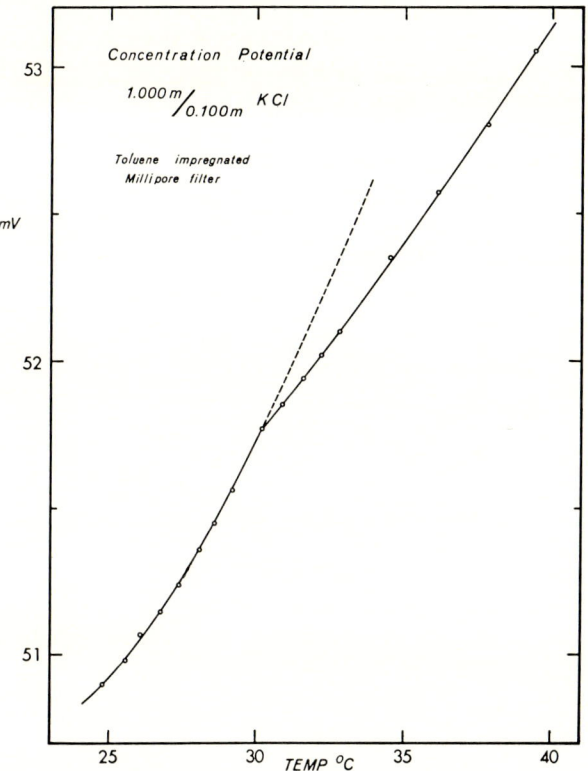

FIGURE 10 Concentration potential across Millipore filter, impregnated with toluene. Drost-Hansen, *et al.*, 1968, unpublished.

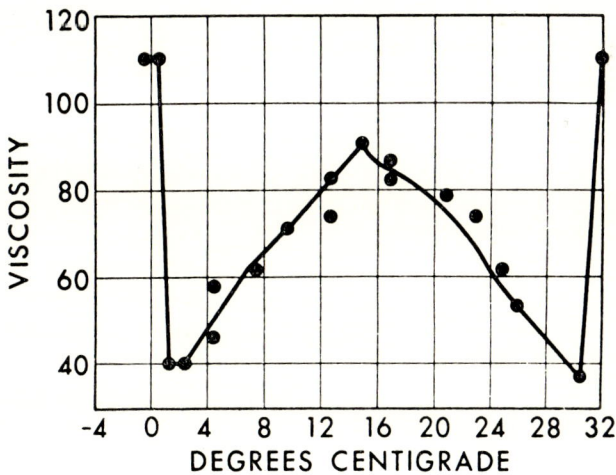

FIGURE 11 Viscosity of a protoplasm from *Cumingia* egg. Data by Heilbrunn, 1924.

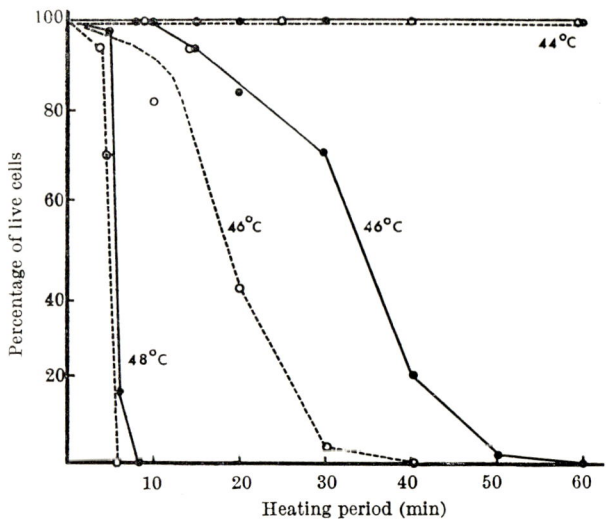

FIGURE 12 Two day survival of cells after heating to temperatures indicated: dotted curves—fibroblasts, solid curve—epithelia cells. Data by Auersperg, 1966.

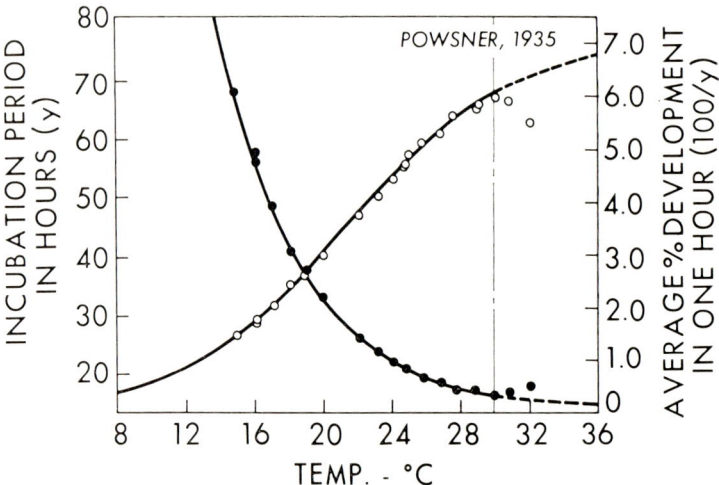

FIGURE 13 Logistics curve for development of the eggs, *Drosophila melanogaster*. See Davidson, 1944.

Role of water structure in various membrane systems 253

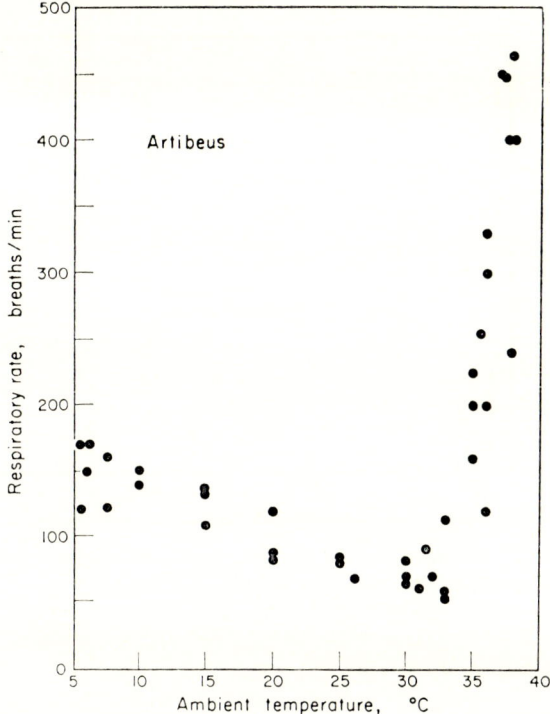

FIGURE 14 Minimal respiration rate of bat, *Artibeus hirsutu* . Data by Carpenter and Graham, 1967.

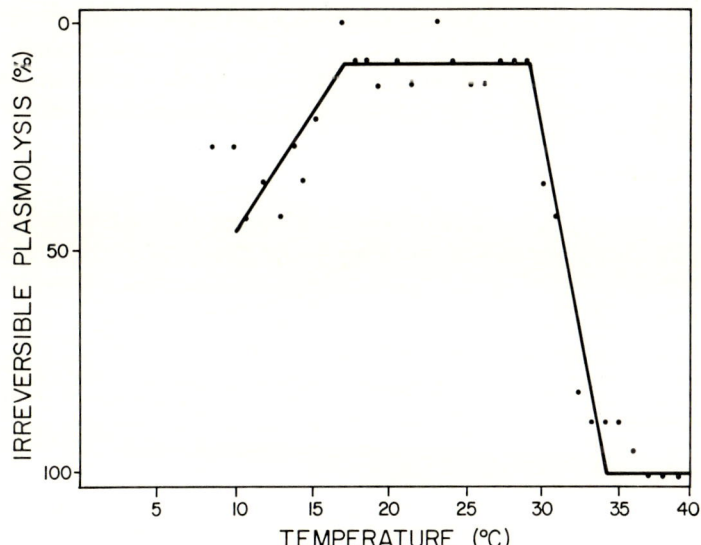

FIGURE 15 Cells surviving three day exposure to temperatures indicated, *Nitella*. Data by Thorhaug, 1968, unpublished thesis.

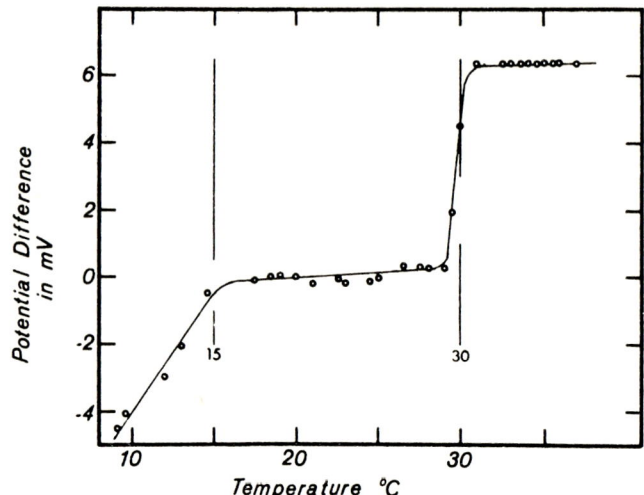

FIGURE 16   Potential across the membrane of *Valonia utricularis*. Data by Thorhaug. See Drost-Hansen and Thorhaug, 1967.

V   DISCUSSIONS AND CONCLUSIONS

Evidence has been presented to demonstrate the existence of abrupt changes in the properties of water near interfaces, in particular, water in membranes and in biologic systems. Previously, it had been proposed that these changes are due to higher-order phase transitions in structures units, stabilized by the proximity to an interface. In order to be consistent with reality, future theoretical work on membranes must allow for and incorporate quantatively such water structure effects in the description of membrane functioning.

Acknowledgement

The author wishes to thank the National Aeronautics and Space Administration for partial support of this study under Grant NGL 10-007-010.

Contribution No. 1116 from Institute of Marine and Atmospheric Sciences, University of Miami.

References

1. ALDFINGER, K. H. 1968, Private communications; see also: PESCHEL, G. and ALDFINGER, K. H. 1967, *Naturwissenschaften*, **54**, 614.
2. AUERSPERG, N. 1966, *Nature*, **209**, 415.
3. DAVIDSON, J., 1944. *J. Anim. Ecol.*, **13**, 26. On relationship between temperature and rate of development of insects at constant temperatures.

4. CARPENTER, R. E. and GRAHAM, J. B. 1967, *Comp. Biochem. Physiol.*, **22**, 709.
5. DROST-HANSEN, W. 1965A. *N. Y. Academy of Science*, Annals 125, Art. 2, 471.
6. DROST-HANSEN, W. 1965B. *Ind. and Engr. Chem.*, April issue p. 18.
7. DROST-HANSEN, W. 1965C, *Proc. First Int. Symp. on Water Desalt.*, **1**, 382, U.S. Gov. Printing office, spring 1967.
8. DROST-HANSEN, W. 1967, *Adv. in Chem. Series*, **67**, 70.
9. DROST-HANSEN, W. 1968, *Chem. Phys. Letters*, **2** (8), 647.
10. DROST-HANSEN, W. 1968B, 'On the Structure of Water Near Solid Interfaces and the possible existence of long-range order'. *Ind. and Chem. Engr.*, Nov.–Dec. 1969.
11. DROST-HANSEN, W. and co-workers, 1968, unpublished.
12. DROST-HANSEN, W., and OPPENHEIMER, C. H. 1960, *J. Bacteriology*, **80**, 21.
13. DROST-HANSEN, W., and THORHAUG, A. K. 1967, *Nature*, **215**, 506.
14. FORSLIND, E. 1966, *Svensk Naturvetenskap*, p. 9.
15. JOHNSON, G. A., LECCHINI, S. M. A., SMITH, E. G., CLIFFORD, J., and PETHICA, B. A. 1966, *Disc. Far. Soc.*, **42**, 120.
16. JOHNSON, J. H., EYRING, H., and POLISSAR, M. J. 1954, 'The Kinetic Basis of Molecular Biology', Wiley & Sons, N. Y.
17. HEILBRUNN, L. V. 1924, 'The viscosity of protoplasm at various temperatures.' *Am. J. Physiol.*, **68**, 645.
18. LUBEZKY, I., BETTELHEIM, F. A., and FOLMAN, M. 1967, *Trans. Far. Soc.* **63**, 1794.
19. NELSON, S. S., and BLEI, I. 1966, 'Model Membrane Studies Related to Ionic Transport in Biological Systems'. Office of Saline Water, Research and Development Progress Report 221, U.S. Gov. Printing Office, Wash. D. C.
20. THORHAUG, A. K. 1968, Thesis unpublished.
21. TIMMERMANS, J., and BODSON, H. 1937, *Compt. Rend.*, **204**, 1804.

DISCUSSION

ROCKSTEIN I was intrigued by the last two of your curves, because a number of enzyme systems which I have studied have optimal temperatures at 35°. In the honeybee, for example, this is more understandable since they live in hives which are about 33°C. But the housefly which does best at about 20° to 25° also shows in some of the enzyme systems concerned with the energizing of flight, optimal temperatures of 35° and one obtains a typical inverted U-shaped curve.

DROST-HANSEN Sometimes one finds enzyme systems that do show abrupt anomalies very close to the temperatures where we have seen them in a variety of systems. In other cases, the changes occur at different temperatures. Firstly, I think we have to recognize that if we are not looking at something where the water structure plays a dominant role, there's no reason for us to expect to see any anomalies at all. Secondly, there are many other things that go on and if you have consecutive and competing reactions with different temperature coefficients—different energies of activation—then obviously you will get something which is a mixture, and may not necessarily sharply correlate with the temperatures we have discussed. In return, could I ask how closely did you measure these things?

ROCKSTEIN These represent data based on at least 10 to 20 animals per determination and done at least 3 to 5 times per temperature determination. As far as 'close' is concerned, our data is reproduceable within 5 percent $\pm$ 5 per cent for biological chemical data. This isn't too bad, I think.

DROST-HANSEN That's beautiful, but did you measure at one degree intervals —or were they five degrees?

ROCKSTEIN No, these were at five degree intervals.

DROST-HANSEN Well, this is one of the problems. You don't quite know what goes on then. That is, at least, my contention.

JAUCH I wonder if you could just elaborate a little bit on how you imagined the structure change in water to occur. What kind of structure is it that you are envisaging?

DROST-HANSEN As politicians in this country say, 'I'm glad you asked that question'. As I mentioned, we do not yet know what the structure of water is. Essentially, there are those who contend that water is on the average a liquid, with no elements of structure at all; I am talking about bulk liquid—water, away from any surface or interface. There are those who contend it's a mixture—mixture between structured units of one sort or another as well as monomers. What type of structures can you have? You can have the Nemethy–Scheraga type of clusters. You can have broken down ice-lattices, which have certainly been popular as models, or you can have a Pauling type of water, a clathrate-like cage structure. My own feelings at this time are that there are some attributes in the properties of water which would suggest that the structure of bulk water, indeed, has 'sites' available, implying the likely existence of discrete voids, as you would have both in clathrate type structures as well as in the large, open spaces between adjacent hexagonal rings of water molecules in an ice lattice. I cannot discriminate between the two possibilities. The important difference is that you have pentagonal symmetry in the clathrates and you have hexagonal symmetry in the ices. For a number of reasons that I can't elaborate on here, I feel at this time slightly more swayed toward the clathrate cage structures than I do toward the broken-down ice-lattice structures. So much for water in general. Now regarding water near an interface, I think what we are seeing is a stabilization, particularly by hydrophobic surfaces, of water structures that are very likely clathrate cage-like entities. I believe what we are seeing in thermal anomalies are transitions from one type of structure—say with 40 molecules—to smaller structured units with perhaps only 20 or 30 molecules. This is the type of structural transitions I'm envisioning at this time. Let me also throw in that, of course, there are 8 or 9 different high pressure ice polymorphs, and who knows, one or more of these types of structures may be what we encounter.

THORHAUG I'd like to report some more recent and unpublished data at very closely-spaced temperature intervals, concerning the giant algal cell *Valonia*. I have an illustration of the mean change in potential difference across *Valonia macrophysa* cells. Each point represents 25 cells which have been measured over the temperature range of 8° to 36° C. The Nernst equation which is being used currently to express the potential difference across algal cells would predict that the potential was a monotone function and that there was no discrete break such as seen here at 15 and 30° C. This is for the species *Valonia macrophysa*. In a similar manner *Valonia ventricosa* shows discrete breaks at 15 and 30 °C, also unpredicted by the Nernst equation. Each one of these points represent 30 cells. This may indicate that it does pay to measure at sufficiently close temperature intervals because if I had measured this only at 5° intervals, it is quite obvious that one would not have been able to predict whether the deviations were due to scatter in my data or from actual anomalies occurring in the potential.

DROST-HANSEN I might add that in addition to these curves, Dr. Thorhaug also has a large number of survivals (or irreversible plasmolysis) experiments. She has further determined the amount of sodium and potassium in the intercellular sap of the organisms and these data also show very abrupt changes at the temperatures of the thermal anomalies.

TASAKI   Is the effect of the abrupt change in potential a reversible effect?

DROST-HANSEN   It is reversible to the following extent: As I understand it, you can heat the cells (as in these experiments) up to, say, 37°, and you can cool them down again and you get the same curve. However, if you maintain the cells, as you saw on the curve for *Nitella*, for three days at 37°, you kill the cells, but for short time exposures, the course is reversible.

PAPER 12

Structural and functional properties of bacterial cell membranes

MILTON R. J. SALTON
*Department of Microbiology, New York University
School of Medicine, New York*

INTRODUCTION

ONE OF THE outstanding features to emerge from the past two decades of intensive work on the comparative anatomy of microorganisms, is the marked difference in the degree of differentiation of the internal structures of bacteria and blue-green algae on the one hand and yeasts, fungi, algae and protozoa on the other (Salton, 1968a). Unlike the higher microorganisms, the bacteria and blue-green algae do not possess organized mitochondrial structures, nuclear membranes enclosing the chromatin, nor do they possess a well-developed membranous endoplasmic reticulum. Although bacteria lack the variety of membrane-bounded organelles seen in thin sections of cells of higher organisms, it is clear that the biochemical functions associated with these structures are performed by the membrane systems of the bacterial cell. Thus the relatively 'undifferentiated' bacterial cell appears to be less complex with respect to the variety of membrane structures and offers some attractions for investigations of membrane structure and function.

Anatomical studies of bacterial cells show that they possess, at the most, two distinct membrane systems or regions, namely, the limiting plasma membrane and the internal ramifications of the invagination of the membrane to form the 'mesosome' membranes (Fitz-James, 1960). The development of the mesosome membranes is generally more conspicuous in Gram-positive bacteria than in the majority of Gram-negative organisms. The thin section of *Micrococcus lysodeikticus* presented in Figure 1, illustrates the typical appearance of the surface structures and internal organization of a Gram-positive organism while the anatomy of the Gram-negative bacterium, *Escherichia coli*, is shown in Figure 2. The latter sections do not show any mesosome structures, but their presence in Gram-negative bacteria, including *E. coli*, is well documented (Ryter, 1968; Salton, 1967a).

Although the membranes of bacterial cells appear to be distinguishable into two distinct systems or regions, it is not known to what extent the plasma and mesosome membranes differ structurally and functionally. Reports on the localization of specific functions in the mesosome have been conflicting. For example, on the basis of reduction of tetrazolium compounds and tellurite, it

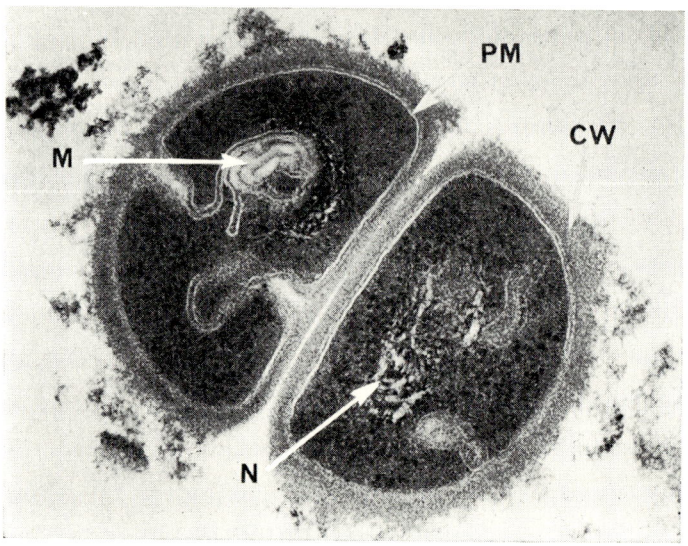

FIGURE 1 Electron migrograph of a thin section of *Micrococcus lysodeikticus* illustrating the thick cell wall (CW), the plasma membrane (PM) and its invagination to form the mesosome (M) and the nuclear region (N). × 137,500. (Salton, 1968c). (Reduced to 0.65 of original).

has been concluded that the mesosomes or 'chondrioids' are the site of the respiratory chain components of the bacterial cell (van Iterson and Leene, 1964; Leene and van Iterson, 1965). In contrast to this work, Sedar and Burde (1965) concluded from the electron microscopy of *Bacillus subtilis* cells stained cytochemically for succinic dehydrogenase, that this enzymatic activity was found on 'membranous organelles associated with the cytoplasmic membrane' (i.e. mesosomes) as well as on the septal plasma membrane, the nuclear area and the plasma membrane.

Further uncertainty has arisen from the observations of Ferrandes, Chaix and Ryter (1966) indicating that the cytochromes of *B. subtilis* were localized in the mesosome fraction and not in the plasma membrane. Attempts to confirm this with a *B. subtilis* strain in our laboratory (Munoz, unpublished data) were unsuccessful. Moreover, Ellar and Freer (unpublished results) succeeded in extruding and separating the mesosomes from *Micrococcus lysodeikticus* and *Sarcina lutea* and with the former organism succinic dehydrogenase and cytochromes were detectable in both mesosome and plasma membrane

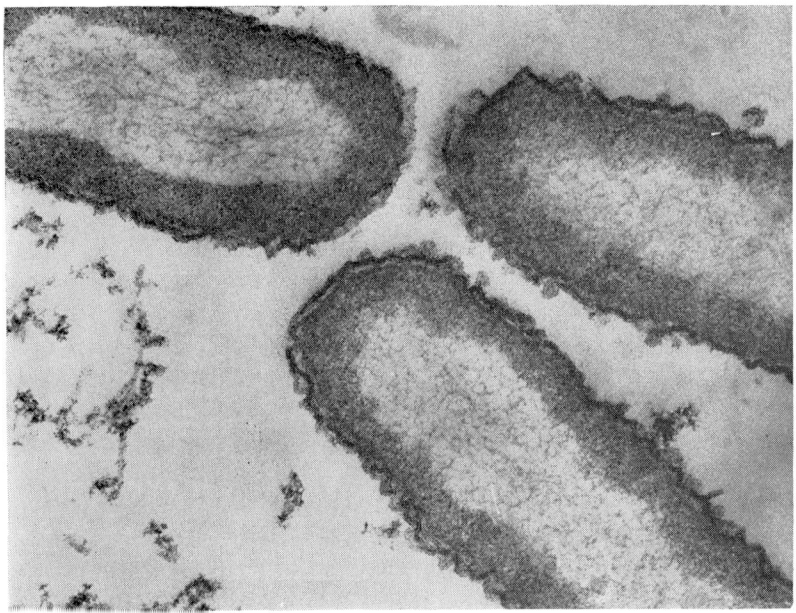

FIGURE 2 Electron micrograph of a thin section of *Escherichia coli* showing the multiple-layered outer envelope structure and the internal organization of the cell. × 105,000. (Salton, 1968c). (Reduced to 0.65 of original).

fractions. From studies of *Bacillus lichenformis* mesosome and plasma membrane fractions, Rogers, *et al.*, (1967) have suggested that succinic dehydrogenase represented 'a true component of the membrane' of the 'ghost' system and that the ferricyanide reductase may well be a component of the mesosome.

It is evident that further work is needed before any firm conclusion can be made about the functional differences between plasma and mesosome membranes in bacteria. Indeed, it may well be that 'mesosomes' represent a class of membranous structures having anatomical similarities but whose functions within a single cell or from one bacterial group to another may vary significantly. It is pertinent in this respect to note that mesosomes appear concomitantly with penicillinase production in the inducible *Bacillus lichenformis* strain used by Ghosh, Sargent and Lampen (1968). That these mesosomes perform a secretory function in this organism seems an inescapable conclusion (Ghosh, *et al.*, 1968). Apart from the involvement of the mesosomes of *B. licheniformis* in the secretion of penicillinase, it is not possible at the present time to conclude, with any degree of confidence, that certain specific functions are assigned to the mesosome membranes and not to the plasma membranes, or *vice versa*.

The development of extensive intracellular membrane systems in Gram-negative bacteria is frequently associated with some special feature of their physiology. Thus photosynthetic bacteria possess vesicular, lamellar or specialized membranes or membranous organelles bearing the biochemical apparatus for photosynthesis (Cohen-Bazire and Kunisawa, 1963; Drews, 1960; Holt, Conti and Fuller, 1966). Other Gram-negative bacteria possessing specialized mechanisms for energy production and well-developed internal membrane systems of a vesicular or stacked lamella type include the nitrogen-fixing Azotobacter (Robrish and Marr, 1962), nitrifying bacteria of the Nitrosocystis, Nitrosomonas and Nitrobacter groups (Murray and Watson, 1965; Remsen, Valois and Watson, 1967). Some of the membrane fragments separated from disrupted cells of *Nitrosocystis oceanus* exhibited a highly ordered, 'crystalline' array of subunits while other fragments showed a striking resemblance to mitochondrial membrane preparations (Remsen, *et al.*, 1967). However, as these authors point out, the present status of the studies of the membrane of this organism do not permit conclusions concerning the functions of the structural subunits.

Thus in the bacterial cell a multiplicity of functions including transport, cell-wall biosynthesis, respiratory activities, chromatin separation and DNA replication (?), and phospholipid synthesis are all mediated by the membrane or membrane systems (see reviews, Salton, 1967a; Rothfield and Finkelstein, 1968; Ryter, 1968, and Fitz-James, 1967). With these diverse functions to be performed by the membranes, it would be surprising if the enzymes and transport carriers involved were uniformly distributed throughout the plasma and mesosome membranes. It thus appears much more likely that bacterial membranes are far from functionally uniform structures and that they may be made up of 'mosaics' or regions in which specific functions are localized. The investigations in the author's laboratory have been directed towards the determination of the structure and molecular architecture of a bacterial cell membrane system (that of *Micrococcus lysodeikticus*) in terms of specific functional 'markers'.

Bacterial cell membranes

Investigations of the chemical composition, structure and biochemical properties of bacterial cell membranes have been carried out in a number of laboratories and no attempt will be made to review these studies here. Much of the work has been confined to bacteria of the Gram-positive groups since there are numerous cell-wall degrading ('muralytic') enzyme systems available for the selective removal of the wall (Salton, 1967a). The membranes from the

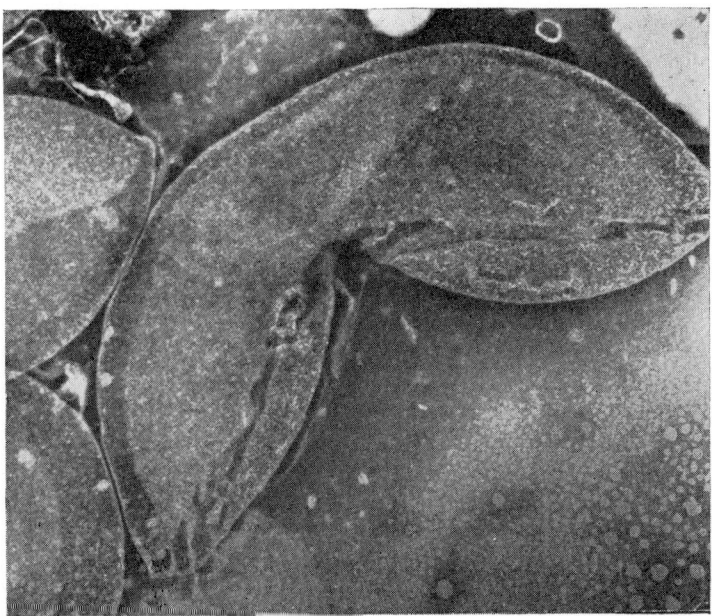

FIGURE 3 Electron micrograph of isolated envelopes of *Proteus vulgaris* negatively stained with phosphotungstate, showing multi-layered structures and 'pitted' appearance of envelope components, × 106,000. (Reduced to 0.65 of original).

Gram-positive bacteria can then be readily isolated by 'osmotic lysis' of the protoplasts formed by enzymatic digestion of the walls or by recovery of the membranes from lysates of the bacterial cells. The 'cleanest' membranes, uncontaminated by wall products, have generally been obtained from Gram-positive organisms whose walls are completely solubilized by wall-degrading enzymes.

The isolation of plasma membranes from Gram-negative bacteria has proved to be much more difficult because of the complexity of the outer envelope structure and the unavailability of suitable enzyme systems for the selective removal of components external to the rigid peptidoglycan structure (Salton, 1967a). Surface structures in preparations obtained by mechanical disruption of Gram-negative bacteria were formerly referred to as 'walls' (Salton and Horne, 1951). However, they are, strictly speaking, 'envelope' structures and not walls, since they possess the outer organized lipopolysaccharide complexes as well as the rigid peptidoglycan layers and inner plasma membrane fragments or sheets (Salton, 1967a). The complexity of

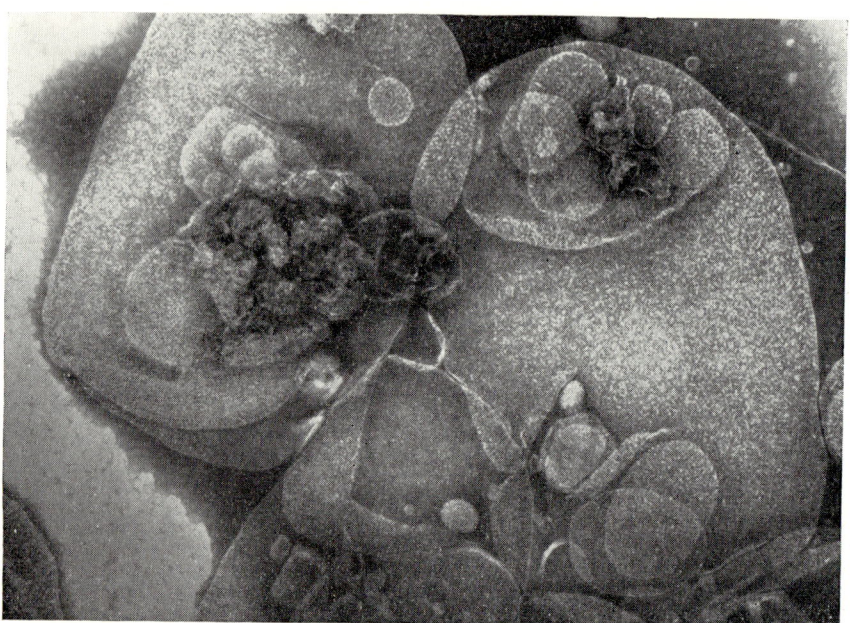

FIGURE 4 Electron micrograph of isolated envelopes of *Escherichia coli* negatively stained with phosphotungstate, illustrating the complexity of envelope fraction. Many of the circular discs may arise from fragmentation of the plasma membrane. Note the 'pitted' appearance, × 106,000. (Reduced to 0.65 of original).

envelopes from the two Gram-negative organisms *Proteus vulgaris* and *Escherichia coli* is illustrated in the electron micrographs of the structures negatively stained with phosphotungstate (Figures 3 and 4). The multilayered character of these envelopes is contrasted with the appearance of membranes isolated from the Gram-positive bacteria, *Bacillus subtilis* and *Staphylococcus aureus* (Figures 5 and 6).

Recent progress has been made in achieving a better resolution of the membrane structures from strains of *E. coli* by Kaback and Stadtman (1966)

Bacterial cell membranes

and by Miura and Mizushima (1968). Although these preparations are relatively free of the peptidoglycan components such as muramic acid and diaminopimelic acid, the extent of contamination with outer envelope components of the lipopolysaccharides has not yet been clearly established. However, it is realized that for some functional studies, such as active transport (Kaback and Stadtman, 1966), complete removal of outer envelope structures

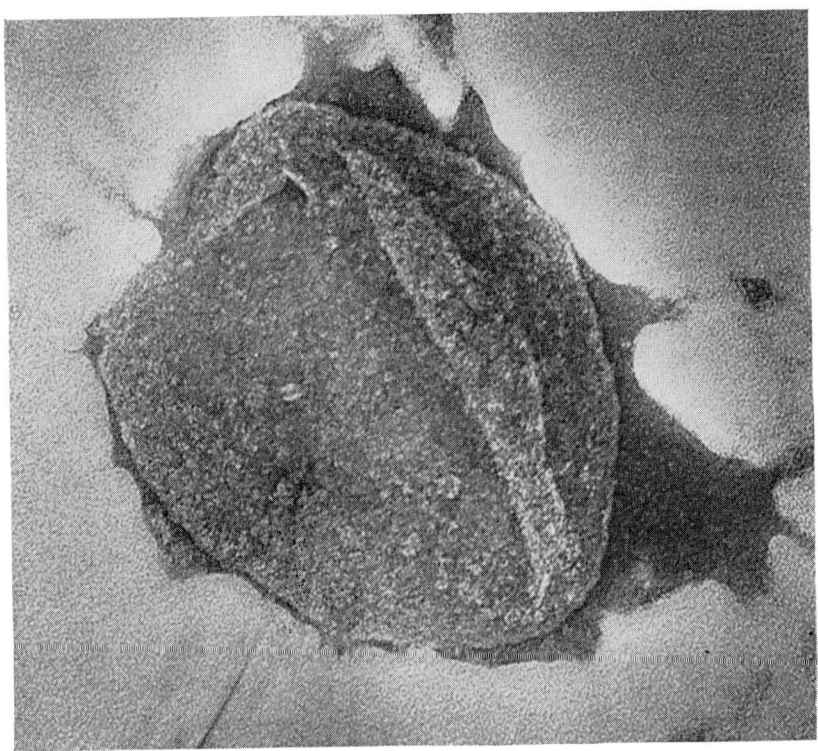

FIGURE 5 Electron micrograph of *Bacillus subtilis* membrane negatively stained with ammonium molybdate, ×300,000. (Reduced to 0.65 of original).

is not absolutely essential. The latter obviously becomes a requirement if specific localization of a structural or functional component in the plasma membrane or outer envelope is to be established.

Thus, at the present time, the Gram-positive bacteria are still the most suitable organisms for the unambiguous isolation of bacterial membrane systems. For the above reasons, we have used the lysozyme-sensitive Gram-positive organism, *Micrococcus lysodeikticus*, as a model system for the investigation of the chemical composition, structural and functional properties of a bacterial membrane. Other investigators, notably Grula and his colleagues (Grula, et al., 1967) have also selected this organism for studies of membrane structure.

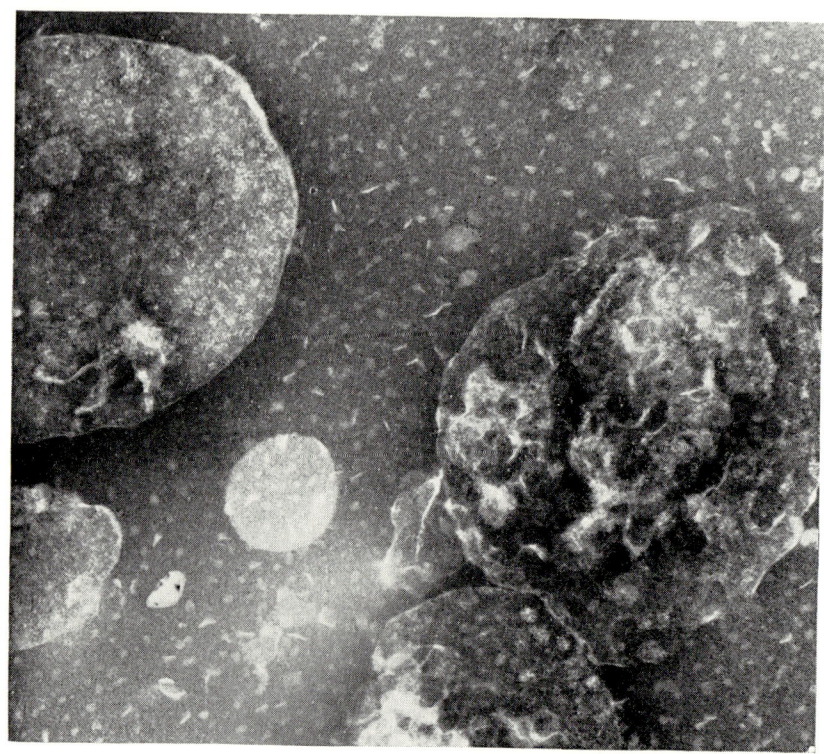

FIGURE 6 Electron micrograph of *Staphylococcus aureus* membrane fraction negatively stained with phosphotungstate, × 300,000.
fraction negatively stained with phosphotungstate, × 300,000. (Reduced to 0.65 of original).

COMPOSITION OF *MICROCOCCUS LYSODEIKTICUS* MEMBRANES

Membranes from *Micrococcus lysodeikticus* have been isolated under the conditions described by Salton and Freer (1965) and carefully washed to ensure freedom from cytoplasmic contamination and subjected to the criteria outlined by Salton (1967b) for establishing their homogeneity. The overall chemical composition of the membranes of *Micrococcus lysodeikticus* is very similar to that of other bacterial membranes and membranes isolated from cells or organelles of higher organisms. The chemical composition of

M. lysodeikticus membranes together with the data for several other Gram-positive bacteria is presented in Table I. The features of the lipid components and the nature of the fatty acid constituents are summarized in Tables II and III. Amino acid analysis (Freer and Salton, unpublished results) is in general accord with the data reported by Grula, *et al.* (1967) confirming the virtual absence or very low contents of cysteine.

Infrared spectroscopy of the membranes of *Micrococcus lysodeikticus* (Green and Salton, 1969) has yielded spectra very similar to those reported for mammalian cell membranes (Wallach and Zahler, 1968; Chapman, Kamat and Levene, 1968). As with animal cell membranes, the dominant absorption bands of *M. lysodeikticus* membranes can be assigned to the two major classes of chemical constituents, the proteins and lipids. The Amide I band of the bacterial membrane in the region of $1650 \, \text{cm}^{-1}$ has been interpreted as being due to an α-helical or random coil conformation or a combination of both. The characteristics of the infrared spectra also suggest the existence of weak interactions between lipid hydrocarbons and hydrophobic regions of the proteins (Green and Salton, 1969).

Thus *Micrococcus lysodeikticus* cell membranes are essentially protein-lipid structures, and evidence from the electrophoretic separation of lipids from proteins by the use of deoxycholate (Salton and Schmitt, 1967a) suggests that little, if any, lipid is covalently linked to protein. The lipids of *M. lysodeikticus* membranes are accounted for largely by the phospholipids (Macfarlane, 1962) and under the growth conditions used in our laboratory the principal phospholipids are cardiolipid (diphosphatidyl glycerol), phosphatidyl glycerol and phosphatidyl inositol. The variety of phospholipids in the membranes under these conditions is thus rather simpler than that found for other membranes (Macfarlane, 1964; Kates, 1966). No sterols are present in these bacterial membranes, but they contain small amounts of carotenoids and menaquinone (Salton and Schmitt, 1967b).

The variety of proteins in the membranes of *Micrococcus lysodeikticus* has been investigated by disc electrophoresis in polyacrylamide gels. Because of their insolubility, their association with lipids and the need for dissociating agents in the polyacrylamide gel systems, it is difficult to reach any final conclusion about the total number of individual proteins associated with the isolated membranes. Our data for various methods of dissociation indicate something of the order of 15–20 individual bands on disc gel systems (Salton, Schmitt and Trefts, 1967). The complexity of protein patterns on disc gel electrophoresis of dissociated membranes is in agreement with reports on other bacterial and animal cell membranes (Schneiderman, 1965; Neville, 1967; Rottem and Razin, 1967; Zahler, Wallach and Lüscher, 1967). Our studies are not in accord with the concept of a 'structural protein' accounting

TABLE I Chemical composition of membranes isolated from gram-positive bacteria.†

	% Protein	% Total lipid
Bacillus spp.	58–75	20–28
M. lysodeikticus	65–68‡	23–26
	49§	
Sarcina lutea	57§	23
Staph. aureus	69‡	30
	73§	

† Data from Salton (1968c)
‡ Biuret method
§ Lowry method.

TABLE II Principal components identified in 'total lipid' of *Micrococcus lysodeikticus* membranes.

Phospholipids (approx. 75–80% of total lipid)
    Cardiolipin (diphosphatidyl glycerol)
    Phosphatidyl glycerol
    Phosphatidyl inositol
Menaquinone–9 (4–5% of total lipid)
Carotenoids (approx. 0.5% of total lipid)
Other components detected (Macfarlane, 1962, 1964; Kates, 1966) glycolipids, lipo-amino acids, fatty acids, glycerides.

TABLE III Fatty acid composition of *Micrococcus lysodeikticus* membrane lipids.

Methyl esters	% Composition
$C_{12:0}$	0.4
$C_{13:0\ br.}$	1.2
$C_{14:0}$	4.4
$C_{15:0\ iso}$	0.4
$C_{15:0\ anteiso}$	85.4
$C_{16:0}$	0.2
$C_{16:0\ br.}$	5.0
$C_{17:0\ iso}$	0.4
$C_{17:0\ anteiso}$	2.6
$C_{18:0}$	trace

for the major portion of the membrane protein. The results obtained so far with *M. lysodeikticus* would indicate that a number of functional proteins can be dissociated by sodium dodecyl sulfate (SDS) into subunits behaving rather like 'structural protein' on disc gel electrophoresis. It is evident that the final interpretation of attempts to assess the variety of membrane proteins must await further characterization of individual functional proteins and a knowledge of the constituent subunits obtained with dissociating agents such as SDS, urea or guanidine hydrochloride.

DISSOCIATION AND REAGGREGATION OF BACTERIAL MEMBRANES

Unlike the rigid bacterial cell wall, the underlying plasma membrane is sensitive to dissociation with surface-active agents. The differential sensitivities of wall and membrane were established after Weibull (1953) developed the methods for preparing stable bacterial protoplasts. On removal of the cell walls by digestion with lysozyme, the protoplasts of Gram-positive bacteria such as *Bacillus megaterium* and *Micrococcus lysodeikticus* exhibited marked susceptibility to lysis with SDS and other surfactants (Gilby and Few, 1957; Salton, 1957). It could be anticipated, therefore, that surface-active compounds would have a direct action on isolated bacterial membranes and this was confirmed in later studies by Salton and Netschey (1965) and Razin, Morowitz and Terry (1965).

The disaggregation of *Micrococcus lysodeikticus* by anionic surfactants of the long chain alcohol sulfate type occurs very rapidly and can be followed conveniently by measuring the decrease in the absorbance (turbidity reduction) of membrane suspensions. The results of the dissociation of *M. lysodeikticus* membranes in 0.05 M tris-HCl buffer (pH 7.4) at 25°C by alcohol sulfates of alkyl chain lengths of C_{10}, C_{12} and C_{14} are illustrated in Figure 7. The speed with which the dissociation occurs is so fast that it is most unlikely that any enzyme action is involved in the breakdown of the membrane structure. Indeed, there is general agreement that surface-active agents owe their lytic and membrane dissociating activities to their ability to disrupt the relatively weak interactions holding the membrane lipids and proteins together. The disaggregation of the membranes of *M. lysodeikticus* by SDS and by a nonionic agent of the polyoxyethylated alkyl phenol type, Nonidet P.40, was more effective than that observed with the cationic compound, dodecyltrimethylammonium bromide (Salton and Netschey, 1965).

Examination in the ultracentrifuge, of the products of dissociation of *M. lysodeikticus* by various surface-active agents (Salton and Netschey, 1965) and *Mycoplasma laidlawii* by SDS (Razin, Morowitz and Terry, 1965), revealed components with sedimentation coefficients of about 2–3 S, possessing a surprising degree of apparent homogeneity. Such components represent lipid-protein-detergent aggregates rather than resolved membrane protein and membrane lipid. Moreover, dissociation of *Micrococcus lysodeikticus* membranes by exposure to ultrasound yielded a uniform major component with a sedimentation coefficient of 4.2 S (Salton and Netschey, 1965) and this migrated as a single, negatively-charged component of protein-lipid and

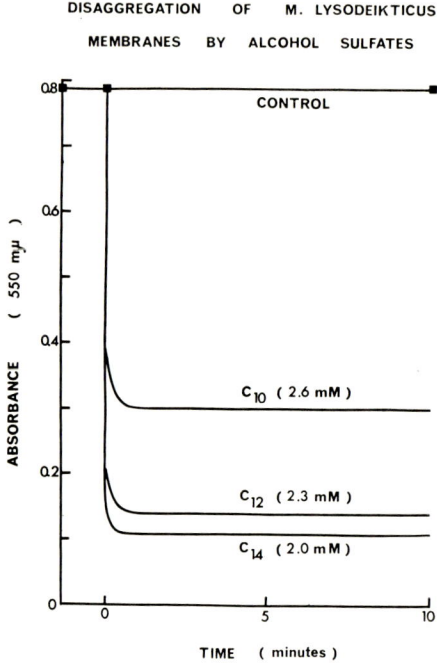

FIGURE 7 Disaggregation of isolated *Micrococcus lysodeikticus* membranes in 0.05M tris-HCl buffer (pH 7.4) at 25 °C by alcohol sulfates of chain lengths C_{10}, C_{12} and C_{14}. The dissociation was followed by recording the absorbance changes at 550 mμ. Data from Salton (1968c).

carotenoid on sucrose gradient electrophoresis (Salton, 1968b). However, Engelman, Terry and Morowitz (1967) did achieve some resolution of SDS dissociated mycoplasma membranes into protein and lipid when these were subjected to gradient centrifugation using a 'sandwich technique'.

The complete dissociation and separation of membranes into their constituent proteins and lipids is difficult to achieve without resorting to organic solvents. A degree of selective separation of *M. lysodeikticus* membranes into

3 distinct fractions by detergent fractionation procedures using ammonium sulfate treatment was reported by Salton, Schmitt and Trefts (1967). 'Fraction 1' was colorless, being devoid of the yellow carotenoid pigments, 'fraction 2' contained the bulk of the membrane lipid and little of the membrane cytochromes and 'fraction 3', which was insoluble, contained only small amounts of lipid and virtually all of the membrane cytochrome complement (Salton, Schmitt and Trefts, 1967). The distribution of protein and lipid (as ^{32}P-labeled phospholipid) is illustrated in Table IV for *M. lysodeikticus* and

TABLE IV Percentage distribution of recovered protein and ^{32} P-labeled lipid in fractions separated by dissociation of bacterial membranes with surface-active agents.

	Protein			^{32} P-lipid
	ML(A)*	ML(B)*	SL*	ML(A)*
Nonidet fraction				
1	23.2	20.0	18.9	15.5
2	46.3	46.0	44.0	80.2
3	30.5	34.0	37.1	4.4
Deoxycholate + triton fraction				
1	21.0	8.8		15.3
2	61.5	70.0		81.0
3	17.5	21.2		2.7
Deoxycholate fraction				
1		10.4	15.1	
2		28.2	46.7	
3		61.4	38.2	

ML(A)*—^{32} P-labeled membranes of *M. lysodeikticus*.
ML(B)*—membranes of *M. lysodeikticus*.
SL*—membranes of *S. lutea*.
(From Salton, Schmitt and Trefts, 1967).

Sarcina lutea membrane fractions. Polyacrylamide disc electrophoresis of fractions 2 and 3 both showed considerable complexity in band patterns, although the resolution in the former fraction was poor, presumably because of the large amount of lipid in these fractions. Fraction 3, which contained cytochromes *a*, *b* and *c*, gave normal cytochrome spectra when this fraction was dissolved in SDS and reduced with dithionite (Salton, Schmitt and Trefts, 1967). However, on examination in the electron microscope this fraction was completely amorphous and devoid of any structural organization.

By far the most efficient method for the removal of lipid from the membrane proteins is the extraction procedure using *n*-butanol which was developed by Maddy (1964) for the examination of erythrocyte membrane proteins. The behavior of the bacterial membrane protein was unlike that of the erythrocyte membrane in that not all of the protein remained in the 'soluble' state in the aqueous phase. Most of the cytochromes were recovered as an insoluble, interfacial layer on centrifugation of the butanol extracted membrane suspension. However, the aqueous phase proteins on examination in the acid-urea polyacrylamide electrophoresis system exhibited a good resolution into a number of discrete protein bands (Salton, Schmitt and Trefts, 1967).

For the isolation of certain functional 'markers' from the membranes of several Gram-positive bacteria we have found that dissociation of the membranes with deoxycholate has provided a very effective method. This aspect of our investigations will be discussed in more detail in a subsequent section of this contribution.

Membranes dissociated with surface-active agents show a marked propensity for reaggregation when the dissociating agent is removed by dialysis. The reassembly into structures possessing membrane profiles on examination of thin sections in the electron microscope has been reported by Razin, Morowitz and Terry (1965) and Terry, Engelman and Morowitz (1967) in their studies of *Mycoplasma laidlawii* membranes dissociated with SDS. Butler, Smith and Grula (1967) have also shown that *M. lysodeikticus* membranes dissociated with SDS will reaggregate into membranous sheets possessing characteristic double-track profiles when viewed in thin sections.

There are very few reports on the reassociation of lipid-free 'soluble' protein and protein-free lipid from cell membranes. Zahler, Wallach and Lüscher (1967) carried out such a recombination experiment with erythrocyte proteins and the stroma lipids and obtained membrane-like structures with apparently 'correctly placed antigenic sites'. However, the profiles of the reassociated material were not as clearly resolved as those of the free lipid. Grula, *et al.* (1967) has shown that 'stripped subunits' of *M. lysodeikticus* membranes are still in the form of insoluble membrane sheets and that they will recombine with native and foreign phospholipids. With aqueous-phase soluble proteins obtained after *n*-butanol extraction of the membranes of *M. lysodeikticus* (residual lipid < 1 per cent by weight) we have found that small vesicles are reformed by reassociation with protein-free lipid (Folch washed) after removal of the solvent by dialysis. The appearance of such vesicles is illustrated in the negatively stained preparation shown in Figure 8. Suspensions of these vesicles could be agglutinated with membrane antisera indicating the 'outside orientation' of at least some of the antigens of the membrane. These results and those of Zahler, *et al.* (1967) suggest that there must be a preferred orientation when the lipids and proteins reassociate.

These dissociation and reassociation studies with the above systems must remain something of a laboratory curiosity until specific functional parameters can be determined, such as those that have been performed with reassembled components of mitochondrial membranes (e.g. McConnell, et al., 1966; Racker, 1967; Kopaczyk, et al., 1968).

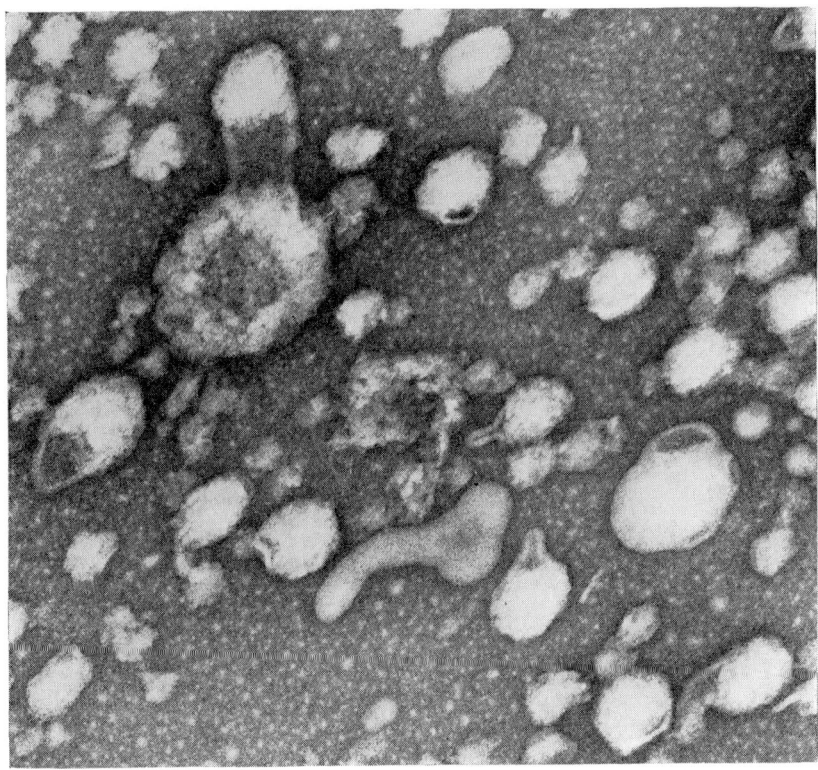

FIGURE 8 Electron micrograph of negatively stained vesicles formed by the reassociation of lipid-free, soluble protein and total lipid from *M. lysodeikticus* membranes. The lipid dissolved in *n*-butanol was homogenized with the protein solution and the solvent removed by dialysis, × 300,000. (Reduced to 0.65 of original).

ISOLATION OF FUNCTIONAL 'MARKERS' FROM *MICROCOCCUS LYSODEIKTICUS* MEMBRANES

Attempts to gain some insight into the molecular architecture of cell membranes would obviously be greatly facilitated by the ability to identify specific functional proteins in the membranes. Cytochemical staining combined with electron microscopy has been employed successfully with bacteria for determining the distribution of activities such as adenosinetriphosphatase (Voelz and Ortigoza, 1968), esterases (Baillie, *et al.*, 1967) and succinic dehydrogenase (Sedar and Burde, 1965). Although these methods are highly specific for a particular enzymatic activity, a method which could be used for the specific localization of a protein (e.g. with ferritin-labeled antibody) would have some advantages. Such an approach as the latter would involve the purification of a membrane enzyme and preparation of antisera to the protein. Adenosinetriphosphatase (ATPase) appeared to be a suitable enzyme since

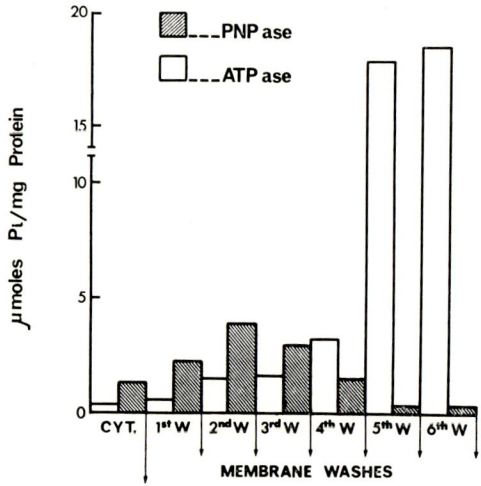

FIGURE 9 The release of *M. lysodeikticus* ATPase and PNPase into the cytoplasmic fraction (CYT) and into consecutive washes (1-4) in 0.03M tris-HCl-0.001M EDTA followed by two washes (5 and 6) in 0.003M tris-HCl (Munoz, *et al*, 1968a).

its localization could be determined by both the cytochemical and ferritin-labeled antibody methods.

Membrane-bound ATPase activities have been reported for a number of Gram-positive bacteria (Georgi, *et al.*, 1955; Abrams, *et al.*, 1960; Weibull, *et al.*, 1962; Ishikawa and Lehninger, 1962). *Micrococcus lysodeikticus* ATPase had been investigated by Ishikawa and Lehninger (1962) and it was later shown to be activated by Ca^{2+} (Ishikawa, 1966). Since membrane-associated ATPases can be 'solubilized' by release from the bacterial membranes (Ishikawa and Lehninger, 1962; Abrams, 1965) we selected this enzyme as a suitable functional protein for further investigations.

Bacterial cell membranes

Investigations by Munoz, et al. (1968a) in our laboratory have shown that the Ca^{2+}-activated ATPase can be selectively released from *M. lysodeikticus* membranes by lowering the ionic strength of the buffer after they had been subjected to a series of washes to remove contaminating cytoplasmic proteins. The *M. lysodeikticus* membrane ATPase is thus released in a similar manner to the Mg^{2+}-activated ATPase of *Streptococcus faecalis* membranes investigated by Abrams (1965) and Abrams and Baron (1968). The pattern of release of the ATPase and polynucleotide phosphorylase (PNPase) from *M. lysodeikticus* is illustrated in Figure 9.

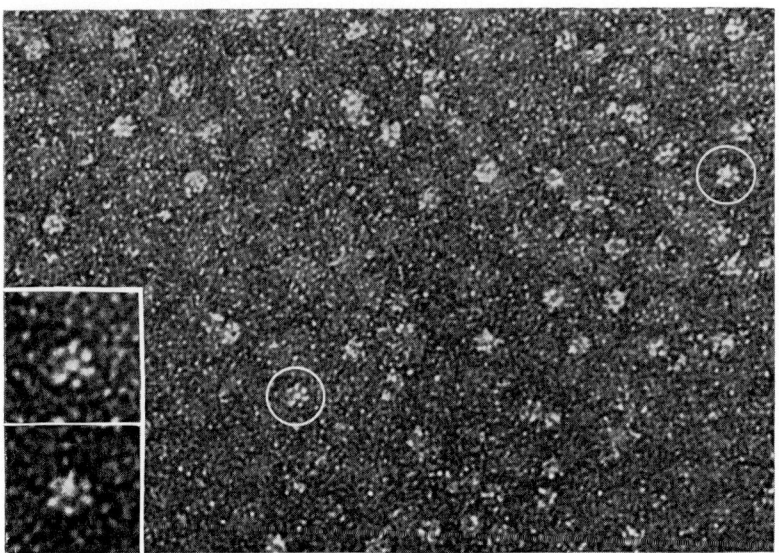

FIGURE 10 Active ATPase enzyme preparation negatively stained for electron microscopy showing spherical particles of approx. 100 Å diameter, × 500,000. Selected particles in circles enlarged in inset to show central subunit surrounded by 6 additional subunits, × 1,000,000. (Munoz, et al., 1968b). (Reduced to 0.65 of original).

Examination of the active ATPase fractions by negative staining with 1 per cent ammonium molybdate in 2.0 per cent ammonium acetate buffer (pH 7.2) revealed the presence of uniform spherical particles of approximately 100 Å diameter as shown in Figure 10 (Munoz, et al., 1968b). These particles appear to be made up of a central subunit surrounded by 6 additional subunits as indicated in Figure 10. Some of the particles possessed rectangular profiles but whether or not they represent pairs of discs of the ATPase enzyme or macromolecular structures bearing other enzymatic activities must await further investigation. Particles similar to those in the purified enzyme preparations were often found associated with membrane fragments as illustrated in Figure 11. Abram (1965) and Biryuzova, *et al.* (1964) have described stalked particles of about the same size associated with bacterial membranes. Similar structures have been observed for the ATPase activity associated with mito-

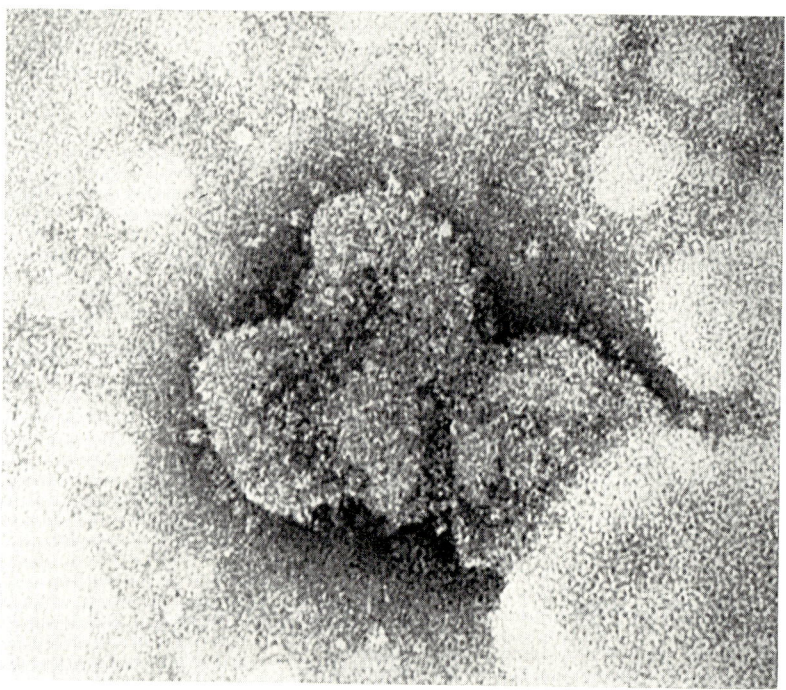

FIGURE 11 Membrane fragment negatively stained to show a large number of attached particles, × 300,000. (Munoz, *et al.*, 1968b).

chondrial membranes (Racker, 1967; Stiles and Crane, 1966) suggesting that there may be a basic type of structural organization common to both mitochondrial and bacterial membrane systems.

The ATPase from *M. lysodeikticus* membranes has been purified by chromatography on Sephadex G-200 and highly purified preparations gave a single band on polyacrylamide disc electrophoresis when stained for protein and ATPase activity (Munoz, *et al.*, 1969). For the specific detection of the enzyme in the gels, the method of Weinbaum and Markman (1966) was used. This enzyme constitutes one of the major proteins of the membranes of

M. lysodeikticus, accounting for approximately 10 per cent of the total membrane protein (Munoz, *et al.*, 1969). Moreover, it has been identified as one of the principal membrane antigens and it gave a single line of precipitate in agar gels when the ATPase band from the polyacrylamide gels was reacted with antisera to washed membranes (Munoz, *et al.*, 1969).

One other interesting feature to emerge from these studies is that in its membrane-associated state, the ATPase protein does not exhibit enzymatic activity unless 'activated' by trypsin (Munoz, *et al.*, 1969). Although we do not have any direct proof as to the function of the ATPase protein in the bound state on the membrane, it is tempting to suggest that it is involved in oxidative phosphorylation. The work of Ishikawa and his colleagues provides strong evidence in favor of this function for the ATPase, since they have reported the necessity of a coupling factor for oxidative phosphorylation in *M. lysodeikticus* (Ishikawa and Lehninger, 1962; Ishikawa, *et al.*, 1965) and presented evidence identifying the coupling factor as an ATPase (Ishikawa, 1966). As suggested by Munoz, *et al.* (1969) the studies with the *M. lysodeikticus* ATPase indicate an 'allotopic' behavior (Racker, 1967) thus reinforcing the basic similarities between mitochondria and the corresponding structures in the bacterial membrane performing these functions.

As already mentioned in the section on the dissociation of membranes, it has been possible to devise suitable fractionation procedures for the selective separation of the *M. lysodeikticus* membrane cytochromes. No organized structures were found in these fractions, so further attempts were made to isolate functional components of the electron transport system in which some structural integrity was still preserved. It was found that treatment of the isolated membranes of *M. lysodeikticus* with 1 per cent sodium deoxycholate (DOC) in 0.05 M tris-HCl buffer (pH 7.5) resulted in the dissociation of membrane lipid, carotenoid and much of the protein from the insoluble structures containing the cytochromes and succinic dehydrogenase activity (Salton, *et al.*, 1968). The release of ^{32}P-labeled lipid, protein and carotenoid on 6 successive extractions of the membranes with 1 per cent DOC in tris-HCl buffer is shown in Figure 12 and the changes in the density of the fractions on sucrose gradient centrifugation are demonstrated in Figure 13.

Examination of the DOC-insoluble residues from *M. lysodeikticus* membranes as negatively-stained preparations in the electron microscope revealed the presence of membranous sheets. The appearance of these DOC-insoluble residues is contrasted with that of untreated membranes in Figure 14. Although the membrane sheets in the DOC-insoluble fractions are very similar to untreated membranes in the variation of sizes of sheets and fragments, their folded, 'collapsed-sac' appearance suggests that they are much more rigid than the original membranes. Analysis of the lipid contents of these insoluble residues established that the total lipid is about 3–5 per cent compared to 23–26 per cent for whole membranes (Salton, et al., 1968). It thus appears likely that these lipid-depleted sheets are held together by weak interactions between hydrophobic regions of the proteins. Indeed, these sheets are completely dissociated by 0.5 per cent SDS, acetic acid-water (1:2, v/v)

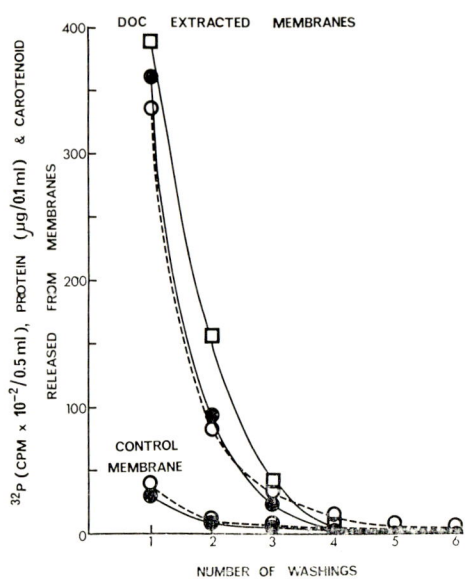

FIGURE 12 Release of ^{32}P-labeled lipid (●), protein (○) and carotenoid (□), absorbance at 475 mμ × 10^2, (1 cm cuvette) from *M. lysodeikticus* membranes subjected to 6 successive washes with 1 per cent DOC in 0.05M tris, pH 7.5 (DOC extracted membranes) and untreated membranes washed × 6 with 0.05M tris (control membrane). Determinations were performed on supernatant washed after centrifugation for 50 min. at 33,000 xg at 0 °C. Initial control and DOC extracted membrane suspensions contained 4.9 mg protein/ml. (Salton, et al., 1968).

and by Triton X-100 and only partially 'solubilized' by 8 M urea. The removal of the dissociating agent by dialysis against buffer without any added cations resulted in reaggregation into less ordered aggregates.

The apparent rigidity of the DOC-insoluble, lipid-depleted sheets suggests that the association with lipid in the native membrane gives the structure a greater flexibility or plasticity. The extent to which lipids may be required for specific enzymatic activities has not yet been fully explored but they may

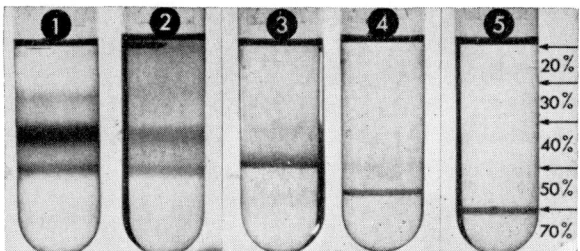

Figure 13. Centrifugation of M. lysodeikticus membrane fractions on discontinuous gradients of 70, 50, 40, 30, and 20 per cent sucrose in 0.05M tris buffer (pH 7.5) at 24,000 rpm for 2 hours at 5 °C in SW 25 rotor. 1 control membrane; 2, membrane in 1 per cent DOC-tris layered on gradient; 3, insoluble residue after 1 wash in 1 per cent DOC-tris; 4, insoluble residue after 2 washes in 1 per cent DOC-tris; 5, DOC-insoluble residue, 6 washes. Fractions in 1 ml tris layered on top of gradients. Interfaces between sucrose solutions indicated at right of gradient 5. (Salton, et al., 1968).

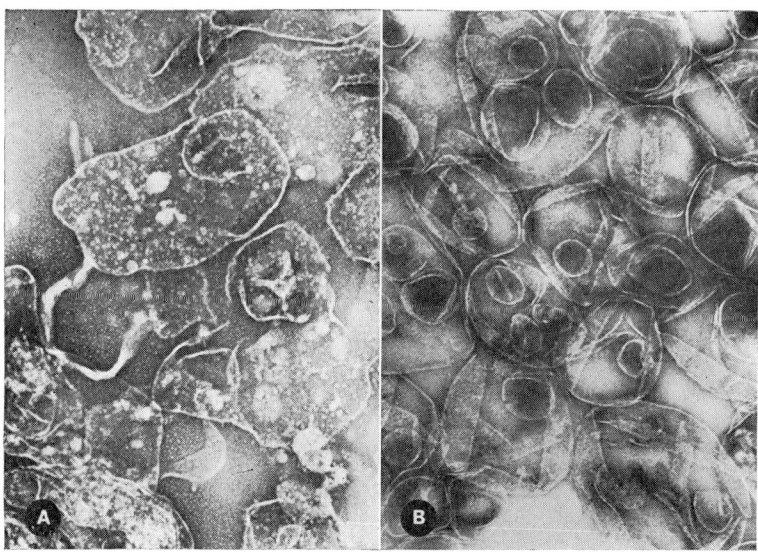

FIGURE 14 Electron micrographs of M. lysodeikticus membrane fractions negatively stained with 2 per cent ammonium molybdate. A, control membrane after 6 washes with 0.05M tris; B, DOC-insoluble residues after 6 successive washes with 1 per cent DOC in 0.5M tris. A and B, × 86,000. Note the disappearance of the spherical particles (ATPase) on extraction with DOC. (Salton, et al., 1968). (Reduced to 0.65 of original).

also have important functional as well as structural properties in the membrane.

The DOC-insoluble residues from *M. lysodeikticus* membranes are composed largely of protein and they account for 10–15 per cent of the initial membrane protein. Membranes of two other Gram-positive bacteria (*Bacillus subtilis* and *Sarcina lutea*) have also been subjected to this fractionation procedure with similar results. The DOC-insoluble residues from *B. subtilis* membranes are compared with untreated membranes in Figure 15 and they show a striking similarity to results obtained with *M. lysodeikticus*. All of the DOC-insoluble fractions examined contained the membrane cytochromes and succinic dehydrogenase activities.

The results suggest that the electron transport system in these bacteria is organized in a well-defined component of the membrane and that its survival as a lipid-depleted sheet is related to the highly insoluble nature of components such as cytochrome *b* and cytochrome oxidase. From the contribution of the DOC-insoluble fraction to the weight of the membrane (about 15 per cent) and to the protein content, it appears unlikely that it would form a completely continuous sheet on the inner face of the entire membrane system. By the use of antisera to specific components of this fraction of the bacterial membrane it is hoped that the determination of the distribution of the electron transfer chain components in the membrane structure will be possible. In this way it is anticipated that we shall be able to build up a more complete picture of the molecular architecture of the membrane by the localization of specific functional proteins.

The concept that membranes are organized as bimolecular leaflets corresponding to a 'unit' double-track structure has certainly undergone considerable revision in the past few years (Korn, 1966; Green and Perdue, 1966). Bacterial membranes seen either *in situ* in thin sections of the intact cell or in the isolated state also possess the traditional, 'unit membrane' profile. This anatomical appearance undoubtedly gives an over-simplified impression of the organization of such membranes. It is evident from our studies with *M. lysodeikticus* membranes that there are regions of the 'unit membrane' system that perform mitochondrial functions and, as with mitochondria, they possess some of the complex particles (ATPase) attached to the membrane structures. There are probably certain areas of the membrane that are completely dissociated by deoxycholate as well as the regions containing the electron-transport system that remain insoluble as the lipid-depleted sheets. By selecting functions other than those responsible for electron transfer and oxidative phosphorylation, it may be possible to 'map' other enzymatic and transport components of the plasma membrane. It appears most unlikely that the bacterial membrane represents a structure in which there is uniformity of

Bacterial cell membranes

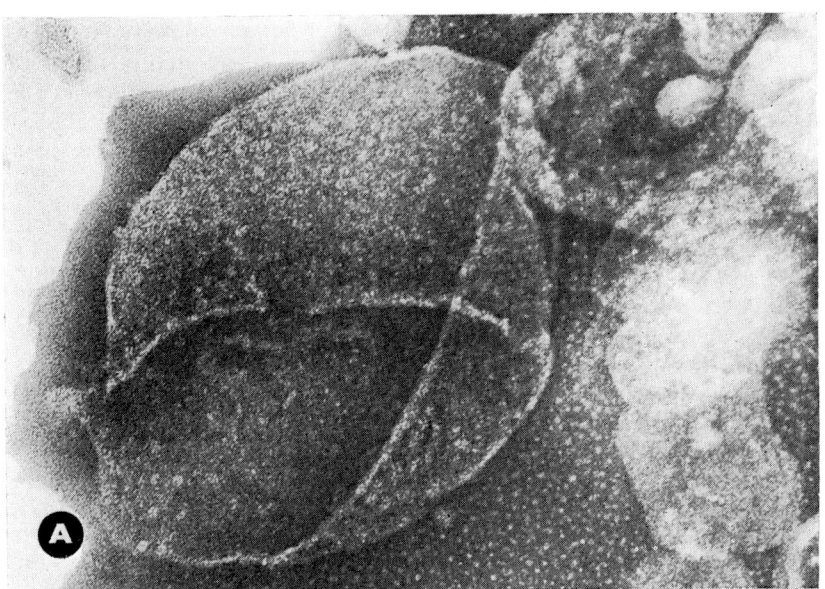

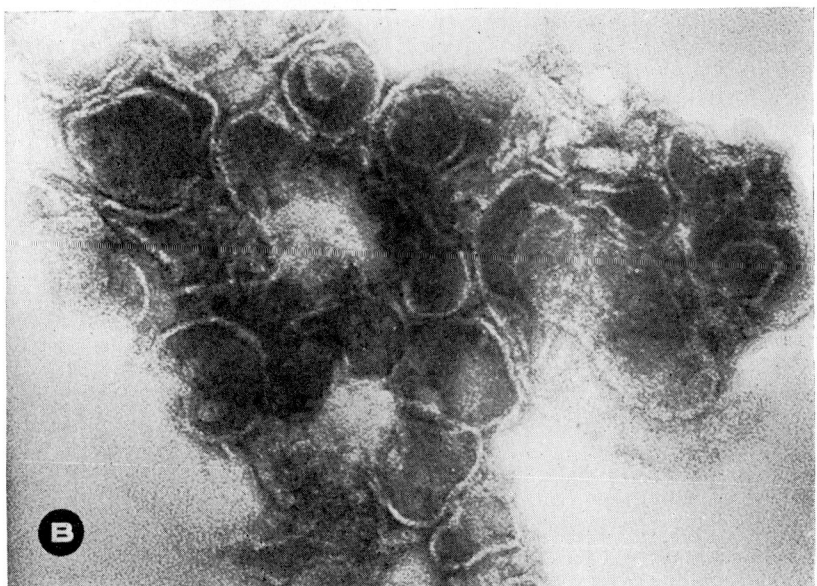

FIGURE 15 Negatively stained preparations of A, untreated membranes of *Bacillus subtilis* and B, DOC-insoluble residues from *B. subtilis* membranes extracted and prepared as for *M. lysodeikticus* membrane fractions in Figure 14, ×300,000. (Reduced to 0.65 of original).

distribution of functional components or uniformity of associations between lipids and proteins or proteins with proteins. The elucidation of the precise architecture and the degrees of stability of interactions between its constituent molecules must, however, await further investigation.

Acknowledgements

The author's work was supported by a National Science Foundation Grant (GB 4603). I wish to thank Dr. John H. Freer for electron micrographs of some of the preparations and Mr. Charles Harman for the preparation of photographs.

References

ABRAM, D., 1965, *J. Bacteriol.*, **89**, 855.
ABRAMS, A., MCNAMARA, P., and JOHNSON, F. 1960, *J. Biol. Chem.*, **235**, 3659.
ABRAMS, A. 1965, *J. Biol. Chem.*, **240**, 3675.
ABRAMS, A., and BARON, C. 1968, *Biochemistry*, **7**, 501.
BAILLIE, A., THOMSON, R. O., BATTY, I., and WALKER, P. D. 1967, *J. Applied Bacteriol.*, **30**, 312.
BIRYUZOVA, V. I., LUKOYANOVA, M. A., GELMAN, N. S., and OPARIN, A. I. 1964, *Dokl. Akad. Nauk, SSSR.*, **156**, 198.
BUTLER, T. F., SMITH, G. L., and GRULA, E. A. 1967, *Canad. J. Microbiol.*, **13**, 1471.
CHAPMAN, D., KAMAT, V. B., and LEVENE, R. J. 1968, *Science*, **160**, 314.
COHEN-BAZIRE, G., and KUNISAWA, R. 1963, *J. Cell Biol.*, **16**, 311.
DREWS, G. 1960, *Arch. Mikrobiol.*, **36**, 99.
ENGELMAN, D. M., TERRY, T. M., and MOROWITZ, H. J. 1967, *Biochim. Biophys. Acta*, **135**, 381.
FERRANDES, B., CHAIX, P., and RYTER, A. 1966, *Compt. Rend.*, **263**, 1632.
FITZ-JAMES, P. C. 1960, *J. Biophys. Biochem. Cytol.*, **8**, 507.
FITZ-JAMES, P. C. 1967, *Protides of the Biological Fluids*, **15**, 289.
GEORGI, C. E., MILITZER, W. E., and DECKER, T. S. 1955, *J. Bacteriol.*, **10**, 716.
GHOSH, B. K., SARGENT, M. G., and LAMPEN, J. O. 1968, *J. Bacteriol.*, **96**, 1314.
GILBY, R., and FEW, A. V. 1957, *Proc. Intern. Congr. Surface Activity*, 2nd, London, **4**, 262.
GREEN, D. E., and PERDUE, J. F. 1966, *Proc. Natl. Acad. Sci., U.S.*, **55**, 1294.
GREEN, D. H., and SALTON, M. R. J. 1969, manuscript submitted, *Biochim. Biophys. Acta*.
GRULA, E. A., BUTLER, T. F., KING, R. D., and SMITH, G. L. 1967, *Canad. J. Microbiol.*, **13**, 1499.
HOLT, S. C., CONTI, S. F., and FULLER, R. C. 1966, *J. Bacteriol.*, **91**, 311.
ISHIKAWA, S., and LEHNINGER, A. 1962, *J. Biol. Chem.*, **237**, 2401.
ISHIKAWA, S., YAMASHITA, S., ARAKI, S., and SHIMAZONO, N. 1965, *J. Biochem. Tokyo*, **57**, 235.
ISHIKAWA, S. 1966, *J. Biochem. Tokyo*, **60**, 598.
ITERSON, VAN W., and LEENE, W. 1964, *J. Cell Biol.*, **20**, 377.
KABACK, H. R., and STADTMAN, E. R. 1966, *Proc. Natl. Acad. Sci., U.S.*, **55**, 920.

KATES, M. 1966, *Ann. Rev. Microbiol.*, **20**, 13.
KOPACZYK, K., ASAI, J., and GREEN, D. E. 1968, *Arch. Biochem. and Biophys.*, **126**, 358.
KORN, E. D. 1966, *Science*, **153**, 1491.
LEENE, W., and ITERSON, VAN W. 1965, *J. Cell Biol.*, **27**, 237.
MACFARLANE, N. G. 1962, *Nature*, **196**, 136.
MACFARLANE, N. G. 1964, *Adv. Lipid Research*, **2**, 91.
MCCONNELL, D., TZAGOLOFF, A., MACLENNAN, D. H., and GREEN, D. E. 1966, *J. Biol. Chem.*, **241**, 2373.
MADDY, A. H. 1964, *Biochim. Biophys. Acta*, **88**, 448.
MIURA, T., and MIZUSHIMA, S. 1968, *Biochim. Biophys. Acta*, **150**, 159.
MUNOZ, E., NACHBAR, M. S., SCHOR, M. T., and SALTON, M. R. J. 1968a, *Biochem. Biophys. Res. Commun.*, **32**, 539.
MUNOZ, E., FREER, J. H., ELLAR, D. J., and SALTON, M. R. J. 1968b, *Biochim. Acta.*, **150**, 531.
MUNOZ, E., SALTON, M. R. J., NG, M. H., and SCHOR, M. T. 1969, *European J. Biochem.*, **1**, 490.
MURRAY, R. G. E., and WATSON, S. W. 1965, *J. Bacteriol.*, **89**, 1594.
NEVILLE, D. M. 1967, *Biochim. Biophys. Acta*, **133**, 168.
RACKER, F. 1967, *Federation Proc.*, **26**, 1335.
RAZIN, S., MOROWITZ, H. J., and TERRY, T. M. 1965, *Proc. Natl. Acad. Sci., U.S.*, **54**, 219.
REMSEN, C. C., VALOIS, F. W., and WATSON, S. W. 1967, *J. Bacteriol.*, **94**, 422.
ROBRISH, S., and MARR, A. G. 1962, *J. Bacteriol.*, **83**, 158.
ROGERS, H. J., REAVELEY, D. A., and BURDETT, I. D. J. 1967, *Protides of the Biological Fluids*, **15**, 303.
ROTHFIELD, L., and FINKELSTEIN, A. 1968, *Ann. Rev. Biochem.*, **37**, 463.
ROTTEM, S., and RAZIN, S. 1967, *J. Bacteriol.*, **94**, 359.
RYTER, A. 1968, *Bacteriol. Rev.*, **32**, 39.
SALTON, M. R. J., and HORNE, R. W. 1951, *Biochim. Biophys. Acta*, **7**, 177.
SALTON, M. R. J. 1957, *Proc. Intern. Congr. Surface Activity, 2nd, London*, **4**, 245.
SALTON, M. R. J., and FREER, J. H. 1965, *Biochim. Biophys. Acta*, **107**, 531.
SALTON, M. R. J., and NETSCHEY, A. 1965, *Biochim. Biophys. Acta*, **107**, 539.
SALTON, M. R. J. 1967a, *Ann. Rev. Microbiol.*, **21**, 417.
SALTON, M. R. J. 1967b, *Trans. N. Y. Acad. Sci., Ser. II.*, **29**, 764.
SALTON, M. R. J., and SCHMITT, M. D. 1967a, *Biochem. Biophys. Res. Commun.*, **27**, 529.
SALTON, M. R. J., and SCHMITT, M. D. 1967b, *Biochim. Biophys. Acta*, **135**, 196.
SALTON, M. R. J., SCHMITT, M. D., and TREFTS, P. E. 1967, *Biochem. Biophys. Res. Commun.*, **29**, 728; 1968, *erratum*, **30**, 446.
SALTON, M. R. J. 1968a, in *Comprehensive Biochemistry*, **23**, 127, Elsevier, Amsterdam.
SALTON, M. R. J. 1968b, in *Microbial Protoplasts Spheroplasts and L-Forms* (Ed. L. B. Guze), 144, Williams and Wilkins, Baltimore.
SALTON, M. R. J. 1968c, *J. Gen. Physiol.*, **52**, 227s.
SALTON, M. R. J., FREER, J. H., and ELLAR, D. J. 1968, *Biochem. Biophys. Res. Commun.*, **33**, 909.
SCHNEIDERMAN, L. J. 1965, *Biochem. Biophys. Res. Commun.*, **20**, 763.
SEDAR, A. W., and BURDE, R. M. 1965, *J. Cell Biol.*, **27**, 53.

STILES, J. W., and CRANE, F. L. 1966, *Biochim. Biophys. Acta*, **126**, 179.
TERRY, T. M., ENGELMAN, D. M., and MOROWITZ, H. J. 1967, *Biochim. Biophys. Acta*, **135**, 391.
VOELZ H., and ORTIGOZA, R. O. 1968, *J. Bacteriol*, **96**, 1357.
WALLACH, D. F. H., and ZAHLER, P. H. 1968, *Biochim. Biophys. Acta*, **150**, 186.
WEIBULL, C. 1953, *J. Bacteriol.*, **66**, 688.
WEIBULL, C., GREENAWALT, J., and LOW, H. 1962, *J. Biol. Chem.*, **237**, 847.
WEINBAUM, G., and MARKMAN, R. 1966, *Biochim. Biophys. Acta*, **124**, 207.
ZAHLER, P. H., WALLACH, D. F. H., and LÜSCHER, E. F. 1967, *Protides of the Biological Fluids*, **15**, 69.

DISCUSSION

GREER I wonder if any of your studies can be correlated with studies on mitochondria to explore the evolutionary relationship that has been postulated between bacteria and mitochondria?

SALTON I think there's every opportunity to do that by investigating the function and structures of those parts of the bacterial membrane responsible for the mitochondrial activities. There are indeed similarities between mitochondrial membranes with the 'stalked particles' and the particles associated with the bacterial membranes. Moreover the ATPase is a calcium activated enzyme; it is not activated by Na^+ or K^+ so is not involved in transport of these ions. Investigations of the protein structure (amino acid termini and sequence) of the electron transport components, the cytochromes, and the ATPase should reveal any evolutionary trend from one bacterial group to another. So I think this sort of approach of studying the protein structure in relation to 'phylogeny' could reveal any evolutionary relationships with mitochondria of higher cells.

MYSELS At the very beginning, you mentioned that the enzymatic attack of the outer wall also removes the internal membrane system. Could you provide some mechanism as to what happens to that bacterial membrane system?

SALTON Yes, I think that if you look at the bacterial cell in 'profile' in thin sections, one sees the outer wall, then the plasma membrane and then the ramification of the membrane inside (i.e. the mesosome membranes). When the cell wall is removed with lysozyme in the presence of an osmotic stabilizer, the internal membrane could either uncoil and become part of the protoplast membrane, or if it is in the form of vesicles some of the internal membrane would be everted or 'extruded'.

MYSELS So that would mean that the internal volume increases.

SALTON Yes, that does indeed happen. The diameter of the fixed stabilizer protoplasts is significantly greater than that of the original, intact cells.

KEDEM It looks as if the ATPase is on the membrane. Furthermore, if you remove it, you're left with a smooth sheet; but functionally I suppose it must be accessible at least to the same degree from the inside, and go right through it. Isn't that so?

SALTON The evidence suggests that the ATPase particles are on the inside of the membrane and this of course would make sense from the viewpoint of its function. As to whether parts of the particle penetrate to the outside, we really don't know, but I really think we have a reasonable chance of approaching this by using different systems of dissociating the purified ATPase to see if any of the different subunits or parts of the stalk could be identified with specific antisera. In this way if any of the ATPase structure penetrated to the outside, the components could be identified with ferritin labeled antibodies.

Structure of the mitochondrial cristael membrane

D. E. GREEN and G. VANDERKOOI
(Presented by D. E. Green)
Institute for Enzyme Research, University of Wisconsin, Madison, Wisconsin 53706

Biological membranes fall into one of two categories—membranes with tripartite repeating units, and smooth membranes, the repeating units of which lack projecting sectors (1). Membranes which are concerned with the linking of electron transfer to synthesis of ATP (e.g., the inner membranes of the mitochondrion and the chloroplast), or with active transport (sarcoplasmic reticulum and the microvilli), or with the invasion of host cells by virus (e.g., the influenza and herpes virus), fall within the first category while all others appear to be of the second category (1). We shall restrict our discussion of the structure of biological membranes to that of the inner membrane of the mitochondrion—a membrane which may be considered as representative of the category of membranes with tripartite repeating units.

The Membrane Systems of the Mitochondrion

The mitochondrion is called an organelle because it is a system of membranes rather than a single membrane. Figure 1† is a diagrammatic representation and interpretation of the relationships between the boundary membranes and the cristae. We are defining the two parallel, spheroidal boundary membranes as the outer membranes and the tubular membranes as the inner (cristae) membranes (2). It is to be noted that the cristal tubes are continuous with the inner boundary membrane. The lumen of each crista extends beyond

---

† Figures 1 to 6 taken from GREEN and GOLDBERGER, *Molecular Insights into the Living Process*, Academic Press, New York (1967).

the junction point with the inner boundary membrane and into the space between the boundary membranes.

There is general agreement about the topological relations of the inner boundary membrane and the cristael membranes but not about the question

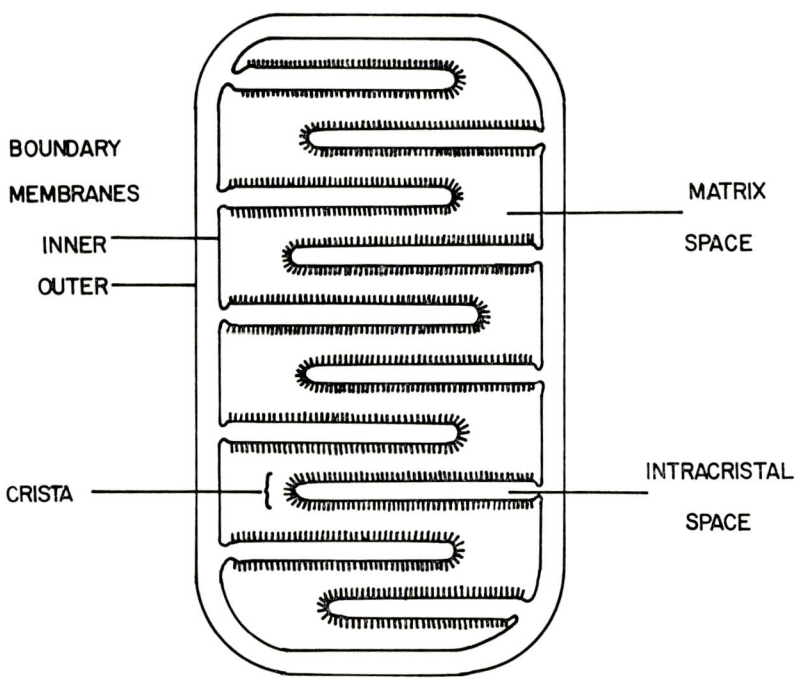

FIGURE 1 The membrane systems of the mitochondrion. The lumen in the cristael membranes is continuous with the lumen that separates the outer and inner boundary membranes. This continuity is achieved by a perforation in the inner boundary membrane wherever the crista is attached. The tripartite repeating units are localized exclusively in the cristael membranes. The enzyme systems associated with the cristael membranes are integral parts of the repeating units whereas the enzyme systems associated with the boundary membranes can be separated from these membranes without loss of membrane structure.

of their chemical identity. Our view, based on electron microscopic and biochemical evidence, is that the cristael membranes have electron transfer activity, whereas the inner boundary membrane does not (2). By virtue of this postulated difference in enzyme function, we treat the inner boundary membrane as a membrane distinct from the cristael membrane even though the two membranes form a continuum. At a certain stage in the biogenesis of yeast mitochondria, the two parallel boundary membranes are formed before cristae are laid down (3, 4). From such evidence we are inclined to make a differentiation between the smooth boundary membranes and the segmented cristael membranes. In our present usage, the term inner mitochondrial membrane is synonymous with the cristael membrane and does not include the inner boundary membrane.

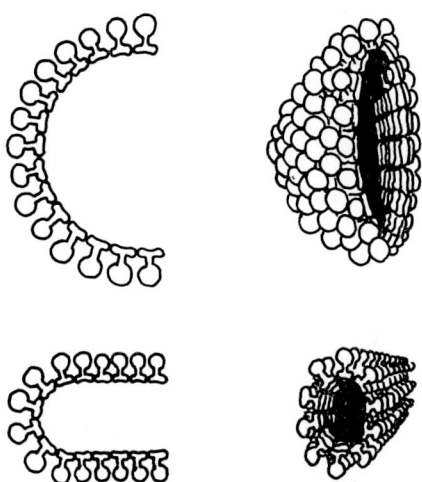

FIGURE 2 Membranes as two dimensional continua of nesting repeating units. Tubular and vesicular membranes are shown in the diagram. The repeating units need not necessarily have a tripartite structure. This is an optional feature. The important notion implicit in this interpretation of membrane structure is that the membrane is only one repeating unit thick and that the building block of membrane structure is the repeating unit. The interior space of the membrane is presumed to be fluid-filled.

The Repeating Units of the Cristael Membrane

The cristael membrane may be considered as a two-dimensional curved continuum of nesting tripartite repeating units (see Figure 2). The membrane is merely the integrated expression of the component repeating units. No structured elements other than the repeating units need be invoked to account for the known properties of the membrane. The lumen of the tubules is usually electron transparent but regular structure is often seen in electron micrographs. The space external to the tubules (the matrix space) is believed by electron microscopists, on the basis of staining properties (5), to contain structured components. Our own view is that the matrix space, at least in beef heart mitochondria, is a fluid-filled space and that the apparent structure derives by fragmentation of the repeating units of the cristael membrane. We shall leave this question open because it is not central to the considerations to which the present communication is addressed.

The repeating unit of the cristael membrane is a composite of three bonded sectors—a spherical headpiece linked to a quasi-cuboidal basepiece via a cylindrical stalk (see Figure 3). The basepiece is the membrane-forming sector of the repeating unit and it can exercise that role even when the headpiece-stalk sectors have been detached by appropriate means (6). The membrane continuum, thus, depends exclusively on the nesting of basepieces. The headpiece-stalk sectors do not nest one with another. They are in fact projections from the membrane-forming basepieces, and are detachable without compromising the principle of the membrane continuum.

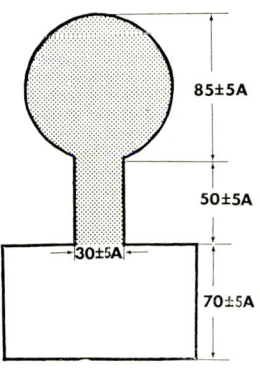

FIGURE 3 The structure and dimensions of the tripartite repeating unit of the mitochondrial cristael membrane. The center to center distance between the headpieces of nesting tripartite repeating units is estimated to be 110-114 Å.

Membrane Formation from Repeating Units

The cristael membrane can be 'depolymerized' into its component repeating units by exposure to bile salts such as cholate in presence of salt (7). Organized membrane structure disappears as the result of such exposure. The solubilized membrane appears in electron micrographs as a suspension of fine particles which we have equated with individual tripartite units stabilized by association with bile salts. When the concentration of bile salts is reduced by dialysis or dilution or by other means, the disaggregated repeating units reassociate to form a reconstituted vesicular membrane (see Figure 4). From this simple experiment an important conclusion may be drawn. A membrane can be self-assembled from its disaggregated repeating units. The information required for the alignment of repeating units that underlies membrane formation is therefore contained within the repeating units.

Structure of the mitochondrial cristael membrane

De novo membrane formation under the conditions specified above also provides clear evidence that the membrane is in fact a composite of nesting repeating units. The conditions for solubilization of the membrane do not

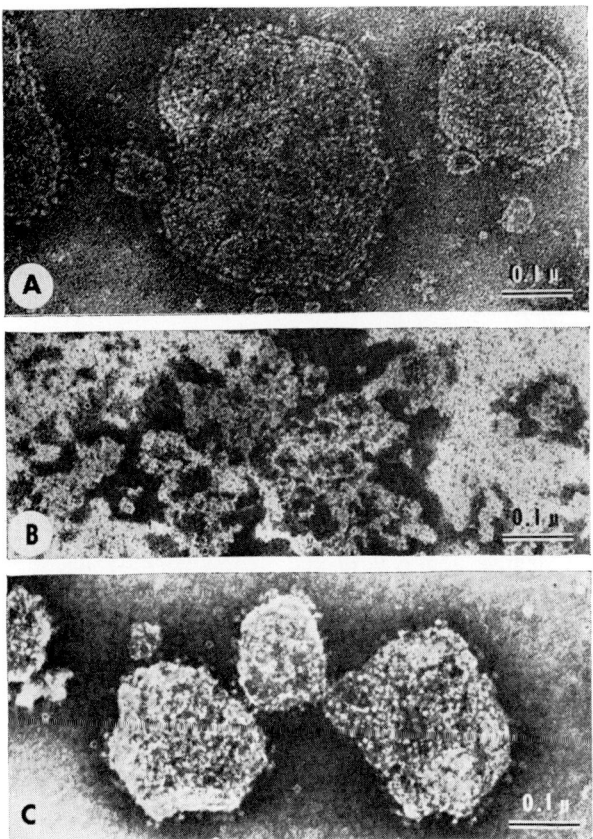

FIGURE 4 Reconstitution of a vesicular membrane from depolymerized repeating units. A: The starting membranous preparation showing tripartite repeating units; B: The same as A after solubilization with bile salts; C: The same as B after removal of bile salts by dilution and sedimentation. Negatively stained with phosphotungstate.

lead to the resolution of the repeating units into their component macromolecules. The forces that hold repeating units together in the continuum are weakened sufficiently to compel disaggregation of the membrane, but the forces that hold together the macromolecules of a repeating unit are largely unaffected.

Tripartite repeating units can be resolved into basepieces and headpiece-stalk sectors (8). Membranes containing such stripped tripartite repeating units can also be depolymerized by bile salts; the basepieces thus monomerized will reaggregate to form a membrane when the bile salt is removed. Thus the headpiece-stalk sectors of tripartite repeating units play no role in membrane formation, but rather are concerned with the mechanism of energy transductions in the cristael membrane.

THE MOLECULAR COMPONENTS OF THE CRISTAEL MEMBRANE

The Proteins

The headpiece is a sphere about 85 Å in diameter (9). Since the headpiece consists entirely of protein, the mass of the headpiece estimated from the volume and the density (1.3 gm per cm^3) would come to a value of about 260,000. The molecular weight of the headpiece obtained by isolation has been determined to be about 284,000 (10). Each headpiece is made up of at least four different species of protein which vary in molecular weight from 25,000–60,000 (10, 11). Thus the headpiece is a complex of some 7 molecules of protein bonded together in a specific configurational pattern (10). The headpiece can be depolymerized at 0° into its component repeating units by appropriate salts in high concentration, e.g., NaBr at 2 M concentration (10, 12). To some degree, removal of the depolymerizing agent will lead to reconstitution of the headpiece both structurally and functionally (12).

There are four chemically distinct basepieces which have been isolated, labeled Complex I, II, III, and IV (13), but all of these have approximately the same size and are the shape of a rectangular block, 114 × 114 × 60 Å (14). The volume of a basepiece is thus considerably larger than that of the headpiece. But whereas the headpiece consists entirely of protein, the basepiece is a composite of both phospholipid and protein. Each basepiece has been found to contain protein in an amount corresponding to 250,000 Daltons—virtually the same amount of protein as is contained in the headpiece (15). The ratio of phospholipid to protein in a basepiece has been estimated to be 1:1 (16). Thus the total mass of a basepiece comes to a value of 500,000, and this mass is consistent with the volume of a parallelogram with the dimensions assigned above and a value for the density intermediate between that of protein and phospholipid. Each basepiece contains at least 5 different species of protein and, thus, like the headpiece, is a multiprotein complex (17). It has been possible to resolve a basepiece into its component protein molecules (17, 18), but reconstitution of the basepiece from the mixture of proteins has not yet been accomplished.

The stalk is a cylindrical unit 50 Å long and 30 Å in diameter (14, 19). The mass of such a unit, given the assigned dimensions and a value of the density for that of protein, is estimated to be about 25,000. A protein corresponding dimensionally to the stalk has been isolated, and on the basis of molecular sieve determinations, the molecular weight has been estimated to be about 18,000 (20). We may conclude, therefore, that the stalk corresponds to a single protein.

Nonprosthetic Proteins

The proteins of the headpieces and basepieces of the tripartite repeating units are of two types: proteins which have a prosthetic group, and therefore fulfill a catalytic or otherwise active functional role, such as the hemoproteins and flavoproteins; and proteins which lack a prosthetic group, and do not have any known catalytic or active functional role (21). These two categories of proteins occur in approximately a 1:1 ratio. The nonprosthetic proteins of the basepieces are known as core proteins (22, 23), while those of the headpieces are called structural proteins. All of the core proteins and structural proteins studied thus far *appear* to have similar amino acid composition, molecular weight, and peptide maps after tryptic digestion (21, 22). The validity or significance of these analyses may be questioned, though, on account of the impure samples used, and the strong aggregating tendencies of these proteins to form water insoluble species. Other features which these proteins have in common are the capability for bonding phospholipid hydrophobically to the extent of about 30 per cent by weight, and the capacity for interacting electrostatically with water soluble catalytic proteins, to form water-soluble adducts. The core proteins are more resistant to digestion by proteolytic enzymes and to solubilization by detergents than are the structural proteins (21)

At present it is not possible to specify the role of the nonprosthetic proteins. They may be truly noncatalytic and serve only in an organizational capacity, as was originally thought when the name 'structural protein' was applied. It is now known, however, that the material originally called structural protein is for the most part no more than disulfide-crosslinked subunits of the headpiece (24). The available evidence does not exclude the possibility that the major electrophoretic band of structural protein in its native form is a nonprosthetic enzyme bearing an active site, the activity of which is lost when the headpiece is depolymerized. There is somewhat more reason to suppose that the core proteins of the basepieces have only an organizational, membrane-structuring, noncatalytic function. The studies of Korman and Vande Zande on Complex IV (25) have shown that it can be clearly split into a core protein sector and a prosthetic-group containing sector, in the weight ratio of about 1:1. Even in this case however, the absence of any detectable catalytic activity

The Phospholipids

The available evidence suggests that each basepiece contains some 300 molecules of phospholipid and that no phospholipid is linked either to headpiece or stalk. The phospholipid is composed essentially of three species—phosphatidyl choline, phosphatidyl ethanolamine, and cardiolipin which are present in the proportions of 41:33:15 per cent of the total phosphorus content (26). Phosphatidyl choline is a strong zwitterion; phosphatidyl ethanolamine is a weaker zwitterion and has weak acidic tendencies; and cardiolipin is a strongly acidic phospholipid. Thus, in the aggregate, mitochondrial phospholipid carries a net negative charge—this largely by virtue of the cardiolipin.

Mitochondrial phospholipids are characterized by a high degree of unsaturation of the fatty acyl residues (there are two such residues in phosphatidyl choline and ethanolamine and four in cardiolipin) (26). The degree of unsaturation of the fatty acyl residues varies from one species of phospholipid to another but on the average there are at least 2–3 such unsaturated bonds per atom of phosphorus.

The fatty chains in the three species of phospholipids are uniformly 16–18 carbon atoms in length. It is the presence of two or four such chains in each molecule of phospholipid that accounts for the insolubility of phospholipid in water.

MEMBRANE STRUCTURE

The cristael membrane may be thought of as a two-dimensional crystal. The basepieces, which are lipoprotein complexes, are the repeating units of the two-dimensional array. The fact that there are at least four chemically distinct species of basepieces does not detract from the gross crystallinity of the structure, since they are all about the same size and shape, and are probably randomly distributed throughout the membrane. An examination of various electron micrographs of negatively-stained specimens of the reconstituted membranes of Complex II and Complex IV indicates that they belong to a square two-dimensional space group. The membrane model to be developed in what follows is quite similar to the model recently proposed by Sjostrand (27). It is illustrated in Figure 5.

It is possible to extract a large portion (80–90 per cent) of the membrane lipid without disrupting the membrane and without making the membrane noticeably thinner (28, 29), although the membrane is weakened when this is done. This shows that the lipid is not essential to the membranous structure, and consequently, protein–protein interactions are the primary bonds which hold the membrane together. If a lipid-extracted membrane is physically disrupted, an amorphous state results, containing three-dimensional clusters, whereas if a membrane containing its normal lipid content is disrupted, vesicles form and the membranous nature of the material is not lost (7). This shows that the lipid is important in making the two-dimensional array of protein molecules more stable than a three-dimensional aggregate.

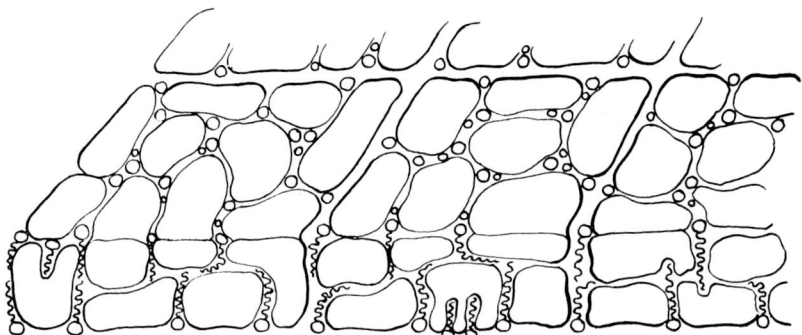

FIGURE 5 A proposed model of channels within the basepieces of repeating units into which phospholipid molecules can fit. The fatty residues of the phospholipid molecules (indicated by zigzag lines) fit into the channels whereas the polar heads (indicated by circles) are external to the channels. The channels are consequences of the imperfect fitting of the set of proteins which form each of the basepieces.

Localization of the Lipid

The protein and phospholipid in the cristael membrane exist in about a 1:1 ratio after the headpieces (which are lipid-free) have been removed. A portion of the lipid apparently sloughs off easily, since the membranes which form from isolated basepieces contain only about 20–25 per cent lipid (6). If the membrane is in fact composed primarily of protein molecules interacting with each other, where is there room for this much lipid? Crystals of water soluble proteins contain large amounts of liquid of crystallization (mostly water) between the macromolecules, with the macromolecules touching each other at relatively few points. Thus, in crystalline ferricytochrome c, 55 per cent of the crystal by volume is liquid (30) in lysozyme crystals, 33.5 per cent by weight of the crystal is liquid (31). Likewise, the oxyhemoglobin crystal contains a considerable amount of space between the hemoglobin complexes which is filled with liquid, but in this case there are also two cavities penetrating into the four-subunit hemoglobin complex, each of these being about 20 Å long, 8–10 Å wide, and 25 Å deep (32). These cavities are lined with polar amino acids and are filled with liquid. There is, therefore, considerable basis for the suggestion that membrane proteins, while touching each other at some points, still can have a considerable amount of space between them, just as the protein crystals do. In the protein crystals, polar amino acid residues line the interprotein spaces, which favors these spaces being filled with water. If the interprotein spaces in a membrane are lined with nonpolar amino acids, the filling of these spaces with the aliphatic tails of phospholipids would be favored over filling them with water. The full amount of lipid could be taken up in this way if the average spacing of proteins in membranes is similar to that in crystals. Penetration of the aliphatic chains into the interprotein spaces should be possible, just as it is possible for small molecules to diffuse into a protein crystal, since the chains are flexible, and a CH_2 group is about the same size as a water molecule.

Several additional observations can be made concerning this model. The thickness of the cristael membrane is about 60 Å. If it is composed of globular proteins of an average molecular weight of 25,000, it would be between one and two protein molecules thick. An aliphatic chain of 18 carbon atoms is roughly 25 Å long. This means that the aliphatic tails of phospholipids, which normally contain 16–18 carbon atoms, could penetrate nearly to the middle of the membrane, while the polar head remains on the surface. All crevices could therefore be filled if the aliphatic chains penetrate the membrane from both sides.

The phospholipid polar heads on the surface of the membrane will give it water solubility, or at least an affinity for water. This feature is closely related with the observation that phospholipids stabilize the two-dimensional nature of a membrane (7), since two polar membrane surfaces are not likely to associate with each other in an aqueous system, whereas two nonpolar surfaces will almost certainly associate. The nonpolar phospholipid tails which penetrate between the proteins, on the other hand, cause the membrane to mimic a lipid bilayer in transport properties.

There are various pieces of evidence that strongly favor the idea that the phospholipid heads are on the surface, and the tails in the interior, in addition to those already presented (33, 34). Phospholipase C (phosphatidyl choline choline-phosphohydrolase) hydrolyzes away the phosphoryl choline groups of lecithin without disrupting the membrane, and apparently without releasing the remaining diglyceride from the membrane. This would indicate that the phosphoryl choline group is on the surface. The aliphatic tails of the phospholipids have a considerable degree of unsaturation, but little peroxidation of ethylenic linkages takes place in the intact membrane. Since these bonds are quite unstable toward peroxidation when the lipids are isolated, it indicates that these linkages are buried in a nonpolar region in the intact membrane.

The theory of membrane formation from repeating units (15) is based on the postulate that the polar heads of the phospholipid molecules are localized on only two of the six faces of the quasi-cuboidal basepiece (these two faces being parallel). If the channels between and in the repeating units have consistent directionality, then the postulate of the asymmetric localization of phospholipid is fully satisfied.

Aqueous Acetone Extraction

It has been pointed out that, since 10 per cent aqueous acetone will only extract about 80 per cent of the membrane lipid (26), there apparently are two types of lipid present, weakly bound and strongly bound. A large proportion (60 per cent) of the lipid which remains after the aqueous acetone extraction is cardiolipin (35), which before extraction accounts for only 16 per cent of the total phospholipid (26). If, on the other hand, the membrane is extracted with chloroform-methanol mixtures, the cardiolipin is quantitatively extracted. This shows that the tightness of the binding of the cardiolipin is not due to its acidic nature, since if that were the case, it should not be extracted so easily by chloroform-methanol. Rather, it is bound more strongly by hydrophobic bonds and nonbonded (London-van der Waals) forces than are the other lipids. Aqueous acetone would not be expected to disrupt these interactions as well as chloroform-methanol would. The cause for the stronger bonding of cardiolipin, as compared to other phospholipids, is immediately evident from its structure. A molecule of cardiolipin has four aliphatic chains, as compared to the two aliphatic chains of lecithin or phosphatidyl-ethanolamine. This means that the bonding energy of the cardiolipin molecule will be twice as great as that of the other lipids. It is quite reasonable to suppose that this additional energy is adequate to keep the cardiolipin bound to the protein in the presence of a marginal lipid solvent such as aqueous acetone.

Basepiece Bonding

The proteins and lipids within the individual basepieces are very firmly bound to each other, but the bonds holding the basepieces together are not as strong, permitting the separation and isolation of the basepieces as intact units. Both polar and nonpolar interactions are involved in the forces which hold the basepieces together. This is evident from the observation that bile acids and salt must both be used to disaggregate the membrane; neither bile acid nor salt by itself is effective. Prolonged sonication under alkaline conditions (pH 8) in the complete absence of salt can also disaggregate the inner membrane (36). In this case sonication leads to the rupture of the membrane, but reconstitution by recoalescing of disaggregated repeating units is interfered with by the alkaline medium. The proteins of the disaggregated repeating units at pH 8 are more negatively charged than at neutral pH, and this change in charge leads to enhancement of the electrostatic repulsion between particles. The barrier to interaction between disaggregated repeating units is, thus, intensified by raising the pH of the medium.

Since the basepieces are all about the same size, no special arrangement of them is dictated. Apparently any basepiece can bond equally well to any other basepiece. This has been demonstrated through the formation of membranes by the various isolated complexes, taken singly or mixed with other complexes (37).

Bonding of the Detachable Sector

The stalk is bonded hydrophobically to the basepiece and both electrostatically and hydrophobically to the headpiece. These conclusions are deduced from the following considerations. The isolated headpiece is a water soluble species that does not aggregate (9), whereas the isolated headpiece-stalk tends readily to aggregate (8). Moreover the isolated stalk readily polymerizes to form water-insoluble aggregates (19). These observations suggest that the stalk has a hydrophobic character, while the headpiece resembles a normal soluble protein with little tendency to polymerize.

Reagents like 2 M NaBr which lead to the depolymerization of the headpiece also rupture the link between headpiece and stalk (12). It is difficult to decide from this fact alone whether salt weakens the link between stalk and headpiece directly, or indirectly by inducing the depolymerization of the headpiece.

Detergents are needed to rupture the link of the stalk to the basepiece and this would suggest a strong hydrophobic component in this interaction. Moreover, headpiece-stalk sectors can interact with phospholipid bilayers in the doublet or hydrophobic form, i.e. the form of the bilayer in which the fatty acyl residues are directed outward (38). Electron microscopy shows that the headpiece-stalk sectors are oriented, after interaction with bilayers in the doublet form, in exactly the same fashion as they are oriented on the membrane (38).

FUNCTIONS OF THE CRISTAEL MEMBRANE

As a preliminary to a discussion of the localization of functions in the three sectors of the tripartite repeating units, we shall have to digress to a listing of the various functions. The inner membrane contains basically two systems—the electron transfer chain and the ATPase complex (39). The electron transfer chain catalyzes the oxidation of succinate or of DPNH by molecular oxygen while the ATPase complex catalyzes the hydrolysis of ATP to ADP and inorganic phosphate. In the functional cristael membrane, the transfer of electrons in the chain is coupled to the union of ADP and P_i to form ATP, and conversely, the hydrolysis of ATP is coupled to the reversal of electron transfer.

The Electron Transfer Chain

The electron transfer chain is a composite of four complexes which collectively catalyze the oxidation of succinate and of DPNH by molecular oxygen (13). These four complexes of the chain are multiprotein units which have been localized in the basepieces of the tripartite repeating units (15) whereas the ATPase complex, also a multiprotein unit, has been identified with the headpiece of the tripartite repeating unit (9).

The sequential transfer of electrons via the four complexes is diagrammatically represented in Figure 6. Complexes I, III and IV catalyze sequentially the oxidation of DPNH by molecular oxygen while Complexes II, III and IV catalyze sequentially the oxidation of succinate by molecular oxygen. The links between complexes are small molecules. Coenzyme Q accepts electrons from Complex I or II in their reduced form and reduced coenzyme Q transfers electrons to Complex III in its oxidized form (40). Ferricytochrome c accepts electrons from Complex III and ferrocytochrome c transfers electrons to Complex IV. Coenzyme Q and cytochrome c are, therefore, viewed as mobile carriers, not associated with a single basepiece, but free to move from one basepiece to another.

Coenzyme Q is a benzenoid derivative with a side chain of 50 carbon atoms. A molecule such as this certainly cannot come out into the aqueous phase, but must stay within the nonpolar part of the membrane. Its ability to move any great distance would therefore be quite limited, but it should be able to act as an electron carrier between neighboring basepieces.

Cytochrome c (a protein of 12,500 in molecular weight) is known to bind electrostatically to phospholipids (41, 42). The addition of salt ruptures this cytochrome c-lipid complex. It is also well known that cytochrome c is relatively easily extracted from mitochondria by the addition of salt, and that, once extracted, it is easily soluble in aqueous media. It is therefore plausible to assume that cytochrome c binds electrostatically to the charged phospholipid heads, which line the surface of the cristael membranes. It could then 'roll' or move about the surface of the membrane quite easily, fulfilling its function as electron carrier. The random Brownian motion of the cytochrome c molecules is adequate to account for movements of the distances which would be required by this model, which are of the order of 100 Å.

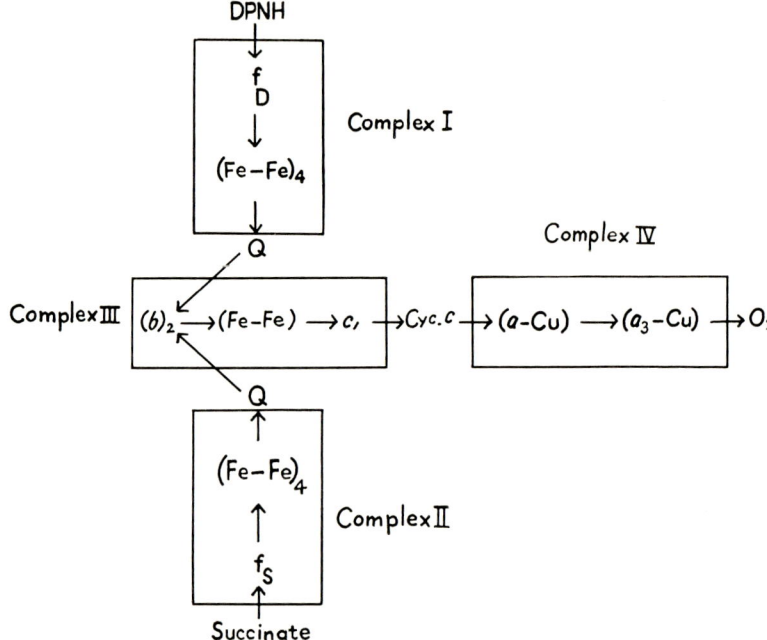

FIGURE 6 The four complexes of the electron transfer chain. Distribution of the oxidation-reduction proteins among the complexes of the electron transfer chain. Fe–Fe represents nonheme iron; f_S, the succinic deyhdrogenase; and f_D, the DPNH dehydrogenase. Coenzyme Q (Q) is the mobile link between Complexes I and III and between Complexes II and III. Cytochrome c (Cyt. c) is the mobile link between Complexes III and IV.

The electron transfer process may be conceived of as a stepwise partitioning of the free energy differential between DPNH and molecular oxygen. Each complex has a set of oxidation-reduction components suitable for conserving the energy yield in a limited segment of the total electron transfer process. Thus, each of the complexes, with its attached headpiece-stalk sectors is a complete operational unit for coupling electron flow in one of the three component segments of the electron transfer chain to the synthesis of ATP (40).

Since there are multiple complexes of the electron transfer chain there must be multiple species of basepieces. Indeed there are at least four species of basepieces, one for each of the four complexes. However all species of basepieces have the same detachable sectors. That is to say the headpiece and stalk sector are identical in all repeating units regardless of the nature of the basepiece (8).

The ATPase Complex

The hydrolysis of ATP catalyzed by the isolated headpiece is a degenerate form of activity. This reaction is completely insensitive to the action of rutamycin—a reagent which completely suppresses hydrolysis of ATP in the cristael membrane. When, however, the hydrolysis of ATP is catalyzed by the headpiece attached to the stalk, the hydrolysis is completely sensitive to the inhibitory action of rutamycin (8). The stalk is thus the determinant of the rutamycin sensitivity of the ATPase reaction. Clearly the link of the stalk to the headpiece has a modulating effect on the ATPase function. But even that link is insufficient to achieve conservation of energy coupled to ATP hydrolysis. Conservation of energy requires the attachment of the headpiece-stalk unit to the basepiece. Only in the intact tripartite repeating unit can electron transfer be coupled to synthesis of ATP or ATP hydrolysis be coupled to the reversal of the electron transfer process.

We have in the tripartite repeating units two coupled systems—the electron transfer chain in the basepiece and the ATPase complex in the headpiece. The stalk is the physical link between these coupled systems, and as we shall see later, the energized state can be generated either in headpiece or in basepiece and presumably transmitted to the other sector via the stalk.

The Work Performances of the Cristael Membrane

The electron transfer process can energize the coupled synthesis of ATP, the coupled translocation of divalent cations and inorganic phosphate, the coupled translocation of monovalent cations and anions, or the coupled transfer of a hydride ion from DPNH to TPN^+. The last three work performances can also be energized by hydrolysis of ATP in complete absence of electron transfer. All of these work performances are aspects of the coupling action of the tripartite repeating units. There is no necessity to invoke elements in the membrane additional to the tripartite repeating units. The transhydrogenation between DPNH and TPN^+ is catalyzed by a complex (43 other than the complexes of the electron transfer chain and this transhydrogenating complex has been tentatively identified as a new basepiece of the tripartite repeating units.

Reconstitution of Tripartite Repeating Units from Sectors

The individual sectors of tripartite repeating units have all been isolated and this has made it possible to reconstitute the tripartite repeating units from the component sectors. Reconstitution of the tripartite repeating unit goes parallel with membrane formation since the tripartite repeating units once formed immediately form membranes.

The following reconstitutions have been achieved: 1, basepieces + headpiece-stalks; and 2, basepiece-stalks + headpieces (8). In both cases membranes with complete tripartite repeating units have been reconstituted. Reconstitution of the tripartite repeating unit from isolated basepieces, headpieces and stalks has yet to be accomplished, but it is only very recently that the stalk sectors have been isolated and characterized.

SUMMARY

The tripartite repeating unit is the molecular instrument of oxidative phosphorylation. Its presence in the inner membrane of the mitochondrion and the chloroplast is dictated by the molecular necessities of the coupling process. A similar kind of repeating unit, though dimensionally smaller, has been found in membranes that carry out active transport and it may be presumed that this microtripartite repeating unit stands in the same relation to active transport as does the macrotripartite repeating unit to oxidative phosphorylation.

Membranes with special transducing needs have specialized tripartite repeating units, and the structure and properties of such membranes reflect the special characteristics of tripartite repeating units. While smooth membranes are also composites of nesting repeating structures, protein in nature, the segmentation of the membrane is less pronounced in the operational sense that it is more difficult to resolve the membrane into the individual repeating structures. As yet no one of the smooth membranes has been examined in such depth as the inner membrane of the mitochondrion.

References

1. GREEN, D. E., ASAI, J., ASBELL, M. A., HARRIS, R. A., and JOLLY, W. in preparation.
2. ALLMANN, D. W., BACHMANN, E., ORME-JOHNSON, N., TAN, W. C., and GREEN, D. E. 1968, *Arch. Biochem. Biophys.*, **122**, 981.
3. LINNANE, A. W., HUANG, M., BIGGS, D. R., and WALKER, G. D. 1965, in 'Aspects of Yeast Metabolism' (R. K. Mills, ed.). Blackwell Science Press, Oxford.
4. LENAZ, G., private communication.
5. HACKENBROCK, C. R. 1968, *Proc. Nat. Acad. Sci. U.S.*, **61**, 598.

6. KOPACZYK, K., ASAI, J., and GREEN, D. E. 1968, *Arch. Biochem. Biophys.*, **126** 358.
7. GREEN, D. E., ALLMANN, D. W., BACHMANN, E., BAUM, H., KOPACZYK, K., KORMAN, E. F., LIPTON, S., MACLENNAN, D., MCCONNELL, D. G., PERDUE, J. F., RIESKE, J. S., and TZAGOLOFF, A., 1967, *Arch. Biochem. Biophys.*, **119**, 312.
8. KOPACZYK, K., ASAI, J., ALLMANN, D. W., ODA, T., and GREEN, D. E. 1968, *Arch. Biochem. Biophys.*, **123**, 602.
9. RACKER, E., 1967, *Federation Proc.*, **26**, 1335.
10. PENEFSKY, H. S., and WARNER, R. C. 1965, *J. Biol. Chem.*, **240**, 4694
11. BLAIR, J. E., LENAZ, G., and HAARD, N. F. 1968, *Arch. Biochem. Biophys.*, **126**, 753.
12. MACLENNAN, D. H., ASAI, J., TZAGOLOFF, A., ODA, T., and VANDERKOOI, G., unpublished observations.
13. HATEFI, Y. 1963, *Adv. Enzymol.* (F. F. Nord, ed.), **25**, 275.
14. FERNANDEZ-MORAN, H., ODA, T., BLAIR, P. V., and GREEN, D. E. 1964, *J. Cell. Biol.*, **22**, 63.
15. GREEN, D. E., and TZAGOLOFF, A. 1966, *Arch. Biochem. Biophys.*, **116**, 293.
16. PENNISTON, J. T., and GREEN, D. E., unpublished observations.
17. BAUM, H., SILMAN, H. I., RIESKE, J. S., and LIPTON, S. H. 1967, *J. Biol. Chem.*, **242**, 4876.
18. RIESKE, J. S., BAUM, H., STONER, C. D., and LIPTON, S. H. 1967, *J. Biol. Chem.*, **242**, 4854.
19. MACLENNAN, D. H., and ASAI, J. 1968, *Biochem. Biophys. Res. Communs.*, **33**, 441.
20. MACLENNAN, D. H., and TZAGOLOFF, A. 1968, *Biochemistry*, **7**, 1603.
21. GREEN, D. E., HAARD, N. F., LENAZ, G., and SILMAN, H. I. 1968, *Proc. Nat. Acad. Sci. U.S.*, **60**, 277.
22. LENAZ, G., HAARD, N. F., LAUWERS, A., ALLMANN, D. W., and GREEN, D. E. 1968, *Arch. Biochem. Biophys.*, **126**, 746.
23. SILMAN, H. I., RIESKE, J. S., LIPTON, S. H., and BAUM, H. 1967, *J. Biol. Chem.*, **242**, 4867.
24. SENIOR, A. E., and VANDERKOOI, G., unpublished observations.
25. KORMAN, E. F., and ZANDE, VANDE H., in preparation.
26. FLEISCHER, S., KLOUWEN, H., and BRIERLEY, G. 1961, *J. Biol. Chem.*, **236**, 2936.
27. SJÖSTRAND, F. S. 1968, in 'Regulatory Functions of Biological Membranes' (J. Jarnefelt, ed.,), Elsevier Pub. Co., Amsterdam, p. 1.
28. FLEISCHER, S., BRIERLEY, G., KLOUWEN, H., and SLAUTTERBACK, D. B. 1962, *J. Biol. Chem.*, **237**, 3264.
29. CRANE, F. L., STILES, J. W., PREZBINDOWSKI, K. S., RUZICKA, F. J., and SUN, F. F. 1968, in 'Regulatory Functions of Biological Membranes' (J. Jarnefelt, ed.), Elsevier Pub. Co., Amsterdam p. 21.
30. DICKERSON, R. E., KOPKA, M. L., WEINZIERL, J., VARNUM, J., and BORDERS, C. L. 1966, in 'Hemes and Hemoproteins' (B. Chance, R. W. Estabrook, and T. Yonetoni, eds.), Academic Press, New York, p. 365.
31. STEINRAUF, L. K. 1959, *Acta Cryst.*, **12**, 77.
32. PERUTZ, M. F., MUIRHEAD, H., COX, J. M., and GOAMAN, L. C. G. 1968, *Nature*, **219**, 131.
33. SENIOR, A. E., HAARD, N. F., VAIL, W. J., and GREEN, D. E., in preparation.
34. LENARD, J., and SINGER, S. J. 1968, *Science*, **159**, 738.

35. Tzagoloff, A., and MacLennan, D. H. 1965, *Biochim. Biophys. Acta*, **99**, 476.
36. Tzagoloff, A., McConnell, D. G., and MacLennan, D. H. 1968, *J. Biol. Chem.*, **243**. 4117.
37. Tzagoloff, A., MacLennan, D. H., McConnell, D. G., and Green, D. E. 1967, *J. Biol. Chem.*, **242**, 2051.
38. Korman, E. F., and Green, D. E., in preparation.
39. Green, D. E., Allmann, D. W., Harris, R. A., and Tan, W. C. 1968, *Biochem. Biophys. Res. Communs.*, **31**, 368.
40. Green, D. E., and Goldberger, R. F. 1967, *Molecular Insights into the Living Process*, Academic Press, New York.
41. Green, D. E., and Fleischer, S. 1963, *Biochim. Biophys. Acta*, **70**, 554.
42. Das, M. L., Hiratsuka, H., Machinist, J. M., and Crane, F. L. 1962, *Biochim. Biophys. Acta*, **60**, 433.
43. MacLennan, D. H., unpublished studies.

DISCUSSION

KURSUNOGLU   Electrostatic coupling or bonding—what do you mean? Are these charges 'sitting' there or do you mean current-current interaction?

GREEN   No, I simply mean the Coulomb interaction of oppositely charged molecular species.

KURSUNOGLU   They are sitting there?

GREEN   Yes, they're sitting there. A negatively and positively charged structure will be held together by just that kind of a bonding.

KURSUNOGLU   Well, what brings about the stability of such a structure?

GREEN   Many biological macromolecular systems are held together by the apposition of opposite charges. Of course, the stability of the system would be greatly reduced in a high salt medium but under physiological conditions this hazard is not very great.

KURSUNOGLU   They have some kind of crystal structure?

GREEN   Yes, if you like.

CHANCE   Dr. Green, your lucid presentation did not indicate in detail the nature of the phospholipid cytochrome c interaction. I have taken the X-ray crystallographic structure of cytochrome c as found by Dickerson as the basis for my model for lipid cytochrome c interaction where the lipid may actually penetrate the hydrophobic spaces of the protein clearly identified in the X-ray structure. I do not believe that interaction of the hydrophilic surfaces of the membrane proteins with the hydrophobic lipids is an adequate explanation for the membrane structure.

GREEN   In reply to a question previously raised, we have the impression now that the number of cavities per repeating unit would depend upon the number of proteins per repeating unit. The greater the number of proteins that are fitted together within a repeating unit, the more cavities will be found. It is not only the number of cavities but also the size of the cavities that will determine the amount of phospholipid that can be accomodated. Calculation shows that each basepiece of the cristal membrane contains about 300 molecules of phospholipid. We will assume that 150 molecules of phospholipid are oriented on each of the two faces of the basepiece that are parallel with the axis of the membrane. Since there are some 8 molecules of protein within a given basepiece, the number of potential cavities would be some multiple of 10. Each cavity would have to accommodate no more than 15 molecules of phospholipid if the multiple is one, and no more than 7–8 molecules of phospholipid if the multiple is 2.

In reply to the second question, Britton, I think you are confused. Cytochrome c is a highly polar molecule which interacts in a polar manner with the polar heads of the phospholipid molecules. The ease of extraction of cytochrome c with 0.15 M KCl from mitochondria proves this point. Cytochromes a, a_3, b, c_1 and the other components of the basepiece complexes have nonpolar regions on their surfaces which interact hydrophobically with the nonpolar parts of the phospholipid molecules. You are well aware that these cytochromes are not extracted in the same manner as cytochrome c, and, therefore, they must be bonded to the lipid in a different manner than cytochrome c.

TASAKI   In the squid axon membrane, the two surfaces are biochemically very different. For example, the external surface is totally insensitive and the inner surface is very sensitive to the action of various proteolytic enzymes. Does your membrane have this kind of property?

GREEN   Now how do you introduce the protease into the inner surface?

TASAKI   By internal perfusion. With 1 mg trypsin dissolved in 1 ml of internal perfusion fluid, for example, axons lose their excitability within 5 minutes; with the same concentration of trypsin dissolved in external fluid medium, axons remain excitable for more than 10 hours.

GREEN   Clearly the two surfaces of the membrane you are working with have different susceptibilities to the action of proteases. Well, let me speak to that point. In the case of the mitochondrial cristael membrane there is no doubt about the asymmetry of the two faces of the basepieces which are parallel with the axis of the membrane. The face to which the headpiece-stalk is attached is obviously different from the opposite face which lacks this projection. But, you are talking about a membrane the repeating units of which lack tripartite structure. I believe I am correct in saying that the membrane of the nerve axon is devoid of projecting sectors. But, this is not to say that the parallel faces in the repeating units of smooth membranes have identical properties. The boundary membranes of the mitochondrion are also smooth membranes but there are profound differences in the properties of the outside and inside faces of the repeating units. Certain enzymes are intimately associated with the inner face but not the outer face. I am inclined to think that the repeating units of all biological membranes have asymmetric character and it is this asymmetry which would account for the different action of protease on the outer and inner surface of the membrane of the nerve axon.

PODALL   What is your evidence for each of the two components?

GREEN   Do you mean the sectors of the tripartite repeating units?

PODALL   Yes. Do you have electron microscopy or any other experimental evidence for each piece? Have you looked at each piece?

GREEN   Each of the three sectors of the tripartite repeating unit has been isolated and characterized chemically—the headpiece by Racker and his group, as well as by MacLennan and Kopaczyk in my laboratory, the stalk by MacLennan, and the basepieces by Hatefi, Rieske, and others in my laboratory. These sectors have been defined biochemically as well as electron microscopically. The dimensions of the isolated sectors correspond to those of the sectors in the native repeating unit. Does that answer your question?

PODALL   More or less.

GREEN   Well, let's pursue the point. The F_1 of Racker has been isolated and shown to have ATPase activity. It has a molecular weight of about 300,000 and is a spherical unit with a diameter of 85 Å—exactly that of the headpiece. This unit can be reintroduced into a membrane from which the headpieces of the repeating units have been stripped away and the reconstituted tripartite unit now shows ATPase activity, which is sensitive to oligomycin—a sensitivity lacking in the original F_1 preparation. Not all the catalytic functions of the tripartite repeating unit can be reconstituted since some are abolished the by exposure of the membrane to the reagents (e.g., bile salts) required to resolve the repeating units.

PODALL   In other words you can reconstitute a mitochondrial membrane.

GREEN   In part, yes. Certain functionalities intrinsic to the membrane can be reconstituted by the reconstitution of the tripartite repeating units from the component sectors. But not all functionalities. We have not reached the stage of complete reconstitution. Until we solve the problem of reversing the structural damage incurred by exposing membranes to detergents, the way to complete reconstitution of enzymic function will still be barred.

WALLACH   Concerning lipid-protein interaction, in many membranes there are at least two classes of phosphotides—those which are very tightly associated with the membrane and whose composition is genetically determined and those which are very loosely associated can exchange rather rapidly with external phosphotides and lipoproteins and show no species specificity. So one can think of protein bound class and one which is perhaps secondarily associated with lipid that is primarily bound to protein.

GREEN   I quite agree that there is evidence of two categories of phospholipid-protein binding—tight binding and loose binding. You are clearly interested in knowing whether the model we have proposed could account for these two types of binding. Depending upon the dimensions and geometry of the cavity as well as on the nature of the amino acid side chains, which line the cavity, the cavity could be one in which the phospholipid molecules can fit snugly or one in which these would fit loosely. There is the additional consideration that there are differences among phospholipids in respect to the number of fatty acid residues per molecule. This number would be a major factor in determining stability of hydrophobic binding of phospholipid to protein. Thus, cardiolipin with four residues per molecule would be bound far more tightly than lecithin or phosphatidyl ethanolamine which only contain two fatty acid residues per molecule. These chemical differences among phospholipid molecules and among cavities may in large measure account for the tight and loose binding of phospholipid to protein which you have pointed out.

In our laboratory Fleischer and his colleagues have demonstrated that the ultrastructure of the cristael membranes of the mitochondrion after essentially complete extraction of phospholipid is indistinguishable from that of the unextracted membranes. We have found the same to be true of the red blood cell ghost membrane. Taken at face value these observations suggest that the geometry of the repeating units which make up the membrane is unaffected by the removal of phospholipid.

JUNGE   I want to ask a question concerning the hierarchy of causality in your model since I believe that that is discriminating between your hypothesis and the hypotheses that have been put forward by others. As I see it, you give the following time sequence of events: electron transport, conformational change, ATP reaction, and the ion transport phenomenon on the side pass. How is the experimental evidence to exclude the following time cause: electron transport, ion transport phenomena which may be coupled with conformational change and the phospholation reaction.

CHANCE   Dr. Junge, please let me answer a portion of the question and then turn the remainder over to Dr. Green. The chief point of verbal presentation to this meeting was to identify a portion of the important reaction sequences that you mention, and although the complete report on our work is appearing shortly (1), I would like to produce here the one figure that calls attention to the essence of our result. By employing two probes of membrane function, one, the well-known pH indicator bromthymol blue which is mainly recognized for its high affinity to hydrophobic spaces of proteins (2) and the fluorochrome ANS which indicates by the position of its emission maximum chiefly the polarity of its environment (3, 4), we are able to note wtih great accuracy the time relations of the increased acidity of fragments of mitochondrial membranes and an increase of polarity of the membrane that accompanies energization by electron flow. This result is indicated in the figure where absorbancy decreases of bromthymol blue (increasing acidity) are recorded simultaneously with fluorescence increases of ANS. The addition of NADH prior to supplement of oligomycin indicates only small and abortive responses, since these membrane fragments (Racker's ASU particles) require a supplement of oligomycin in order to be effective in energy coupling. Thus, following a supplement of 2 μg per ml of oligomycin NADH causes practically simultaneous response of the two probes suggesting in this case that the phenomenon of hydrogen ion binding by the membrane is nearly simultaneous with the state of the membrane related to the energy coupling reaction. While

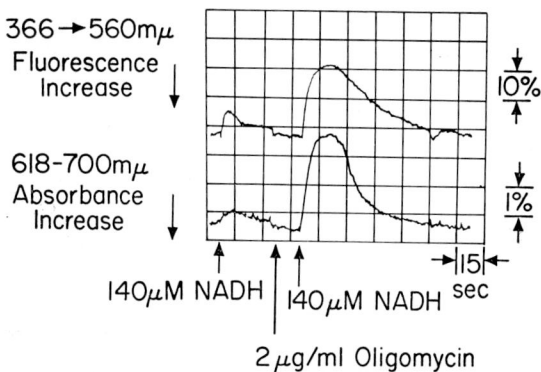

FIGURE 1 The energy dependence of the ANS response. 1 mg ASU protein per ml in 0.3M mannitol-sucrose, 20 mM Tris-HCl, pH 7.4, supplemented with 10 μM BTB and 120 μM ANS.

we can not now distinguish the kinetics of formation of the conformational state of the membrane as indicated by the fluorescence probe from the binding of hydrogen ions to the membrane as indicated by a pH probe, both these processes are considerably slower than the electron transport processes which are activated by NADH addition (5). Interestingly enough, other experiments (1) show that both the bromthymol blue response and the ANS response reach their steady state prior to the function of the membrane in energy requiring reactions, or as Dr. Green calls them, 'work performances', such as the reduction of NAD at a low potential by the higher potential electron donor, succinate. Thus, we find the sequence

$$\text{electron flow} \rightarrow \text{'membrane state'} \rightarrow \text{'work performance'}$$
$$\text{'H}^+\text{ binding'}$$

In one case, sonicated thalicoid membranes, McCarty (6) believes he has established a priority of energy conservation over hydrogen ion movement.

Now, Dr. Green would you like to indicate the relationship of the conformation changes that you observed to the sequence suggested by Dr. Junge?

GREEN   I believe one must make a distinction between small movements of ions which are intrinsic to the conformational change, and massive movements of ions such as take place during translocation of calcium phosphate The massive movements are what we may call work performance, i.e. expressions of cyclical deenergizing of the membrane leading to translocation of ions. With respect to these massive ion movements there is no question which comes first. The membrane can first be energized and then later discharged by addition of divalent ions such as calcium. This stepwise sequence of charge and discharge demonstrates unambiguously that the energizing of the membrane by conformational change precedes the discharge of the membrane by calcium ions. However, the small changes in ion concentration and in pH which accompany the energizing of the membrane are another matter. These are all, we think on the basis of the available evidence, to be expressions of the conformational change in the repeating units during the energizing process. The small pH and ion changes take place simultaneously with the conformational change.

SCHULTZ   In the 19th century, microscopic anatomists had themselves a field day interpreting what they saw in their optical microscopes, and it took 60 years of developing quantitative biochemistry to the point where they could explain some of the imaginative findings. Now, today we see that Dr. Green is a mitochondrial anatomist. What I am worried about is perhaps unjustified, but I would like to ask Dr. Green whether he has such great confidence that the outcome of some of his experiments with the electron microscope might not be as a result of the methods used in preparing the material for the electron microscope?

GREEN   Julius Schultz like many other biochemists who have not used electron microscopy as a tool are fearful of the pitfalls attending the use of an unfamiliar tool. The artifacts he speaks of are easily recognized by the time honored method of controls and I can assure him that we and others have satisfied ourselves by virtue of many controls that the configurational changes we see by electron microscopy are real and independent of the nature either of the fixative or the stain. Moreover, we have used other physical methods—measurement of light scattering changes and changes in pH—and found complete agreement with respect to the fact of configurational change by these various kinds of physical measurements. It needs pointing out that the electron microscope allows the investigator to see directly molecular patterns.

The other physical tools make it possible to deduce molecular processes but not to see them. It is primarily the electron microscope that has made possible the rapid progress of the past two years in recognizing the conformational basis of energy transduction in membrane systems.

MYSELS If I understood your microscope slides of the energizer not in a dry state, there is a great difference in the distribution of volumes between the outside of the membranes and within the membranes.

GREEN What you say is true, but note that there is no change in the volume of the mitochondrion during the energizing process.

MYSELS No. But during the distribution water has been transported.

GREEN Isolated mitochondria (suspended in 0.25 M sucrose) undergo the following configurational cycle: nonenergized (aggregated)→energized→ energized-twisted→non-energized. Mitochondria *in situ* undergo an identical configurational cycle except that the nonenergized configuration is orthodox rather than aggregated. These configurational transitions can take place without any increase in the volume of the mitochondrion. But intrinsic to the cycle, there is a redistribution of water between the space within the crista (intracristal space) and the space outside the crista (matrix space). The orthodox to aggregated configurational transition leads to a movement of water from intracristal to matrix space while the nonenergized (aggregated) to energized or energized to energized-twisted configurational change leads to water movement in the opposite direction. A medium of high osmotic pressure (0.5 M sucrose) can completely suppress the transition from the energized to the energized-twisted configuration by providing a counterforce to the driving force of the configurational transition. Under these conditions neither configurational changes nor water movements take place. The energizing process involves conformational changes in the repeating units and these conformational changes lead to configurational changes in the membranes and accompanying expansion of one mitochondrial space and proportionate contraction of the other.

MYSELS Well, it involves the transport of water from one side of the membrane to the other.

GREEN The membrane is completely permeable to water. There is no need to transport water. When conformational change compels the membrane to expand, the movement of water is sufficiently rapid to fill the space as quickly as it is formed.

MYSELS There still must be a driving force.

GREEN It is the conformational change that compels the redistribution of water by changing the relative osmotic pressure on the two sides of the membrane. The osmotic pressure is the driving force.

MYSELS Well, I wonder whether one might not consider the difficulties of diffusion and viscous resistances, and the driving forces that are involved.

GREEN I must repeat that the available evidence points to extremely rapid and extensive movements of water in the mitochondrion and to the secondary nature of water movements in the energy cycle.

CHANCE It is very important to draw a distinction between the electron microscopic changes described by Dr. Green, and the fluorescence and pH probes that we observe. First, Dr. Green is working with intact mitochondria which have the possibility of calcium uptake which he has stated is 'intrinsic to the conformation change'. The membrane fragments that we used have lost their capacity to translocate calcium and one can imagine that we are, therefore, dealing with some precursor of the effects that Dr. Green discusses.

A related question is just where in the membrane do our fluorochromes indicate structural changes; the location of the electron transport components or the location of the energy coupling components? The latter point can be evaluated by a second membrane-bound fluorochrome called aurovertin; this is one of the antibiotics which like rutamicin or oligomycin that block energy conservation at the ADT–ATP stage and do not effect electron transport (7). These energy coupling reactions are located in the projecting subunits of the membrane and are identified with the coupling factor F_1 and F_0. The antibiotic aurovertin shows a great enhancement of its fluorescence when it binds to the F_1 factor *in vitro* and on the membranes as well. Since the location of this fluorochrome is known, it is of interest to know whether the state of this portion of the membrane is affected by energy coupling. Experiments similar to those reported on ANS show (Figure 1) no observable change of the fluorescence of bound aurovertin on activating electron transport, and thus we can conclude that the changes in the state of the membrane which increases ANS fluorescence in the energy coupling reaction do not measurably effect the state of aurovertin bound to the coupling factor F_1. This leaves us with membrane state changes reported by ANS and BTB that have to do with the 'core' of the membrane if one pleases to identify the cytochrome region by this name. We believe that the association of the energy coupling reaction with a change in the state of the membrane, and a location of the energy coupling site with the 'core' of the membrane is one of the more exciting developments of energy coupling studies. I hope that it will be productive of a better definition of the structure and function relationships of this interesting membrane system of mitochondria.

GREEN Britton Chance has pointed out the apparent paradox that the fluorescence shown by aurovertin in consequence of binding to the headpiece is unaffected by the energizing process. Since aurovertin reacts with the site at which Pi and ADP combine to form ATP, it was hoped for by Britton Chance that the fluorescence of aurovertin would reflect events in the energy cycle. There was no such reflection. It should be pointed out that the interaction of Pi and ADP takes place *after* the repeating unit is energized. Theory would not support the expectation that aurovertin fluorescence would necessarily reflect conformational change in the repeating unit. The fact of conformational change is undisputed. What is uncertain is which probe is close enough to the action to measure this change. Where the probe is located is crucial in this matter. Apparently aurovertin was in the wrong location to be useful and apparently this applies with equal force to ANS.

Britton Chance raised yet another question that needs a reply—the question whether submitochondrial particles show configurational changes comparable to those observed in mitochondria. The answer is yes. The geometry of ETP_H is significantly different from that of the mitochondrial cristae and hence the form of the configurational changes in the membrane would necessarily have to be different. We have been able to demonstrate configurational changes in ETP_H by light scattering measures. Also we have recently succeeded in demonstrating that the repeating units of ETP_H undergo the same conformational cycle as that observed for the repeating units of the mitochondrial cristael membrane.

References to Dr. Chance's Remarks

1. AZZI, A., CHANCE, B., RADDA, G., and LEE, C. P., *PNAS*, February, 1969 (in press).
2. CHANCE, B., and MELA, L. 1966, *J. Biol. Chem.*, **241**, 4588.
3. WEBER, G., and LAURENCE, D. J. R. 1954, *Biochem. J.*, **56**, 31–P.
4. STRYER, L. 1968, *Science*, **162**, 526.
5. CHANCE, B., and MELA, L. 1967, *J. Biol. Chem.*, **242**, 830.
6. MCCARTY, R. E. 1968, *Biochem. Biophys. Res. Commun.*, **32**, 37.
7. LARDY, H. A., CONNELLY, J. L. 1964, and JOHNSON, D., in *Inhibitors of Phosphoryl Transfer*, **3**, 1961.

PAPER 14

Conformational basis of energy transductions in biological membranes†

DAVID E. GREEN and ROBERT A. HARRIS‡
*Institute for Enzyme Research, University of Wisconsin,
Madison, Wisconsin 53706*

OXIDATIVE PHOSPHORYLATION is in essence the coupling in the mitochondrion of an oxidative process to the synthesis of ATP.§ How this coupling is achieved has baffled three generations of biochemists. Progress in the understanding of the coupling problem has had to await the ultrastructural analysis of the mitochondrion because intrinsic to coupling are profound conformational changes in the mitochondrial inner membrane. In the companion to the present communication, we have already considered the ultrastructural features of the inner mitochondrial membrane that are relevant to the coupling problem. The present communication will deal with with various experimental developments that have led to the concept of the primacy of conformational change in coupled reactions catalyzed by the inner mitochondrial membrane.

Discontinuous Nature of the Electron Transfer Process

An important milestone was reached in our understanding of the electron transfer process in mitochondria with the discovery in 1967 by Rieske, Baum, Silman and Lipton that this process is discontinuous (1–4). They were studying the cleavage of Complex III, one of the four complexes of the electron transfer chain, into a soluble fraction containing cytochrome c_1 and a

† This work was supported in part by National Institute of General Medical Sciences (USPHS) Program Project Grant No. GM–12847.
‡ Postdoctoral fellow of the National Multiple Sclerosis Society.
§ Abbreviations used: Pi, inorganic phosphate; mCl–CCP, carbonyl cyanide m-chlorophenylhydrazone; ETPH, phosphorylating submitochondrial electron transfer particles; NE, nonenergized configuration; E, energized configuration; ET, energized-twisted configuration.

particulate fraction containing cytochrome *b*. Antimycin, a reagent which specifically inhibits electron transfer in Complex III, locked the complex in a form which completely resisted the cleavage reaction. The detailed analysis of the mode of action of antimycin led them to the recognition that a confor-

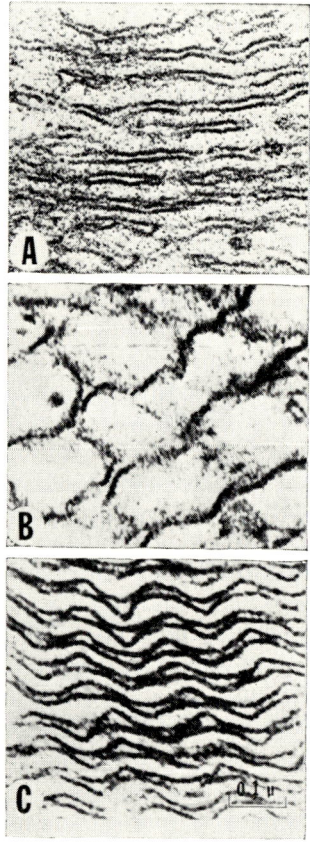

FIGURE 1 The cristal membrane of canary heart mitochondria *in situ* (courtesy of Dr. D. B. Slautterback): A, classical configuration; B, vesicular configuration; and C, zigzag configuration.

mational change is intrinsic to the electron transfer process in Complex III and that antimycin specifically prevents this conformational change. The conformational change is built into the mechanism of electron transfer. Unless this change can take place, electron transfer is frozen. It is sufficient for our present purposes merely to stress that electrons are not transferred continuously from one oxidation-reduction component to another in the electron transfer chain of a complex, and that conformational change is intrinsic to the mechanism of mitochondrial electron transfer.

Three Configurational States of the Cristal Membrane In Situ

David Slautterback in 1965 (5) showed that mitochondria of canary heart muscle examined *in situ* showed domains in which the cristal membrane assumed one of three characteristic configurations (see Figure 1). These configurations were so strikingly different that the thought occurred to us that these were ultrastructural expressions of the energy cycle that underlies coupling reactions in mitochondria. Thereupon, we turned our efforts to exploring the configurational changes in the cristal membrane of isolated mitochondria under conditions in which the energy state of the mitochondrial population could be rigorously controlled.

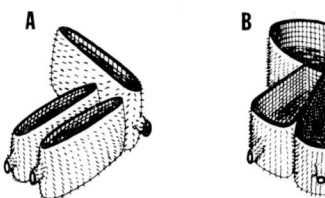

FIGURE 2. The nonenergized configurations of beef heart mitochondria: A, orthodox mode observed *in situ*; B, aggregated mode observed in isolated mitochondria.

Configurational States of the Cristal Membrane in Isolated Mitochondria

Our initial attempts to recognize and interpret the configurational changes in the cristal membrane were frustrated because of an unanticipated complication. The ultrastructure of the cristal membrane of isolated beef heart mitochondria was found to be profoundly different from that of the same membrane in mitochondria *in situ*. What accounted for the difference in geometry of the cristal membrane during the isolation of the mitochondrion? It turned out that the medium used for the isolation of mitochondria compels the expansion of the intracristal space and that the apposition of expanded cristae leads to the emergence of a fusion 'membrane' made up of two membranes pressed together with interleaving of the repeating units of these two membranes (see Figure 2). The cristae of the mitochondrion *in situ* are generally in the orthodox mode. The cristae of the isolated mitochondrion are generally in the aggregated mode. In Figure 3 the three configurational states of the cristal membrane in the aggregated mode are shown (6). Later on we shall present the evidence by which a decision could be reached as to the energy state of each of the three configurations.

Our initial impression was that the crista in the orthodox mode and the crista in the aggregated mode have equivalent coupling capability and that the transition from one mode to the other during isolation was merely a consequence of imposing conditions which expanded the intracristal space. This may be the case but it may also be too simple an interpretation. Recent studies of Williams, *et al.* (7) have shown that the cristae of heart mitochondria *in situ*

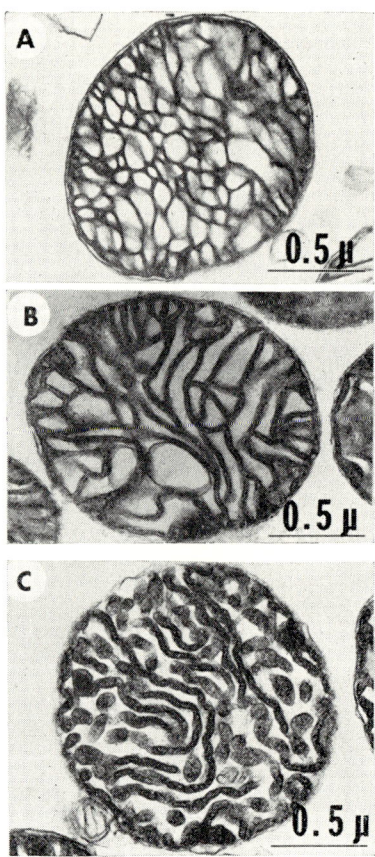

FIGURE 3   Configurations of isolated beef heart mitochondria: A, nonenergized; B, energized; C, energized-twisted.

which are in the energized state are almost indistinguishable in geometry from that of the cristae of isolated mitochondria which are in the energized state. The difference in geometry, i.e., the modal difference applies only to the cristae in the nonenergized state. In the isolated mitochondrion the cristae are in the aggregated mode; in mitochondria *in situ* the cristae are in the orthodox mode. This transition from one mode to another occurs during the energizing process *in situ*.

The Energy Diagram

The energized state of the mitochondrion can be generated either by electron transfer or by hydrolysis of ATP (see Figure 4). Moreover the energized state can be discharged by uncouplers or by reagents which lead to coupled reactions such as translocation of ions or synthesis of ATP. Our problem was to demonstrate that certain configurational states of the cristal membrane were expressions of the energized state of the mitochondrion. The energy diagram makes it clear how such a correlation can be established. All reagents which prevent the generation of the energized state should likewise prevent the formation of the appropriate configurational states of the cristal membrane. The same would be true for reagents that uncouple or lead to the dis-

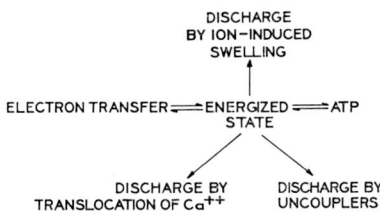

FIGURE 4   The energy diagram.

charge of the energized state. It is obvious that there are a very large number of ways in which the correlation can be tested and we have attempted to establish the correlation by several different methods.

Study of Configurational Change by Electron Microscopy

The first problem was to specify which configurational state of the cristal membrane was nonenergized and which states were energized. The configurational state in which electron transfer and ATP hydrolysis is interdicted would be of course the nonenergized state. This should have the same form as the configurational state obtaining in presence of uncoupler. Once the nonenergized configuration was specified, it remained to characterize the two energized configurations. One of the configurations was generated by substrate or ATP in absence of inorganic phosphate; the other by substrate or ATP only in presence of inorganic phosphate. The configuration generated by substrate or ATP was designated the energized configuration whereas the configuration generated by substrate or ATP in presence of inorganic phosphate was designated the energized-twisted configuration (6, 9, 10). Inspection of the micrographs in Figure 3 will show that there are multiple criteria by which the three configurational states can be distinguished. The transition from the energized-twisted to the nonenergized configuration is by far the more readily recognizable and we have studied this particular transition whenever possible.

The success of the electron microscopic method of studying configurational

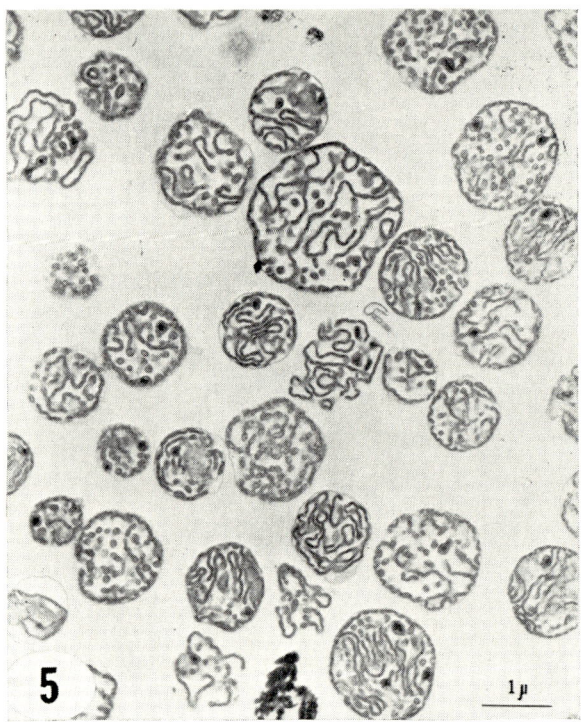

FIGURE 5 A large field of mitochondria in the energized-twisted configuration.

changes in the cristal membrane requires that the mitochondrial population should be uniformly in the same state and that the same configurational change should be noted in all intact mitochondria population. Figures 5 and 6 show fields of mitochondria first in the energized-twisted configuration induced by substrate in presence of inorganic phosphate, and then in the nonenergized configuration induced by addition of respiratory inhibitors to the same mitochondrial suspension. The transition in configuration is general for the entire mitochondrial population. We have used experimental conditions which compel results of a black and white character in respect to configurational change and thus eliminated ambiguity of interpretation of electron micrographs.

There is a second point that has to be underscored before considering the body of experimental evidence bearing on the correlation. During the energy

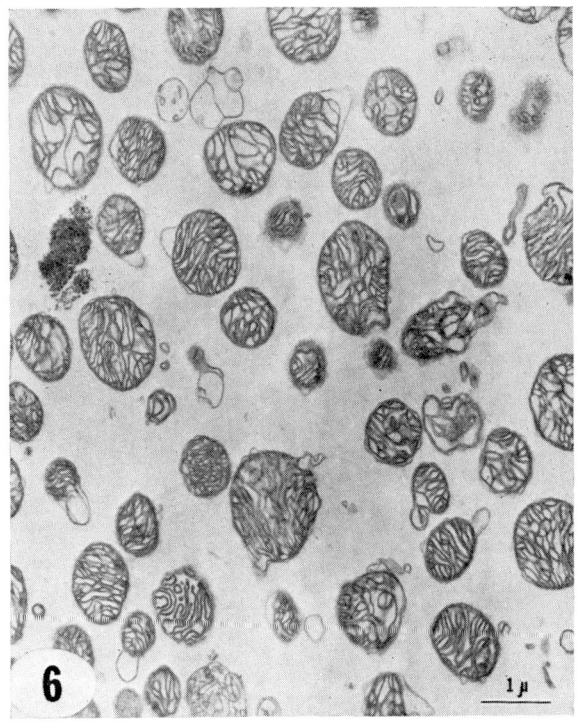

FIGURE 6 A large field of mitochondria in the nonenergized configuration.

cycle each repeating unit undergoes a cycle of conformational change. The configuration of the membrane depends upon the conformation of the individual repeating units. If all the repeating units undergo a transition from the nonenergized to the energized-twisted conformation, a modest though significant conformational change in the repeating units can be translated into a relatively large configurational change in the membrane because of a cooperative effect. This intensification of small conformational changes into large configurational changes of the membrane makes the electron microscopic method ideally suited for study of the energy cycle in membrane systems. At high magnification it is possible to study both configurational and conformational transitions, at lower magnifications only configurational changes in the membrane are recognizable.

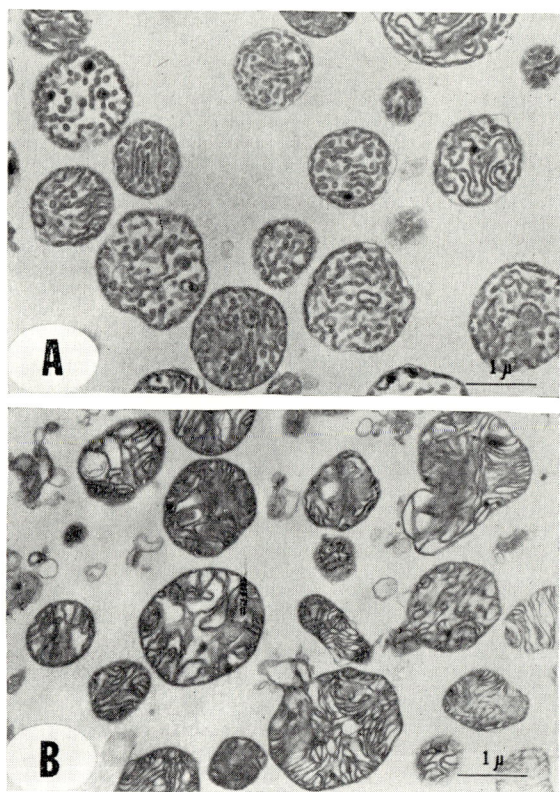

FIGURE 7 Discharge of the energized-twisted configuration by mCl-CCP: A, energized-twisted configuration obtaining before the addition of uncoupler; B, non-energized configuration obtaining after the addition of uncoupler.

To summarize a large body of experimental evidence (6, 9, 10), we can say that a sharp parallelism has been found between the energized state of the mitochondrion and the energized configurations of the cristal membranes. The energized configurations are generated by substrate or by hydrolysis of ATP and discharged by uncouplers. Reagents which prevent electron transfer prevent the generation by substrate; reagents which prevent hydrolysis of

ATP prevent the generation by ATP. Moreover the energized-twisted configuration of the cristal membrane can be discharged by ADP (with synthesis of ATP), by Ca^{++} [with translocation of $Ca_3(PO_4)_2$], or by monovalent salts (with translocation of monovalent salts). A representative set of typical experiments is shown in Figures 7–9.

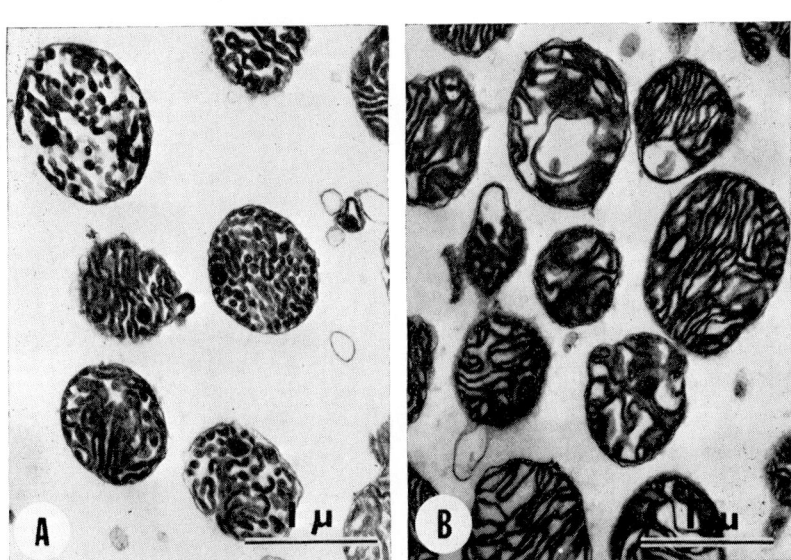

FIGURE 8 Discharge of the energized-twisted configuration by induction of divalent cation uptake: A, energized-twisted configuration obtaining before the addition of Sr^{++}; B, nonenergized configuration obtaining after the addition of Sr^{++}.

A word about the experimental procedure may be in order. The mitochondrial population can be treated in an appropriate manner to induce or discharge a particular configurational state and then a mixture of acrolein and glutaraldehyde is added to the suspension to freeze the mitochondria in the configurational state obtaining at the time. The fixation takes place within seconds of the addition and then the mitochondrial suspension can be processed for microscopic examination at leisure.

Study of Configurational Change by Measurement of Light Scattering

The surface of the mitochondrial particle acts as a reflecting surface for scatter of incident light. When the cristal membrane undergoes configurational

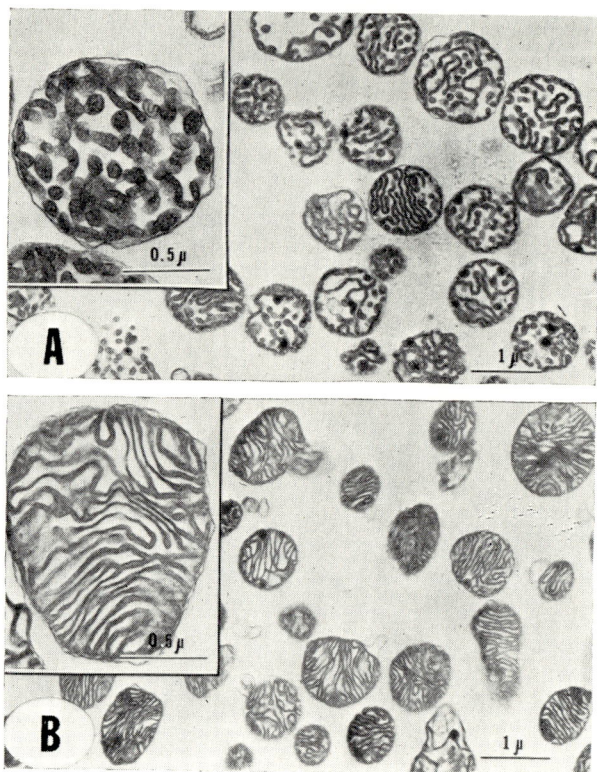

FIGURE 9 Discharge of the energized-twisted configuration by initiation of oxidative phosphorylation: A, energized-twisted configuration obtaining before initiation of oxidative phosphorylation by the addition of ADP; B, mixed states of nonenergized and energized configurations obtaining after the addition of ADP.

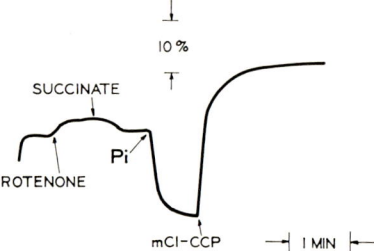

FIGURE 10 Correlation of light-scattering changes with the functional state of beef heart mitochondria. Light scattering was measured 90° to the incident beam with a Brice-Phoenix light-scattering photometer at a wavelength of 546 mμ.

change, the geometry of the scattering envelope of the mitochondrion undergoes a corresponding change and this change can be detected and measured continuously by light scattering at 90° (11–14). Figure 10 shows a record of a typical experiment with a suspension of beef heart mitochondria. The suspension starts in the nonenergized configuration. Addition of substrate leads to a small light-scattering change but further addition of inorganic phosphate induces a major change in light scattering. What this means is that the transition from the nonenergized to the energized configuration involves a small geometric change relative to the transition from the energized to the energized-twisted configuration. Of course with increase in the sensitivity of the instrument for measurement of light scattering both transitions might be measured but we have been obliged by the limitation in sensitivity of the instrument available to us to restrict our study to the study of only one of the two configurational transitions.

We have studied the correlation between the energy state of the mitochondrion and the configurational changes in the cristal membrane by the method of light-scattering changes (13, 14). An exact correlation was again found in a large series of tests. Succinate could generate the energized-twisted configuration but not in presence of antimycin. Uncouplers discharged the energized-twisted configuration and also prevented the generation of this configuration either by substrate or ATP. Moreover ADP or divalent metal ions discharged the energized-twisted configuration of the cristal membrane.

The configurational changes in mitochondria that are so readily recognized by electron microscopy are more difficult to visualize in submitochondrial particles by the same method. This is basically a geometric problem. The submitochondrial particles are considerably smaller than the cristae from which they arise by fragmentation, and the intensification factor, so essential for visualization, is correspondingly smaller. However, the method of measurement of light-scattering change is highly effective in the study of configurational changes in coupled submitochondrial particles (see Figure 11). These studies have shown that the cristal membrane of submitochondrial particles can assume an energized configuration and that the same basic relations apply to the configurational changes as apply to the corresponding configurational changes in mitochondria (13, 14). A new phenomenon was discovered in the course of these studies. As prepared, submitochondrial particles are in a state which resists configurational change. When the particles are first exposed to any one of a series of reagents including ATP then the appropriate light-scattering change can be induced. We interpret the priming of submitochondrial particles by ATP in terms of a change in geometry which makes possible membrane interactions that are intrinsic to the coupling process.

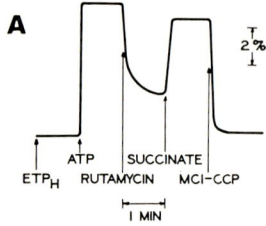

FIGURE 11 Correlation of light-scattering changes with the functional state of ETPH.

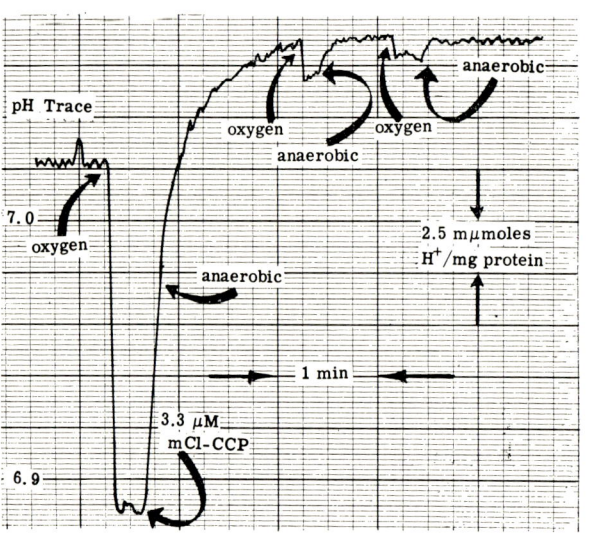

FIGURE 12 Correlation of pH changes with the energy state of beef heart mitochondria. Mitochondria incubated in a closed system in the presence of oxidizable substrate (succinate) assume the nonenergized state (and configuration) shortly after becoming anaerobic. The energized state can then be generated by the addition of oxygen. A pH change is observed upon induction of the energized state.

Study of Configurational Change by Measurement of pH

A conformational change involving as it does a geometric rearrangement of polypeptide chains could be expected to involve either exposure of ionizable groups previously buried in the hydrophobic interior or burial in the hydrophobic interior of ionizable groups previously exposed to the aqueous milieu. Such changes are not necessarily intrinsic to all conformational change but there is a good probability that they will be observed particularly when the magnitude of the conformational change is high. We have attempted to follow conformational changes by changes in pH and the results we have obtained suggest that the measurement of pH change is a most satisfactory and valid way of studying the energized state in mitochondrial and submitochondrial particles (15).

Figure 12 is a continuous tracing of the pH of the medium during an experimental cycle. The transition from the nonenergized to energized conformation is the transition which is most readily studied by this method. The tracing in Figure 12 shows that the pH change induced by addition of oxygen is reversed when uncoupler is added to the system. Inhibitors of electron transfer will suppress the pH change when substrate is the energizing source, and inhibitors of ATP hydrolysis will suppress the pH change when ATP is the energizing source. Thus, by several criteria, measurement of pH as a guide to induction of the energized state leads to exactly the same picture as obtains by measurement of light-scattering changes or by electron microscopic examination.

During the transition of the mitochondrion from the nonenergized to the energized configuration, 15–50 mμmoles of H^+ are released per mg protein. This would correspond to the exposure within one repeating unit of 75–250 groups capable of releasing a hydrogen ion. The number of ionizable groups exposed is surprisingly high. It could be a token of the magnitude of the conformational change during the nonenergized to energized transition of the cristal membrane. However, it most likely means that the steady state differential in pH only in part reflects the pH change intrinsic to conformational change and that the balance of the pH differential is a measure of a cation-proton exchange mechanism operative in the energized state.

Mitochondrial particles, such as EP_1 have lost the capacity for coupling oxidation to phosphorylation and do not show changes by the light-scattering or electron microscopic methods. Nonetheless, such particles show a pH change when energized by substrate and this pH change is reversed on addition of uncoupler (15). It would appear from such experiments that loss of coupling does not necessarily mean the loss of the capacity for conformational change. This capacity is probably intrinsic to the electron transfer process. But the expression of the conformational change as a change in the configuration of the cristae can be drastically altered by chemical alteration of the particle.

When mitochondria are treated to remove Ca^{++} the pH change induced by substrate largely disappears. It can be restored by adding Ca^{++} back to the particle (15). It can be shown that the uptake of Ca^{++} parallels the release of H^+. We interpret this phenomenon as follows. The ionizable groups which are exposed by conformational change in the system do not release a proton unless an acceptor is present which can penetrate into the interior of the particle where the conformational change has taken place. Calcium is assumed to be such an acceptor. Calcium is taken up when the energized state is generated and is released when the energized state is discharged. The movement of H^+ is opposite and equal. We recently have observed that certain preparations of Ca^{++}-free heart mitochondria as well as Ca^{++}-free EP_1 will show H^+ changes during electron transfer. This suggests that under the proper conditions of membrane permeability other cations besides Ca^{++} will penetrate into the interior of the particle in exchange for H^+.

Other Methods of Studying Conformational Changes

We and many others are in the process of exploring additional methods of detecting and measuring conformational changes in the cristal membrane. Several groups (16–18) have independently used the change in fluorescence of the probe 8-anilino-1-naphthalene sulfonic acid during the energy cycle as a method for monitoring conformational changes. Changes in fluorescence of tyrosine or tryptophane residues in proteins within the repeating units could also serve as measures of changes in the immediate environment in which these fluorescing groups are localized. Cytochrome *b* has been known to undergo spectral shifts under energizing conditions which hitherto have been difficult to interpret (19). These spectral shifts could be a reflection of the change of the absorption spectrum of cytochrome *b* with change in the molecular environment induced by conformational change. McConnell has devised reagents (20) with intrinsic electron spin resonance which can be introduced into a membrane and in the membrane these spin label reagents record changes in spin resonance resulting from change in the immediate molecular environment. The tritium exchange technique (21) should also be a useful probe of conformational changes intrinsic to energy transduction.

The introduction into the cristal membrane of reagents which can serve as probes of conformational change is not without its complications. If the reagent finds its way to the immediate vicinity of the molecules participating in the reactions which lead to the conformational change then it is likely that the reagent will interfere with or modulate the energizing reactions. If the reagent is localized at a distance from the critical chemical events, then it may not necessarily reflect the full extent of the conformational change. The combination of the methods of electron microscopy, light scattering, and pH transitions has advantages which no one of the fluorescent or spin-label probes used so far can match.

Fluctuations in Configuration of the Cristal Membrane

The individual repeating units of the cristal membrane go through rapid cycles of conformational change during the energy cycle but experiment has shown that the configuration of the membrane can remain constant for an extended period during which the repeating units are actively cycling (14). How can the constancy of configuration of the membrane be rationalized with the rapid fluctuations of the repeating units? It would appear that the configuration of the membrane is determined by the steady state balance between the rate of energizing and the rate of discharging. Above a critical point the membrane is in, let us say, the energized configuration; below that point it assumes the nonenergized configuration. This means that a membrane in the energized configuration can tolerate a certain proportion of repeating units in the nonenergized conformation without flipping over; and conversely, a membrane in the nonenergized configuration can be stable even though a certain proportion of repeating units are in the energized conformation. This principle regulating the configurational state of the membrane has profound significance for the problem of membrane permeability, cooperativity in membrane systems and the transmission of the energized state throughout a membrane from repeating unit to repeating unit.

The Conformational Cycle

We have postulated that the configurational changes observed in the cristal membrane during the energy cycle reflect corresponding conformational changes in the repeating units (9, 10). Conformational change of the repeating units is postulated to be the primary energy conserving step which initiates a sequence of events which lead to the energized configurational changes of the inner membrane. It has become apparent that conformational change must lead to changes in membrane permeability and to a change in the osmotic pressure exerted by the inner membrane compartment. These changes, executed in the sequence discussed, must establish the configurational changes of the inner membrane.

There is now electron microscopic evidence (22) as well as geometric evidence (23) for a cycle of conformational change of the repeating units during the energy cycle. The dimensional analysis (23) of the cristal membranes in the nonenergized configuration (150 Å in cross-sectional width) has led us to the conclusion that the only way in which two apposed repeating units, each 185 Å long when fully extended, can nest together in a membrane only 150 Å thick, would be by the collapse of the headpiece-stalk sector to a disc 114 × 114 × 25 Å (see Figure 13). Each basepiece would contribute 50 Å to the thickness and each of the two collapsed headpiece-stalks would form discs 25 Å in thickness. The total computed thickness for the apposition of two repeating units with collapsed headpiece-stalks (2 × 50 plus 2 × 25) would account for the observed 150 Å width of the cristal membrane in the non-energized configuration of the aggregated mode. The observed width of the cristal membrane in the energized configuration is about 250 Å. As we shall discuss later, the headpiece-stalk sectors appear to be fully extended in this configurational state but interleaved (see Figure 13). The two basepieces account for 100 Å and, thus, 150 Å is available for one extended headpiece-stalk. The length of the stalk is 50 Å; that leaves 100 Å for the headpiece.

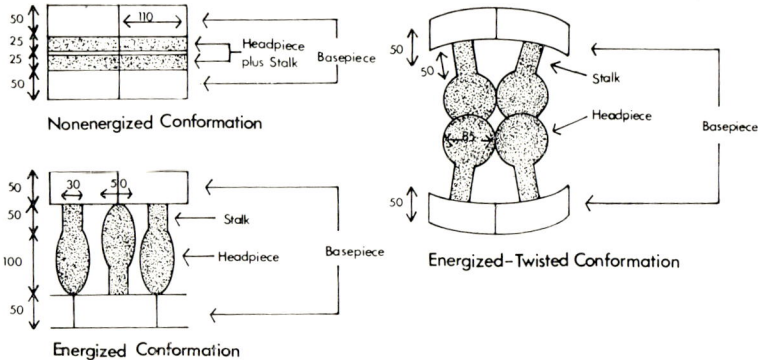

FIGURE 13 The conformational cycle of the tripartite repeating units of the cristal membrane during the energy cycle.

The headpiece is usually seen as a sphere 85 Å in diameter. In the energized conformation, the headpiece appears to have a more ellipsoid shape which accounts for the greater length (100 versus 85 Å). The observed width of the cristal membrane in the energized-twisted configuration is 300 Å. After the contribution of the basepieces to the width is subtracted (2 × 50), a gap of 200 Å between the basepieces is available for the headpiece-stalks. In this configuration, apposed headpiece-stalks are also interleaved but the interleaving involves sets of adjacent headpiece-stalks (probably three headpieces in each set). Each headpiece-stalk unit in this configurational state has a length of 135 Å (50 Å for the stalk and 85 Å for the headpiece). There is thus a good deal of space between the headpieces of one set of repeating units and the basepieces of the apposed repeating units. The geometric necessities of interleaving of headpiece-stalks by sets may compel the helical character of the cristal membrane in the energized-twisted configuration. The interleaving of headpiece-stalks by sets is postulated to lead to periodic changes in curvature of the basepieces and, thus, to a helical configuration of the membrane.

The electron microscopic examination of the cristal membranes in the three configurational states by classical procedures proved inconclusive with respect to the specification of the geometry of the repeating units. By classical staining procedures, the sectors of the repeating units were not readily visualized. Vail, et al., have found that in mitochondria first fixed with glutaraldehyde and then exposed to the action of phospholipase c—an enzyme which hydrolyzes phosphoryl choline from membrane-bound lecithin—the headpiece-stalk units become visualizable (22). Moreover, the domain of the basepieces is specified by a trilaminar structure of the cristal membrane (without the phospholipase c treatment, the trilaminar structure does not show up as frequently). We have interpreted this emergence of ultrastructural detail in terms of the thinning out of the phospholipid polar heads which cover the membrane continuum. With this thinning out, the channels in which the phospholipid is oriented becomes accessible to the staining reagents, and, thus, reagents like osmium tetroxide can readily penetrate to the interior of the membrane and react with the internal structures, namely the headpiece-stalk sectors. The trilaminar appearance of the membrane is an indication of the structure of a basepiece—a quasicuboidal polymeric, protein system with the polar portions of the phospholipid on both of the surfaces which are parallel to the axis of the membrane. The phospholipid domain (plus the exposed edge of the protein) accounts for the dark line on each side of the membrane; the protein domain (i.e., the interior of the basepieces) accounts for the electron-transparent area between the two dark-staining lines. The important point at issue is that we hold that the width of the trilaminar zone defines the width of the basepieces of the repeating units.

The application of the phospholipase c digestion method of Vail, et al. (22) to mitochondria in each of the three configurations has fully confirmed the indications from the geometric analysis that the headpiece-stalk sector must undergo a cycle of conformational change—being apparently collapsed in the nonenergized conformation, and extended in the energized-twisted conformation. In the energized conformation the headpiece of one repeating unit extends directly to the basepiece of the apposed repeating unit. In other words, each repeating unit extends, via its headpiece-stalk, from one side of the membrane to the other side. In the energized-twisted conformation, the link of headpiece with apposed basepiece does not appear to be intact and the headpiece takes up a position about two thirds of the distance from one side of the crista to the other.

The geometric details of the conformational cycle have been considered in some detail not as an exercise in ultrastructural analysis but rather to provide background information required to appreciate what we believe may be the molecular strategy of the conformational cycle. As we shall discuss later, the details of the cycle appear to provide powerful clues to the mechanism of coupling.

The Two Conformational Transitions

The transition from the nonenergized to the energized conformations of the repeating unit (induced by electron transfer or hydrolysis of ATP) is accompanied by a change in pH, is suppressed in presence of uncouplers, and is insensitive to a counter osmotic force such as is imposed by a suspending medium 0.5 M to 0.88 M in sucrose (15, 24, 25). The transition from the energized to the energized-twisted configuration (induced by inorganic phosphate) is highly sensitive to the imposition of a counter osmotic force (15, 24, 25). In fact, in media containing sucrose at concentrations greater than 0.5 M, the second configurational transition is completely suppressed as are all coupled reactions involving inorganic phosphate. The first conformational change involves major redistribution of charged groups expressed as a change in pH and this redistribution is a function of transfer of electrons or hydrolysis of ATP. That is to say, the transduction of redox energy or the energy of hydrolysis of ATP to conformational energy is accompanied by extensive rearrangement of charged groups which leads to a H^+ change and to a configurational change. Uncouplers completely prevent this transduction. The volume change in the first conformational and configurational transition is not large relatively (the width of the cristae increases from 150 Å to 250 Å). The volume change in the second configurational transition is relatively large. This may not be immediately apparent from the increase in width (from 250 to 300 Å), but it must be remembered that the crista in the energized-twisted configuration is tubular and that the transition from a 'pillow case' type crista to multiple tubular cristae involves a considerable augmentation in volume because of the $\frac{4}{3}\pi r^3$ relation and the fact that the diameter of the tubes is 1.2 times greater than that of the original disc-like crista.

Transitions in Configurational Mode

During the configurational cycle of mitochondria *in situ* the repeating units of apposed membranes of separate cristae in the orthodox mode are forced to interact with one another (7). Two factors—both intrinsic to the energy cycle —appear to play a role in establishing aggregation of apposed cristae. They are the initial energy conserving conformational change of the repeating units plus the establishment of a counter osmotic pressure which drives the membrane together. This transition is readily observed in mitochondria *in situ* because the cristae in the nonenergized configuration are in the orthodox mode. In isolated beef heart mitochondria the cristae in the nonenergized configuration are already in the aggregated mode because of the external osmotic pressure of the sucrose suspending medium. In such mitochondria, configurational changes still occur but no transition in mode attends the energizing process.

Once it is appreciated that aggregation of cristae is a consequence of and an expression of the conformational cycle, then aggregation in all its many manifestations can be equated with the energy cycle. Aggregation can take place in a regular way in heart mitochondria where the cristae are arranged in regular, parallel arrays or it can take place in a random fashion in liver or

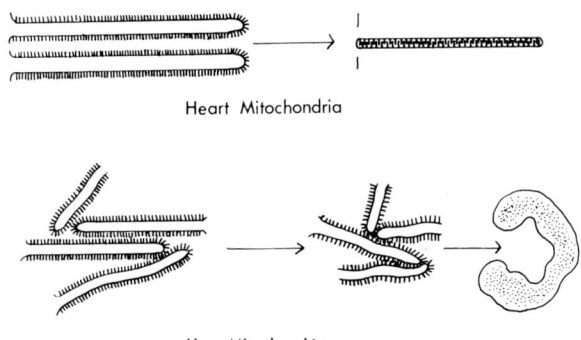

FIGURE 14 The aggregation phenomenon in heart and liver mitochondria.

adrenal mitochondria where the cristae are arranged in a less ordered fashion (see Figure 14). By virtue of the irregularity of the way in which cristae aggregate in liver mitochondria, the distinction between the energized and energized-twisted configuration, so readily made in heart mitochondria, is blurred. It is this technical difficulty that has delayed the recognition that the cristae of liver mitochondria exist in the same basic configurational states as do the cristae of heart mitochondria.

Pseudoenergizing of the Cristael Membrane

A medium which contains an appropriate salt such as NaAc, or a medium made alkaline to pH 10, or even a medium made hypo-osmolar will induce the same configurational changes as does electron transfer or hydrolysis of

ATP without any active process going on (26, 27). In fact, this pseudo-energizing of the cristal membrane by any of the means specified above, is completely insensitive to the action of uncouplers or other inhibitors of active processes (26, 27). The available evidence suggests that pseudoenergizing conditions are those which so alter the aqueous environment which interacts with the cristal membrane that the repeating units subjected to this new environment are spontaneously forced to undergo the same conformational transitions as take place during electron transfer or hydrolysis of ATP. The fact that these transitions are spontaneous, means that the energy difference in the transition between nonenergized and energized-twisted configurations is eliminated. That is to say, conditions can be found in which the energized configuration becomes more stable than the nonenergized configuration and the repeating unit will spontaneously undergo a change in conformation. If mitochondria containing cristae were pseudoenergized to the energized-twisted configuration by any number of means but were resuspended in a different suspension medium, it might be possible to use the pseudoenergized state of the membrane to do work. This would require that the pseudo-energized state would be stable for the duration of the period required to change the composition of the suspending medium. Therefore, utilization of the conformational energy potentially available in pseudoenergized membranes for the synthesis of ATP appears entirely feasible to us. The feasibility of this suggestion is currently being subjected to experimental verification.

Model of Oxidative Phosphorylation

Enough of the pieces of the puzzle of oxidative phosphorylation have been identified to make it profitable to experiment with model-building. The value of a tentative model is that it will compel the integration of the separate events and it will set the scene for prediction and testing of prediction. The model necessarily has to go beyond the available facts; hence, it must be looked upon as a device for visualization and integration of the known facts and for anticipation of features yet to emerge. We fully realize that this model may be wrong in some of its detail, but we maintain that it must be basically right in principle.

In the phenomenon of oxidative phosphorylation we recognize that five sequential events occur: (*a*) the electron transfer process; (*b*) the accompanying energy-conserving conformational change; (*c*) the stabilization of the conformational change by a mechanism of cooperativity between repeating units; (*d*) the interaction of inorganic phosphate with the conformationally perturbed system; and (*e*) the redistribution of energy in the conformationally perturbed system leading to the formation of a high energy phosphoryl

Conformational basis of energy transductions 335

derivative. Before considering the actual model (see Figure 15), let us examine the important facets of each of these five sequential events.

The unit of coupling is one of the four complexes of the electron transfer chain. Each complex is also a basepiece of the cristal membrane. We may consider the complex as a set of four oxidation-reduction proteins or groups which interact with one another in a special way. Two of the four groups are in the reduced form at the beginning of the energizing cycle while two are in the oxidized form. We will arbitrarily designate the components initially in the reduced form as the A group and the components initially in the oxidized form as the B group. Electron transfer takes place between the components of the A and B group whereby two electrons are transferred from the A to the B set of components. The A group becomes oxidized; the B group becomes reduced. As soon as this transfer occurs, a conformational change freezes the system with respect to the electron transfer process (see Figure 16). We can only speculate as to what prevents electron transfer when the system is conformationally perturbed. It could be a problem of accessibility—components which can reduce the A group or oxidize the B group are not accessible to the

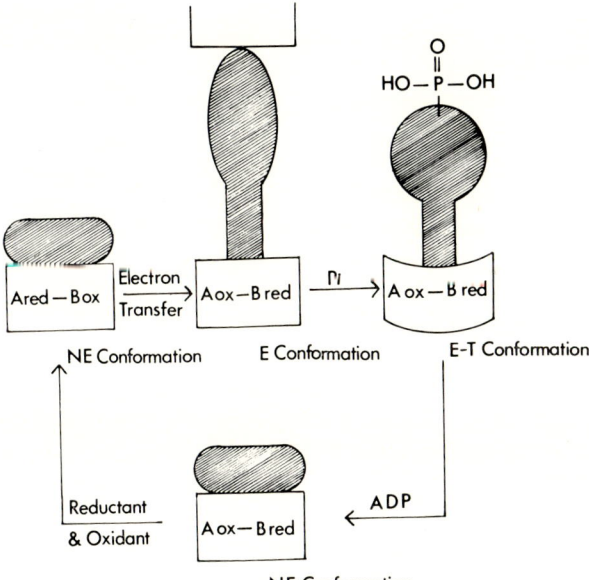

FIGURE 15   A model of oxidative phosphorylation. The basepiece is identified as a complex made up of two sets of oxidation-reduction components (the A and the B sets, respectively). NE abbreviates non-energized; E, energized; and E-T, energized-twisted.

perturbed complex. It could also be a problem of potential. The oxidation-reduction components of the A group in the conformationally perturbed system no longer are reducible by the natural reductant of the unperturbed complex because of an unfavorable potential difference. Whatever the reason for the freezing of the electron transfer process, the fact is that electron transfer does essentially stop in a system with good respiratory control, and resumes only when the system is discharged from the energized to the nonenergized conformation.

The conformational change which accompanies electron transfer in the complex appears to take the form of the extension of the collapsed headpiece-stalk sectors away from the basepiece. If we were to assume that the stalk is in the random coil state in the nonenergized conformation and helical in the energized state, then the extension of the headpiece-stalk sector would be easily rationalized. The helicizing of the stalk would compel the extension of the stalk and similarly the extension of the stalk would compel the transition of the headpiece from a collapsed disc to an extended ellipsoid. What would

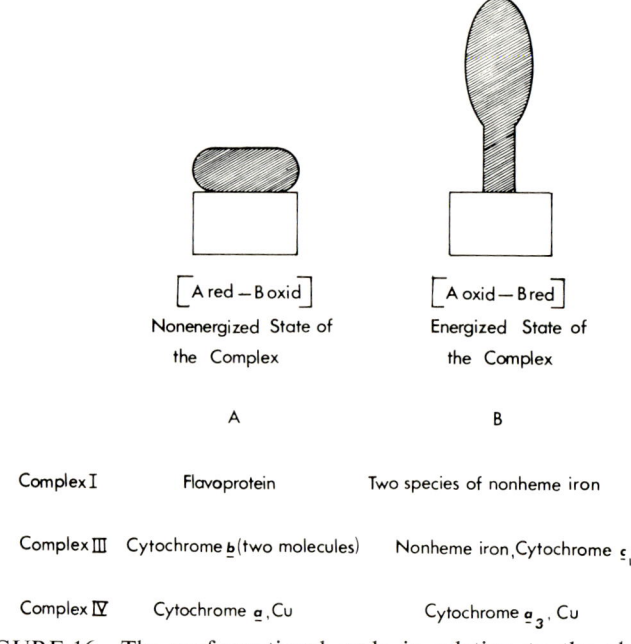

FIGURE 16 The conformational cycle in relation to the electron transfer process. The A and B oxidation-reduction components of the three complexes of the electron transfer chain which are involved in the coupling of electron transfer to synthesis of ATP are listed. Each complex is localized in a separate basepiece.

remain to be explained would be how the electron transfer process could actuate the helicizing of the stalk. Helix-random coil transitions can be controlled by the interpolation of charged groups (inducing the random coil) and by the removal of charged groups (inducing helix formation). All that would be necessary to account for the postulated helicizing of the stalk, as a consequence of the electron transfer process, would be the postulate that the electron transfer process results in the removal of charged groups that prevent helix formation.

The conformational change intrinsic to the extension of the headpiece-stalk sector could be stabilized internally, or, alternatively, stabilized by the formation of a low energy bond which would link together extended stalk to extended headpiece and extended headpiece to apposed basepiece. These two links might be part of the same bonding system. As an example of such a bonding system, we may cite the sulfhydryl-substituted aldimine bond suggested for the visual pigment by Heller (28). Bond formation provides a tactic for conserving the conformational change and freezing the system. If no 'hooking' bonds were formed, the energized repeating unit might spontaneously return to the nonenergized conformation. Presumably the extended state of the headpiece-stalk sectors would be unstable and unless stabilized by some sort of mechanism involving cooperativity between repeating units, the conformationally perturbed system would spontaneously return to the ground state, i.e., the collapsed state.

We can postulate that inorganic phosphate displaces one of the three groups participating in the hooking bond. This entry of phosphate into the system leads to a profound conformational rearrangement of the system. If the link between the headpiece of one membrane and basepiece of the other membrane were severed, a new link between headpiece and phosphate would be established. We may describe the complex thus formed as a low energy phosphate adduct of a conformationally perturbed system. The phosphate group in this complex would not yet be capable of phosphorylating ADP but there would be sufficient energy in the system that if properly channelled could achieve the activation of the bound phosphate to a phosphoryl group. How this could be done is at present unknown. We could postulate that the approach of ADP could lead to the rearrangement of the system with transfer of the phosphate group to a nitrogen atom participating in the hooking bond (assuming a substituted aldimine type of link for the hooking bond). In this way conformational energy could be channelled into bond energy.

Electron transfer can drive the synthesis of ATP and the hydrolysis of ATP can drive reversed electron transfer. In other words, the system is reversible. The energized states can be generated either by electron transfer or by hydrolysis of ATP. How would the model shown in Figure 15 account for the reversibility? If we postulate that the helicizing of the stalk can be activated either by charge changes in the basepiece (electron transfer) or by charge changes in the headpiece (ATP hydrolysis), then the energized state could be generated either by electron transfer or by hydrolysis of ATP via the same basic mechanism.

Predictions of the Model of Oxidative Phosphorylation

It is implicit in the model that we propose, that there are three ways in which uncoupling could be achieved—first by blocking the charged groups (in the basepiece and headpiece) that can actuate the random coil to helix transition of the stalk—second by blocking those groups that participate in the hooking reaction—and third by blocking or displacing the group which has to be displaced by inorganic phosphate. Uncouplers like mCl-CCP and dinitrophenol would block the reaction which leads to the energy-conserving conformational change (these reagents prevent the proton shift which is a hallmark of energy conservation). Other uncoupler-like reagents, e.g. Cd^{++}, may prevent the expression of cooperativity between repeating units and thus may prevent stabilization of the conformational change. Another group of compounds, e.g. arsenate, may lead to uncoupling by displacing the group normally displaced by phosphate. All compounds which 'uncouple' would fall into one of these three categories.

The model of oxidative phosphorylation also serves as a model of active transport. Since the postulated conformational cycle involves the cyclical extension and collapse of the headpiece-stalk sector, a mechanism is suggested for the energized transport of ions from the external medium to the fluid in the interior of the crista. Let us assume that there are binding sites for appropriate ions in the stalks or in the headpiece-stalks and that in the nonenergized conformations the headpiece-stalks and basepiece are all part of the same continuum. Lipid channels provide a device whereby ions from the external medium could equilibrate with the binding sites in either the headpiece or stalk. Facilitating molecules like valinomycin are probably needed to potentiate the penetration of ions into the membrane. The conformational cycle then leads to the extension of the headpiece-stalk from the basepiece and the dissociation of the ions previously bound. In this way, a precise number of ions would be transported during each turn of the conformational cycle.

The model of oxidative phosphorylation presented above points to respiratory control as a geometric problem. When the geometry of apposed cristae would facilitate cooperativity, e.g. hooking or internal stabilization by bond formation, respiratory control would be developed. When the geometry became incompatible with efficient hooking, respiratory control would be poor. Efficient hooking would require parallel, regular, and flat membrane surfaces. Any deviation from these optimal conditions would tend to lower the efficiency of respiratory control. If all of the repeating units became hooked during the energizing transition, respiratory control would be perfect. As the percentage of unhooked repeating units increased, the degree of respiratory control would be reduced since each unhooked repeating unit would act as an energy sink by destabilizing the hooked repeating units. When the percentage of unhooked repeating units reached a critical value, all respiratory control would be lost even though coupling could still be possible. This would be true because respiratory control depends upon the sustained stabilization of the energized state, whereas coupling requires only the instantaneous generation and utilization of the energized state.

Cooperative Phenomena in the Energizing of Membranes

When a few of the repeating units of the cristal membranes are damaged ultrastructurally as happens for example when mitochondria are lyophilized, the coupling activity of the mitochondrion is lost although the capability is still present (29). When lyophilized mitochondria are subjected to sonic irradiation, the fragmented vesicles of ETPH produced thereby, show full coupling capability. Ultrastructural dislocation even of a few repeating units in a membrane is assumed to have an effect far beyond the numbers of units involved because of the phenomenon of cooperativity (30). Dislocation is often accompanied by loss of lipid and detachment of the headpiece—stalk sectors of the repeating units. It is as if the damaged units were uncoupled and behaved as energy sinks. A good many instances of loss of coupling capability by virtue of ultrastructural dislocations of a relatively small number of repeating units have been experimentally verified (29, 30). The cooperative character of membrane phenomena magnifies the effect of damaged repeating units and gives an all or none quality to energy coupling.

It can be shown that the energized state of a basepiece must be transmitted through the membrane by some sort of exchange mechanism (9, 10). By virtue of this capability for transmission of the energized state, the distribution within the membrane of energized repeating units is unrelated to the actual sites where electrons were introduced into the repeating units by electron donors. The pattern of distribution may be uniform rather than bunched and, in fact, may determine the ease with which the membrane can flip from one configuration to another.

The relation between the conformation of the repeating units and the configuration of the membrane has already been referred to previously but some further comment is in order. The repeating units go through rapid conformational cycles but the membrane remains poised in one or another configurational state. The proportion of repeating units in energized *vs* nonenergized conformation will determine which of the three configurations the membrane will assume.

Experimentally the emphasis is placed on compelling the membrane to reflect the dominant conformational state of the repeating units. This is experimentally accomplished by avoiding the steady state situation. The energizing process is separated from the discharge. This leads to synchronization in the conformation of all the repeating units and by this means the experimenter determines the configuration of the membrane. Under steady state conditions which after all are usually the conditions *in vivo*, the membrane may remain permanently in, let us say, the nonenergized configuration simply because the rate of energizing is relatively slow and the rate of discharge is exactly balanced to the rate of energizing. This accounts for the fact that the cristal membranes of mitochondria *in situ* almost invariably are found to be in the nonenergized configuration. The energized configurations of mitochondria *in situ* are only consistently demonstrable when the tissue is exposed to conditions which compel a shift from the steady state situation (7).

Configurational Changes in the Energy Cycles of Membranes Other than the Cristal Membranes

The chloroplast like the mitochondrion couples electron transfer to the synthesis of ATP by union of ADP and Pi. It differs from the mitochondrion in respect to the source of the electrons which are donated to the electron

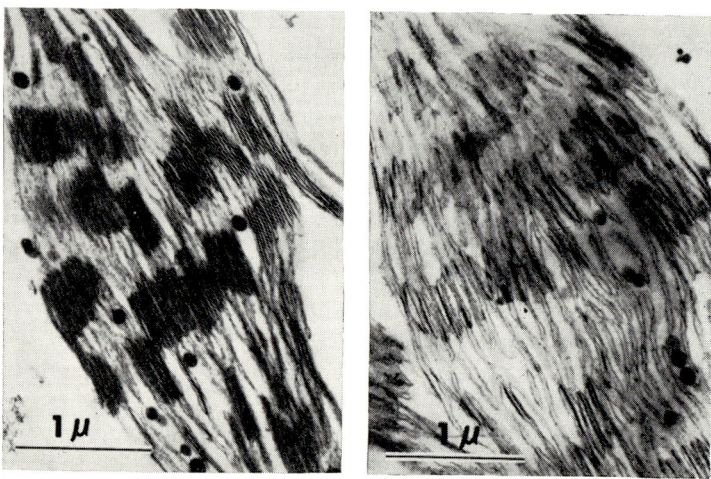

FIGURE 17 The configurational states of spinach chloroplast: left, the energized configuration generated by exposure to light; right, the nonenergized configuration obtaining in the dark.

transfer chain. In the chloroplast these electrons arise by the interaction of light with chlorophyll. In the mitochondrion these electrons are provided by succinate and DPNH. Apart from the source of the electrons the basic mechanism of coupling appears to be the same. Electron microscopic studies in our laboratory as well as in several other laboratories, e.g. Packer (31), have established that in the dark to light transition the inner membranes of the chloroplast undergo profound ultrastructural changes (cf. Figure 17). Moreover, there is a close correlation between these ultrastructural changes and the energy state of the chloroplast.

The plasma membrane of the red blood corpuscle can exist in either of two configurations depending whether ATP is present or not (32). In the intact red blood corpuscle, the shape of a biconcave disc is assumed when ATP is present in the medium; and a spherical shape is assumed when the level of ATP is

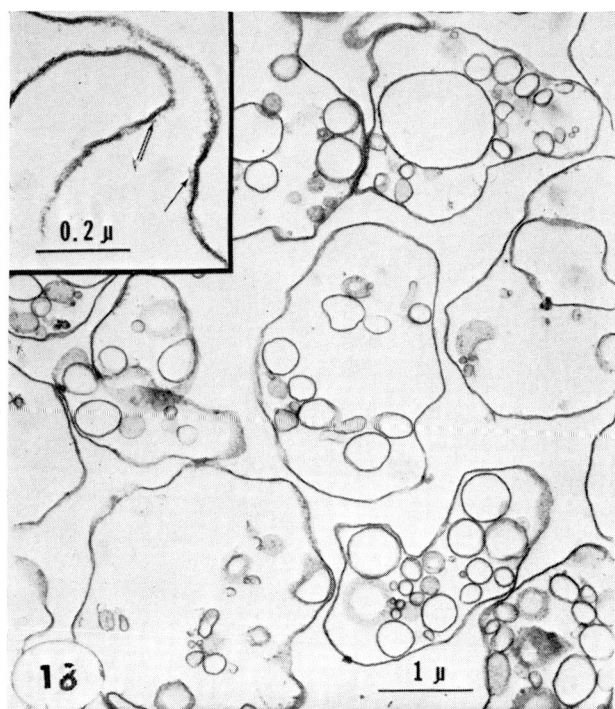

FIGURE 18 Energized configuration of cow erythrocyte ghosts. The vesicles are in the interior of the erythrocyte ghosts. Note that the filamentous material of the boundary membrane faces inward (single arrow), while the filamentous material of the vesicle faces outward (double arrow).

reduced to a negligible concentration. In the ghost membrane of red blood corpuscles (i.e., the membrane after hemoglobin has been allowed to leak out), the cycle of energizing by ATP and deenergizing by removal of ATP leads to an alternation of invagination of the membrane and evagination of the membrane (cf. Figures 18 and 19). The phenomenon of pinocytosis has been shown to be an expression of this configurational cycle. Conditions can be selected in which the energizing of the erythrocyte membrane can be uncomplicated

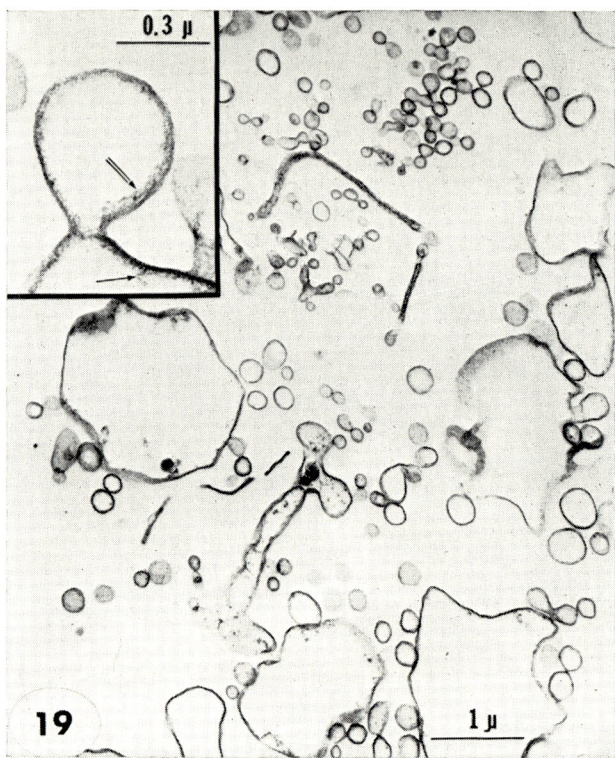

FIGURE 19   Nonenergized configuration of cow erythrocyte ghosts. The vesicles are exterior to the erythrocyte ghosts. Note that the filamentous material of both the boundary membrane (single arrow) and of the vesicle (double arrow) faces inward.

by the discharge of the energized membrane for work purposes. In other words once energized, the membranes remain energized until the level of ATP falls below a critical level. This simple feature of the energy cycle in the red blood corpuscle has made it possible to demonstrate in a very elegant fashion the primacy of conformational change in the erythrocyte membrane (32).

References

1. RIESKE, J. S., BAUM, H., STONER, C. D., and LIPTON, S. H. 1967, *J. Biol. Chem.*, **242**, 4854.
2. BAUM, H., SILMAN, H. I., RIESKE, J. S., and LIPTON, S. H. 1967, *J. Biol. Chem.*, **242**, 4876.
3. RIESKE, J. S., LIPTON, S. H., BAUM, H., and SILMAN, H. I. 1967, *J. Biol. Chem.*, **242**, 4888.
4. BAUM, H., RIESKE, J. S., SILMAN, H. I., and LIPTON, S. H. 1967, *Proc. Nat. Acad. Sci. U.S.*, **57**, 798.
5. SLAUTTERBACK, D. B. 1965, *J. Cell Biol.*, **24**, 1.
6. PENNISTON, J. T., HARRIS, R. A., ASAI, J., and GREEN, D. E. 1968, *Proc. Nat. Acad. Sci. U.S.*, **59**, 624.
7. WILLIAMS, C. H., HARRIS, R. A., VAIL, W. J., CALDWELL, M., VALDIVIA, E., and GREEN, D. E., unpublished studies.
8. HARRIS, R. A., HARRIS, D. L., and GREEN, D. E. 1968, *Arch. Biochem. Biophys.*, **128**, 219.
9. HARRIS, R. A., PENNISTON, J. T., ASAI, J., and GREEN, D. E. 1968, *Proc. Nat. Acad. Sci. U.S.*, **59**, 830.
10. GREEN, D. E., ASAI, J., HARRIS, R. A., and PENNISTON, J. T. 1968, *Arch. Biochem. Biophys.*, **125**, 684.
11. PACKER, L., in 'Methods in Enzymology,' Vol. X, R. W. Estabrook and M. E. Pullman, Eds. (Academic Press, New York, 1967), p. 685.
12. PACKER, L., UTSUMI, K., and MUSTAFA, M. G. 1966, *Arch. Biochem. Biophys.*, **117**, 381.
13. HARRIS, R. A., ASBELL, M. A., and GREEN, D. E. 1969, *Arch. Biochem. Biophys.*, **131**, 316.
14. HARRIS, R. A., ASBELL, M. A., ASAI, J., JOLLY, W. W., and GREEN, D. E. 1969, **132**, 545.
15. HARRIS, R. A., and WILLIAMS, C. 1969, *Federation Proc.*, **29**, 2266.
16. CHANCE, B., PRING, M., AZZI, A., LEE, C. P., and MELA, L. 1969, *Biophys. J.*, **9**, A-90.
17. PACKER, L., and WRIGGLESWORTH, J. M. 1969, *Biophys. J.*, **9**, A-144.
18. PENEFSKY, H. S., and DATTA, A. 1969, *Federation Proc.*, **29**, 2261.
19. CHANCE, B., LEE, C. P., and SCHOENER, B. 1966, *J. Biol. Chem.*, **241**, 4574.
20. STONE, T./J., BUCKMAN, T., NORDIO, P. L., and MCCONNELL, H. M. 1965, *Proc. Nat. Acad. Sci. U.S.*, **54**, 1010.
21. DELUCA, M., and MARSH, M. 1967, *Arch. Biochem. Biophys.*, **121**, 233.
22. VAIL, W. J., SENIOR, A. E., and GREEN, D. E., unpublished studies.
23. PENNISTON, J. T., and GREEN, D. E., unpublished studies.
24. HARRIS, R. A., WILLIAMS, C. H., JOLLY, W. W., ASAI, J., and GREEN, D. E., manuscript in preparation.
25. HARRIS, R. A., and ASBELL, M. A. 1969, *Biophys. J.*, **9**, A-141.
26. BLONDIN, G. A., VAIL, W. J., and GREEN, D. E. *Arch. Biochem. Biophys.*, **129**, 158 (1969).
27. ASAI, J., BLONDIN, G. A., VAIL, W. J., and GREEN, D. E. *Arch. Biochem. Biophys.*, **132**, 524, (1969).
28. HELLER, J. 1968, *Biochemistry*, **7**, 2914.

29. LENAZ, G., JOLLY, W. W., and GREEN, D. E. 1968, *Arch. Biochem. Biophys.*, **126**, 67.
30. JOLLY, W. W., HARRIS, R. A., ASAI, J., and GREEN, D. E. 1969, *Arch. Biochem. Biophys.*, **130**, 191.
31. DEAMER, D. W., CROFTS, A. R., and PACKER, L. 1967, *Biochem. Biophys. Acta.*, **131**, 81.
32. PENNISTON, J. T., and GREEN, D. E. 1968, *Arch. Biochem. Biophys.*, **128**, 339.

PAPER 15

Contrasting protein architectures of plasma and mitochondrial membranes

DONALD F. H. WALLACH, ADRIENNE S. GORDON, JOHN M. GRAHAM and BARRY R. FERNBACH†

INTRODUCTION

IT HAS recently become possible to examine the structure of cellular membranes by spectroscopic techniques. In particular, infrared spectroscopy (IR)‡ in the regions of amide absorption (1–5) and measurements of the optical rotatory dispersion (ORD) and circular dichroism (CD) in the regions of peptide absorption, reflect the architecture of membrane proteins (1, 6–9), while measurements of proton magnetic resonance (PMR) indicate the physical state of lipids in membranes (10, 11). In the following we shall review relevant features of the IR, CD and ORD spectra of the plasma membranes of erythrocytes and Ehrlich ascites carcinoma. We shall also present IR spectra and new optical activity data for mitochondria and mitochondrial membrane components.

INFRARED SPECTROSCOPY

Infrared spectroscopy has been employed to detect the presence of β-conformation in cellular membranes (1–5). In this, use has been made of the fact that the $C=O$ stretching frequencies (Amide I bands) or peptides in helical or 'unordered' conformations lie at 1650–1660 cm^{-1}, whereas the β-conformations have this band near 1630 cm^{-1} (12, 13).

† From the Department of Biological Chemistry, Harvard Medical School, and the Biochemical Research Laboratory, Massachusetts General Hospital, Boston, Massachusetts 02114.

‡ Abbreviations used: CD, circular dichroism; EAC, Ehrlich ascites carcinoma; IR, infrared spectroscopy; ORD, optical rotatory dispersion; PM, plasma membrane; PMR, proton magnetic resonance.

Plasma Membranes

Film IR spectra of the plasma membranes of erythrocytes (Figure 1) and of Ehrlich ascites carcinoma cells (Figure 2) lack the Amide I band of β-conformation, except after heating (1), exposure to acid pH (3–5), and treatment with certain organic solvents (3, 5).

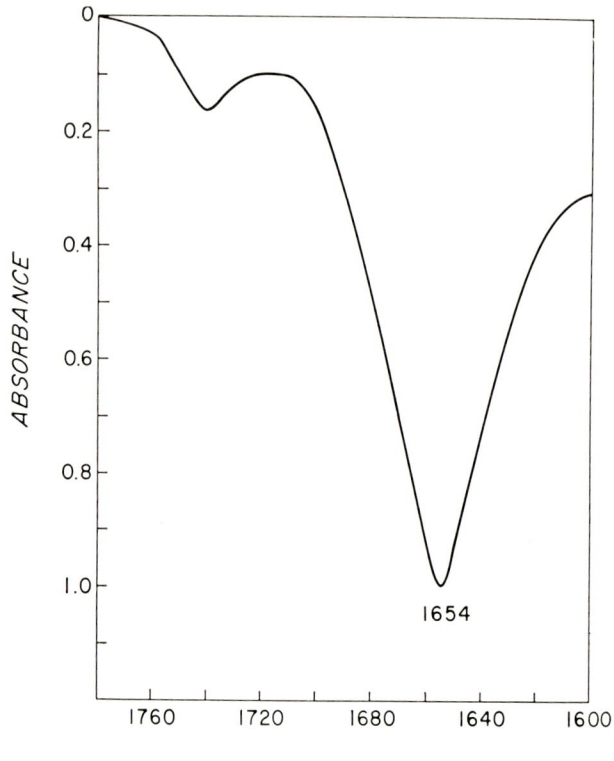

FIGURE 1   Amide I region of the IR spectrum of a film of erythrocyte ghosts on an AgCl plate. Ghosts were applied in 0.007M phosphate buffer, pH 7.4, to plate held at $-76\,°C$, and the frozen layer lyophilized. Spectrum obtained at $\sim 25\,°C$ and 36 per cent relative humidity.

Mitochondria

The film IR spectra of whole rat liver mitochondria (14), of isolated 'inner' membranes (15), of phosphorylating electron transport particles (16) and of 'mitochondrial structural protein' (17), obtained with a Perkin–Elmer spectrophotometer model 221, all have a prominent shoulder near $1630\,cm^{-1}$

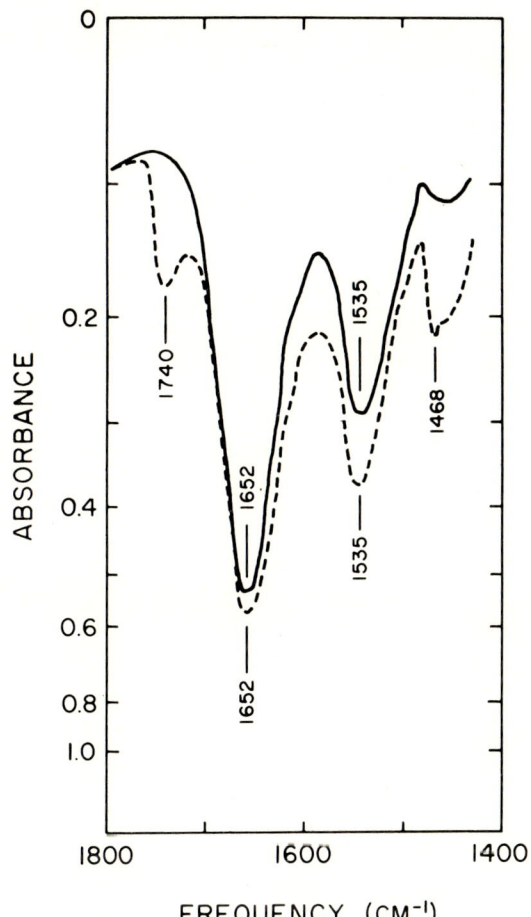

FIGURE 2 Infrared spectrum of a plasma membrane film cast from aqueous suspension: -----, original film; ———, after extraction with 2:1 chloroform:methanol. Figure reprinted by permission from *Proc. Natl. Acad. Sci., U.S.*, **56**, 1552 (1966).

indicative of β-conformation (Figures 3–5). The additional shoulder near 1690 cm^{-1} suggests that the β-structure may be anti-parallel 'pleated-sheet' (12, 13). On the basis of specific succinic dehydrogenase activity, our preparations of 'outer' membranes contained not more than 10 per cent of 'inner' membranes. IR spectra of 'smooth' microsomal membranes give no evidence of β-conformation.

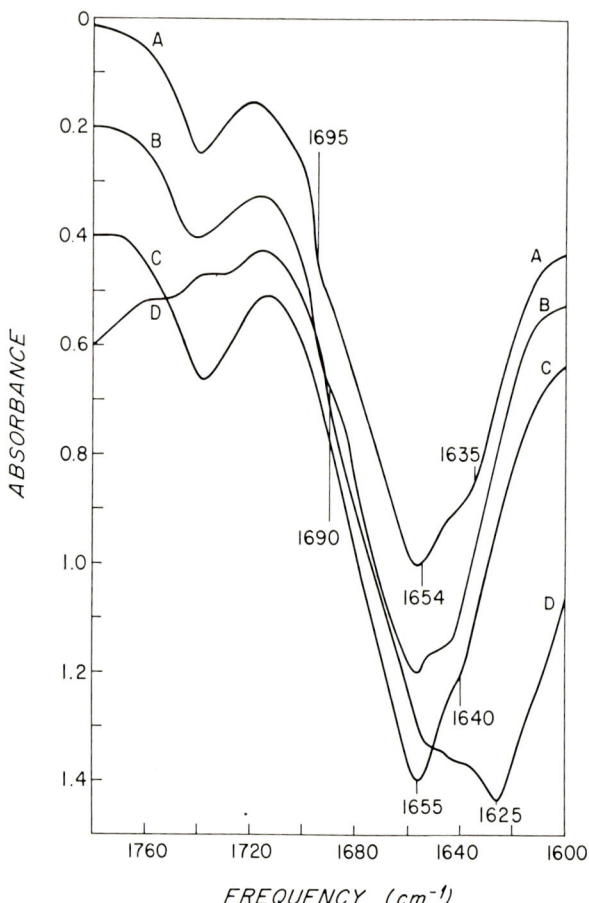

FIGURE 3 Amide I region of the IR spectrum of films of (A) whole rat liver mitochondria; (B) 'inner' mitochondrial membranes; (C) 'outer' mitochondrial membranes; (D) mitochondrial 'structural protein'[17]. Buffer 0.007M phosphate, pH 7.4. Films prepared as in legend for Figure 1. Individual spectra displaced by 0.2 absorbance units.

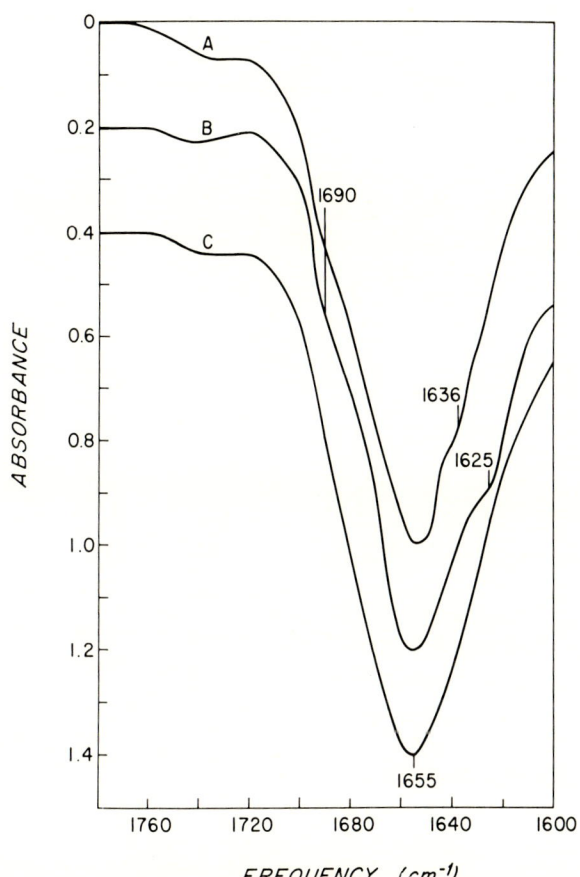

FIGURE 4 Amide I region of the IR spectra of films of (A) mitochondria, (B) 'inner' membranes, and (C) 'outer' membranes after lipid removal. The films of Figure 3 were extracted by dipping in 2:1 chloroform:methanol as in Refs. 3 and 4.

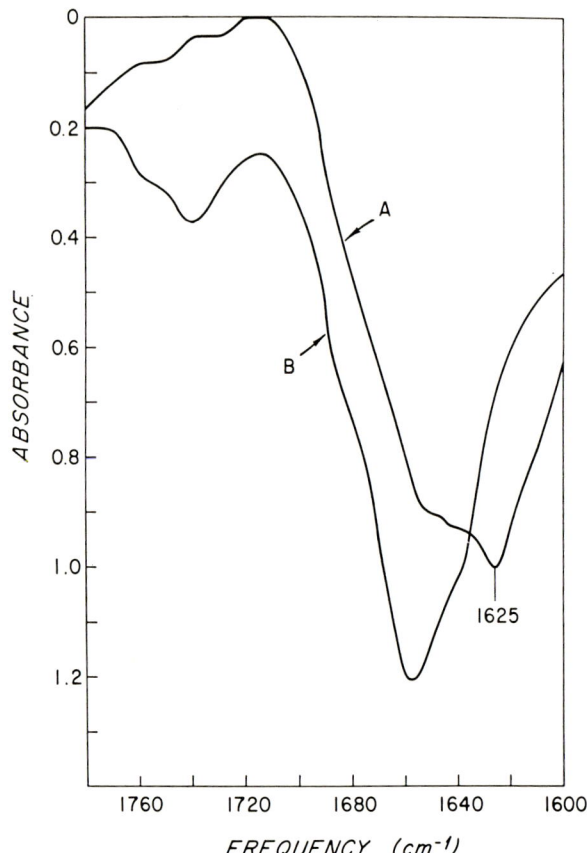

FIGURE 5 Amide I region of the IR spectra of films of (A) mitochondrial 'structural protein'[17] and (B) phosphorylating respiratory particles.[16] The films were prepared as described in the legend for Figure 1.

It is conceivable that the observed β-bands are an artefact of film preparation, although other membranes studied do not show β-conformation in dried films (1, 3–5); this matter will have to be settled by IR spectroscopy in D_2O. It is also possible that the proportion of β-structure in mitochondrial membranes depends upon their metabolic state; this question is under study.

OPTICAL ACTIVITY

CD and ORD are two closely related manifestations of optical activity. CD, expressed in terms of $[\theta]_\lambda$, the ellipticity per optically active residue at wavelength λ, is proportional to $(\varepsilon_L - \varepsilon_R)$, the difference in absorbance for left and right circularly polarized light. Optical rotation, expressed here as $[m]_\lambda$, the mean residue rotation at wavelength λ, is proportional to $(n_L - n_R)$, the difference in refractive index for left and right circularly polarized light.

Since the optical activity of peptide bonds depends strongly upon the secondary structure, and perhaps even the tertiary and quaternary structure of the peptide chain, its measurement can therefore give some clues to protein architecture. In the spectral region of peptide absorption, the CD and ORD spectra of polypeptides and proteins are summations of the contributions of the various individual peptide chromophores in the several conformations present. Since infrared spectroscopy has shown that the plasma membranes (PM) of human erythrocytes and Ehrlich ascites carcinoma (EAC) have negligible β-conformation, discussion of their optical activity can be limited to mixtures of helix and 'unordered' conformation (but this does not hold for mitochondria). Such a mixture is shown in Figure 6, the CD spectrum of poly-L-glutamic acid with half of its peptide bonds in α-helical conformation and half unordered. The figure also shows computed individual spectral contributions of the various peptide chromophores in the two conformations, namely: (a) the large, negative 'helical' $n-\pi^-$ band at 225 mμ ($\alpha, n-\pi^-$); (b) the small, positive 'unordered' band at 217 mμ (U, 217); (c) the large, negative, 'helical' $\pi^0-\pi^-$ band, polarized parallel to the helix axis at 206 mμ ($\alpha, \parallel, \pi^0-\pi^-$); (d) the large, negative, 'unordered' $\pi^0-\pi^-$ band at 198 mμ (U, $\pi^0-\pi^-$); (e) the large, positive, 'helical' $\pi^0-\pi^-$ band, polarized perpendicular to the helix axis at 191 mμ ($\alpha, \perp, \pi^0-\pi^-$).

Figure 7 is the corresponding ORD spectrum. For each CD band there is a symmetric, bimodal ORD curve, centered on the CD band and with the extrema occurring at + and − one halfwidth of the CD band. It is clear that optical rotation occurs well beyond the absorption band and that the position of ORD spectra is very sensitive to the width of the optically active bands. The negative ORD extremum at 233 mμ is due almost entirely to the helical $n-\pi^-$ band at 223 mμ.

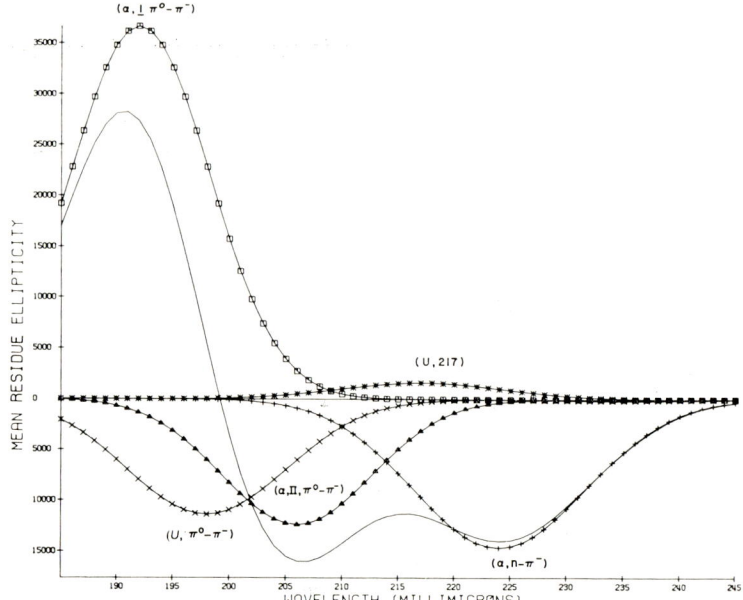

FIGURE 6 Calculated CD spectra of poly-L-glutamic acid and the individual bands contributing to the spectrum. (—) 50 per cent α-helix–50 per cent 'unordered' coil.

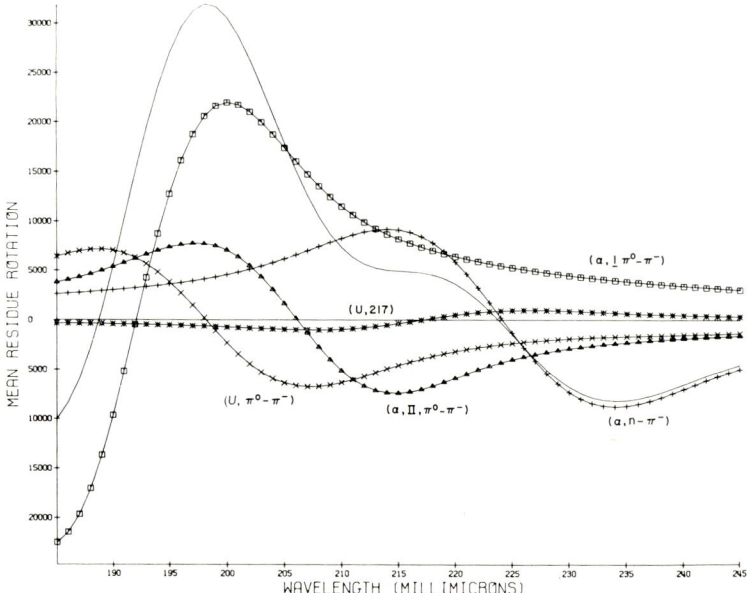

FIGURE 7 Calculated ORD spectra of poly-L-glutamic acid and the individual cotton effects contributing to the spectrum. (—) 50 per cent α-helix–50 per cent 'unordered' coil.

Plasma Membranes

The CD spectra of erythrocyte ghosts and plasma membranes (PM) of Ehrlich ascites carcinoma are presented in Figures 8 and 9 and are rather similar. Both show a prominent negative band at 223 mµ, indicative of considerable helical conformation. The positive band at 195 mµ also indicates

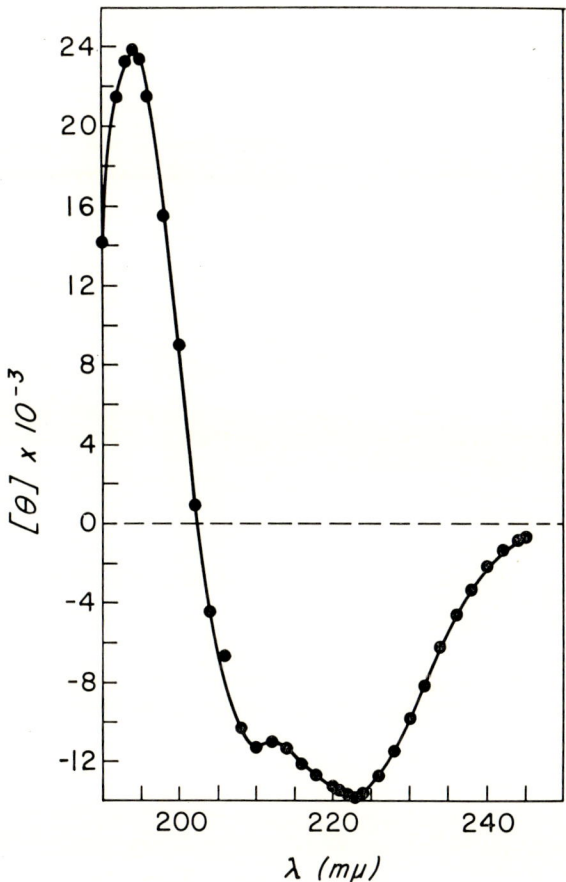

FIGURE 8 Experimental CD spectrum of erythrocyte ghosts. Solvent 0.007M phosphate, pH 7.4.

helix, but it is displaced to the red of the position found in synthetic polypeptides. This displacement, which occurs to a somewhat lesser degree in helix-containing globular proteins, is thought to reflect localization of some of the peptide bonds in an apolar environment of high refractive index (18). The intensities of the membrane CD spectra are small compared with those of synthetic polypeptides. This may be due to the presence of short helical segments, helices other than α-helix, localization of peptide in regions of low polarity, the non-'random' character of peptide regions lacking periodic structure, or a combination of any of these. The picture is only a somewhat more extreme case of what is found in globular proteins (18). The ORD

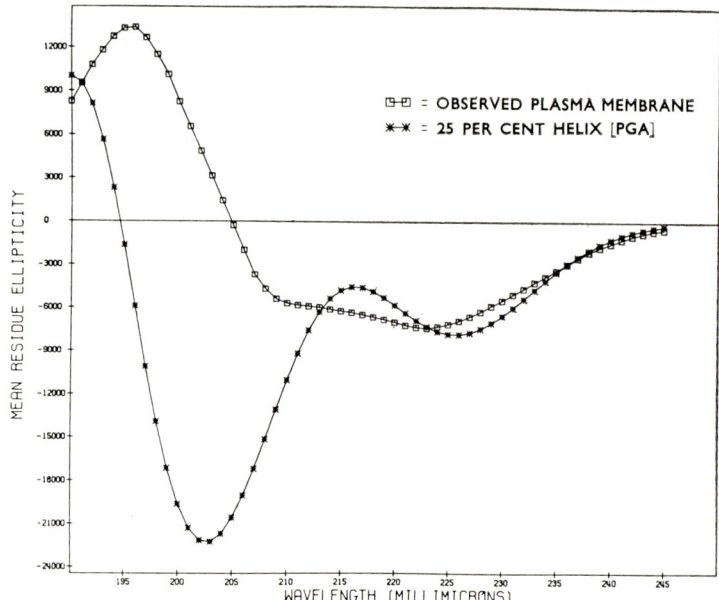

FIGURE 9 Experimental CD spectrum of plasma membrane of Ehrlich ascites carcinoma cells (□) and calculated CD spectrum of poly-L-glutamic acid, 25 per cent α-helix–75 per cent 'unordered' coil (*). Solvent 0.001M Tris-HCl, pH 8.2.

spectra of ghosts and PM, shown in Figures 10 and 11, exhibit a phenomenon characteristic of membrane spectra, namely, displacement of the negative extremum several mμ to the red of its position in α-helical polypeptides. This is due to the large width of the helical $n - \pi^-$ band.

Mitochondria

The optical activity of mitochondrial membranes differ considerably from that of other membranes studied so far. CD spectra (down to 200 mμ) of whole mitochondria, 'inner' mitochondrial membranes and 'outer' mito-

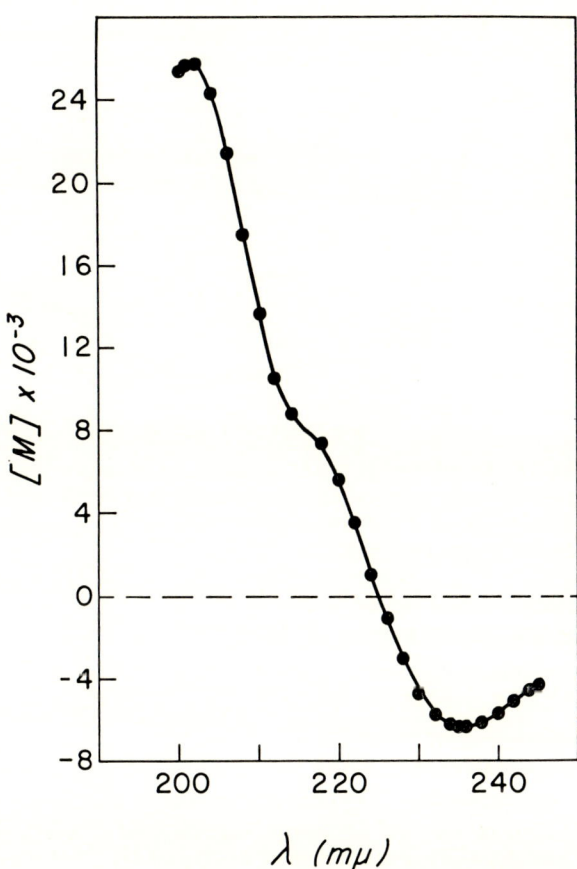

FIGURE 10 Experimental ORD spectrum of erythrocyte ghosts. Solvent 0.007M phosphate, pH 7.4.

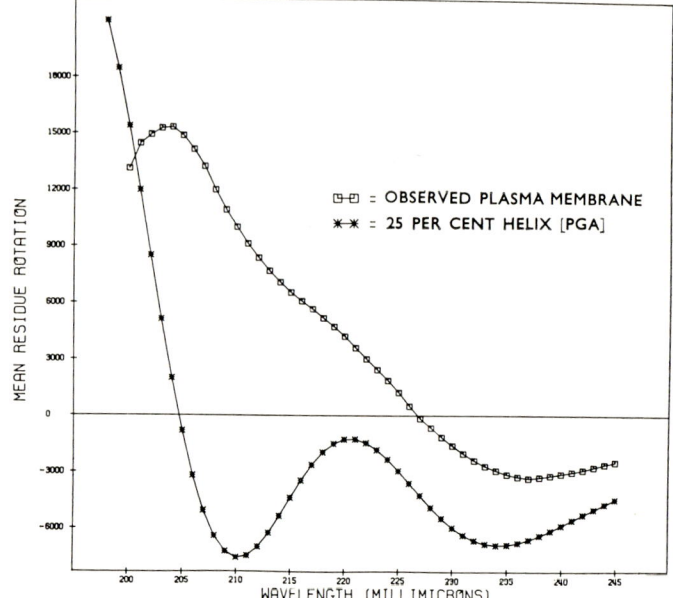

FIGURE 11 Experimental ORD spectrum of plasma membrane of Ehrlich ascites carcinoma cells (□) and calculated ORD spectrum of poly-L-glutamic acid, 25 per cent α-helix-75 per cent 'unordered' coil (*). Solvent 0.001 M Tris-HCl, pH 8.2.

chondrial membranes are shown in Figure 12. All exhibit a prominent minimum ($\Theta \sim 8000$) near 224 mμ; but, unlike that of other membranes, the position of this minimum varies from one mitochondrial preparation to the next, between 223.5 and 225 mμ. There is a very small shoulder near 210 mμ, and crossing-over to positive ellipticity occurs between 205 and 209 mμ. CD peaks were not obtained with these preparations because of poor signal-to-noise ratios. The crossovers of whole mitochondria and 'inner' membranes always lie 1–2 mμ to the red of 'outer' membranes. Corresponding ORD spectra have troughs lying at 236–237 mμ.

The CD spectrum of mitochondrial digitonin fragments (16), Figure 13a is essentially identical to that of 'inner' membranes. A maximum ($\Theta \sim 17,000$) lies at 196 mμ. The ORD spectrum (Figure 13b) has a trough at 237 mμ, a crossover at 226.5 mμ, a shoulder near 220 mμ, and a peak at 204 mμ. Except for our larger ellipticities and rotations, our data are in substantial accord with those of Urry, *et al.* (8).

The presence of β-structure makes interpretation of the optical activity of mitochondrial membranes more difficult than that of other membranes. Two

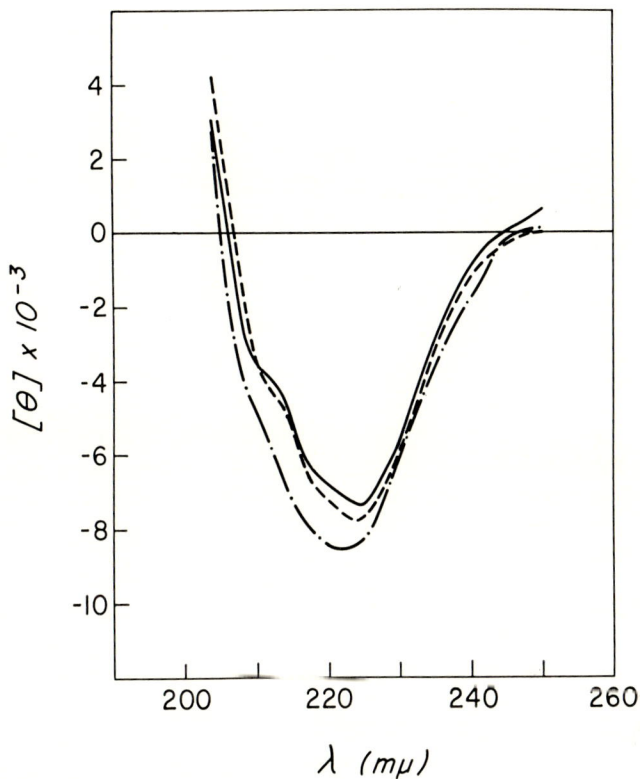

FIGURE 12 CD spectra in peptide absorption region of (a) whole mitochondria (———); (b) 'inner' mitochondrial membranes (– – – –); (c) 'outer' mitochondrial membranes (—·—·—). Buffer 0.007M phosphate, pH 7.4; protein concentration 0.2–0.4 mg/ml; path lengths 1 and 5 mm; ~ 26 °C. Data were calculated assuming a mean residue weight of 130 (see reference 4).

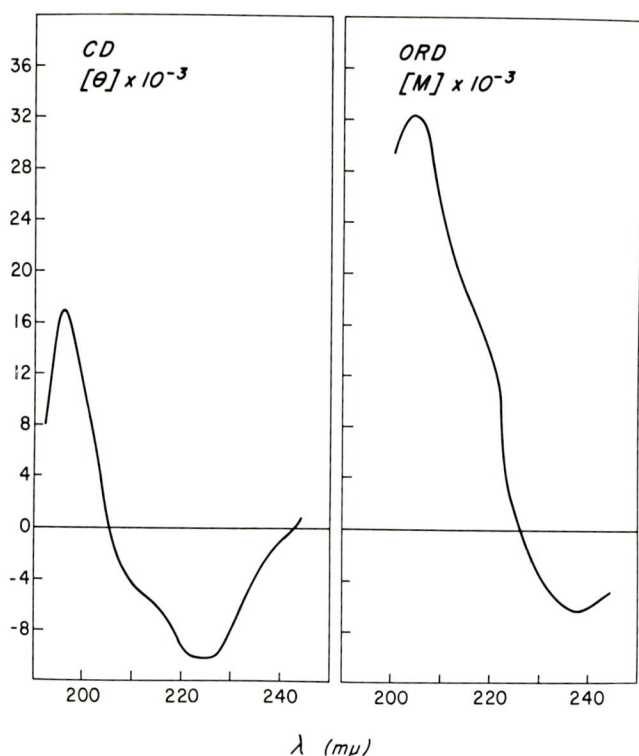

FIGURE 13 Optical activity spectra of phosphorylating respiratory particles.[16] (a) CD; (b) ORD. Conditions as for Figure 12.

recent reports lead us to conclude that it is the presence of this conformation which leads to the peculiarities of the mitochondrial spectra *vis-a-vis* those of other membranes. The first study (19) shows that the CD and ORD spectra of certain β-structured polypeptides with bulky side chains lie at much higher wavelengths than ordinarily expected for this conformation (Table I). The other study (20) shows that the aggregated β-structure formed by 1:1 mixtures of poly-L-glutamic acid and poly-L-lysine at neutral pH has the CD characteristics of β-conformation except for (*a*) very low ellipticity, (*b*) displacement of the minimum to ~ 223 mμ, and (*c*) displacement of the crossover to ~ 210 mμ. (According to these authors hydrophobic interactions are important in stabilization of these aggregates.) It is clear that optical activity spectra of membranes, and of proteins in general, cannot be interpreted even qualitatively without access to IR data.

TABLE I CD spectral positions of β-forming polypeptides†

Polypeptide	Positions (mμ)		
	Minimum	Crossover	Maximum
Poly-L-lysine (β)	219	207	—
Poly-O-acetyl-L-serine	216	203	196
Polyvaline	217	203	198
Poly-O-t-butoxyl-L-serine	220	212	200
Poly-S-carboxymethyl-L-cysteine	228	217	203

† From Stevens, *et al.*[19]

Since physical state and purity of preparations of mitochondrial 'structural' protein are difficult to control, we have examined the optical activity of such preparations only to test the contention of Steim and Fleischer (21) that the spectra indicate aggregation of α-helices. Our ORD data (Figure 14b) are in substantial agreement with those of Steim and Fleischer (21). The CD spectrum in 0.1 N NaOH (Figure 14a) shows a minimum at 222 mμ. At pH 7.4 there is a large decrease in ellipticity and a displacement of the minimum to 228 mμ, with a crossover at 209 mμ. However, it is our view that these spectra might well arise from mixtures of 'unordered conformation' and hydrophobically aggregated β-structure and that they do not warrant conclusions as to presence of α-helix, helical content or helical packing.

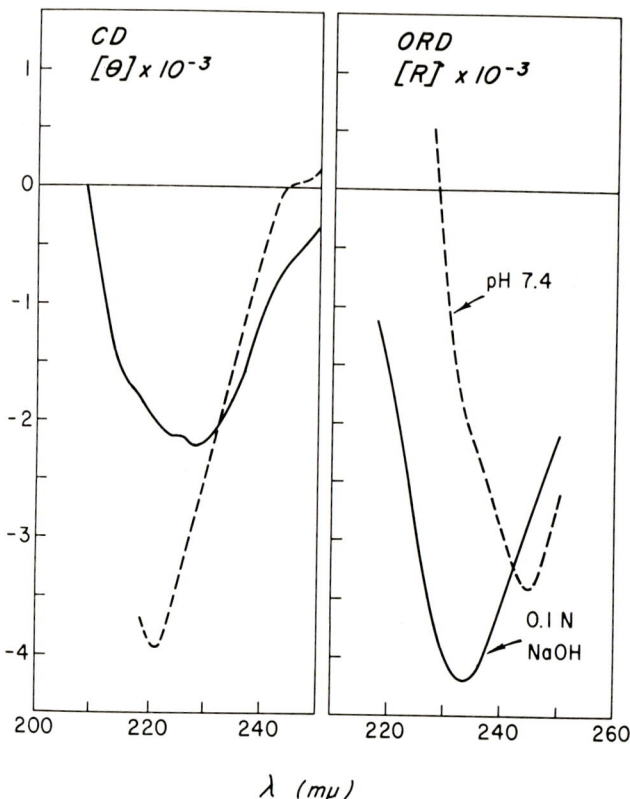

FIGURE 14 Optical activity spectra of mitochondrial 'structural' protein. (a) CD; (b) ORD: (———) in 0.1 N NaOH; (------) same, titrated to pH 7.4 with 1 N HCl.

CONCLUSIONS

The results presented here suggest that the proteins of plasma membranes have a substantial helical content and that they have a globular tertiary structure. However, since IR spectra demonstrate the presence of considerable β-structure in mitochondria, 'inner' mitochondrial membranes, digitonin fragments and mitochondrial 'structural' protein, the optical activities of these structures cannot be simply interpreted. The presence of α-helix is not established, and the spectra could originate from mixtures of 'unordered' and β-conformations. Because of the difficulties of interpretation introduced by the presence of β-structure in mitochondria, we cannot at present comment on the dependence of protein conformation on lipids in these membranes.

ACKNOWLEDGEMENTS

These investigations were supported in part by a research grant (CA-07382) from the National Cancer Institute of the National Institutes of Health, U.S. Public Health Service, through Harvard Medical School, and in part by an award from the Andres Soriano Cancer Research Fund to the Massachusetts General Hospital.

References

1. MADDY, A. H., and MALCOLM, B. R. 1965, Protein conformations in the plasma membrane, *Science*, **150**, 1616.
2. MADDY, A. H., and MALCOLM, B. R. 1966, Protein conformations in biological membranes, *Science*, **153**, 213.
3. WALLACH, D. F. H., and ZAHLER, P. H. 1966, Protein conformations in cellular membranes, *Proc. Nat. Acad. Sci., U.S.*, **56**, 1552.
4. WALLACH, D. F. H., and ZAHLER, P. H. 1968, Infrared spectra of plasma membrane and endoplasmic reticulum of Ehrlich ascites carcinoma, *Biochim. Biophys. Acta*, **150**, 186.
5. CHAPMAN, D., KAMAT, V. B., and LEVENE, R. J. 1968, Infrared spectra and the chain organization of erythrocyte membranes, *Science*, **160**, 314.
6. LENARD, J., and SINGER, S. J. 1966, Protein conformation in cell membrane preparations as studied by optical rotatory dispersion and circular dichroism, *Proc. Nat. Acad. Sci., U.S.*, **56**, 1828.
7. WALLACH, D. F. H. 1968, Optical rotatory dispersion studies *in* Chapter 4, Biological Membranes: Physical Facts and Function, (H. Chapman, editor), London, Academic Press, 176–189.
8. URRY, D. W., MEDNIEKS, M., and BEJNAROWICZ, E. 1967, Optical rotation of mitochondrial membranes, *Proc. Nat. Acad. Sci., U.S.*, **57**, 1043.
9. GORDON, A., WALLACH, D. F. H., and STRAUS, J. H., to be published.
10. CHAPMAN, D., KAMAT, V. B., DEGIER, J., and PENKETT, S. A. 1968, Nuclear magnetic resonance studies of erythrocyte membranes, *J. Mol. Biol.*, **31**, 101.
11. CHAPMAN, D., and KAMAT, V. B. 1968, Nuclear magnetic resonance studies of biological membranes, *in* Regulatory Functions of Biological Membranes, Proceedings of a Sigrid Juselius Symposium held at Helsinki, Finland, Nov. 6–9, 1967, (J. Jarnefelt, editor), Amsterdam, Elsevier Publishing Co., 100.
12. SUSI, H., TIMASHEFF, S. N., and STEVENS, L. 1967, Infrared spectra and protein conformations in aqueous solutions. I. The amide I band in H_2O and D_2O solutions, *J. Biol. Chem.*, **242**, 5460.
13. TIMASHEFF, S. N., SUSI, H., and STEVENS, L. 1967, Infrared spectra and protein conformations in aqueous solutions. II. Survey of globular proteins, *J. Biol. Chem.*, **242**, 5467.
14. HOGEBOOM, G. H. 1954, Fractionation of cell components of animal tissues, *in* Methods in Enzymology Vol. I, Preparation and Assay of Enzymes (S. Colowick and N. Kaplan, editors), New York, Academic Press, 16.
15. VIGNAIS, P. M., and NACHBAUR, J., 1968, A critical evaluation of the contamination, by lysosomes, of preparations of outer membrane of mitochondria, *Biochem. Biophys. Res. Com.*, **33**, 307.

16. HOPPEL, C., and COOPER, C. 1968, The action of digitonin on rat liver mitochondria, *Biochem. J.*, **107**, 367.
17. ALLMANN, D. W., LAUWERS, A., and LENAZ, G. 1967, Preparation of mitochondrial structural protein *in* Methods in Enzymology Vol. X, Oxidation and Phosphorylation (R. Estabrook and M. E. Pullman, editors), New York, Academic Press, 435.
18. STRAUS, J. H., GORDON, A. S., and WALLACH, D. F. H. The influence of tertiary structure upon the optical activity of three globular proteins: myoglobin, hemoglobin and lysozyme, submitted to *J. Biol. Chem.*
19. STEVENS, L., TOWNEND, R., TIMASHEFF, S. N., FASMAN, G. D., and POTTER, J. 1968, The circular dichroism of polypeptide films, *Biochem.*, **7**, 3717.
20. HAMMES, G. G., and SCHULLERY, S. E. 1968, Structure of macromolecular aggregates. I. Aggregation-induced conformational changes in polypeptides. *Biochem.*, **7**, 3882.
21. STEIM, J. M., and FLEISCHER, S. 1967, Aggregation-induced red shift of the Cotton effect of mitochondrial structural protein, *Proc. Nat. Acad. Sci., U.S.*, **58**, 1292.

DISCUSSION

GREEN   Dr. Wallach, you referred to the resolution of the mitochondrion into the inner membrane and the outer membrane. Would you kindly tell us how you achieved this resolution?

WALLACH   We did it by a number of techniques. In particular, a method used recently by Vignais and Nachbaur, involving swelling of the mitochondria followed by discontinuous centrifugation on a sucrose gradient.

GREEN   What I want to know specifically is how you decided that you had isolated the outer mitochondrial membrane?

WALLACH   The outer membrane fraction was almost devoid of NADH oxidase and succinate cytochrome C reductase activity. We estimate that contamination of outer membrane by inner was less than 10 per cent.

GREEN   I'm rather inclined to think on the basis of our own experience that your inner membrane preparation would be mitochondrial in origin and probably free of outer membrane. However, most of the outer membrane preparations reported in the literature have usually been heavily contaminated with microsomes. Are the great differences in the spectra of outer and inner membrane preparations which you have reported possibly due to the fact that your outer membrane preparations are predominantly nonmitochondrial in origin?

WALLACH   I would hesitate to say that the $1630\,\text{cm}^{-1}$ shoulder in the spectrum of the outer membrane could not be due to some contamination. However, smooth endoplasmic reticulum membranes of the liver don't show this, nor do plasma membranes. It is a peculiarity not entirely explicable by contamination. I think the important point is, that unlike the plasma membranes, which have been studied quite a bit, the inner membrane of the mitochondrion contains a significant proportion of its peptide linkages in the antiparallel β-conformation.

GREEN   Yes, I agree that your inner membrane preparations are very likely mitochondrial in origin and contain predominantly components of the inner membrane. I am satisfied with the conclusions you have drawn from the study of the inner membrane preparations. It is the authenticity of the outer membrane preparation which I am questioning. Until this authenticity is established, it may be hazardous to conclude anything about the possible structural differences between outer and inner mitochondrial membranes.

PAPER 16

Cell and photoreceptor membranes†

JEROME J. WOLKEN
*Biophysical Research Laboratory, Carnegie Institute of Technology,
Carnegie-Mellon University, Pittsburgh, Pennsylvania 15213*

INTRODUCTION

THE LIVING CELL is a dynamic system concerned with the processes of energy capture, transfer, storage and conversion—processes vital to its growth, maintenance and reproduction. The cell enclosed by its cell membrane possesses organelles, differentiated structures, to carry on its life processes. These structures include a nucleus, mitochondria, endoplasmic reticulum, ribosomes, golgi system and chromatophores (Figure 1). Cilia and flagella are found in some cells to facilitate cell movement. There are more specialized cell organelles, photoreceptors to capture light energy for phototropisms, photosynthesis and vision. Other sensory receptors to detect changes in the physical and chemical environment have also evolved.

The importance of the cell membrane to the concept of the cell goes back to the beginning of microscopy; for it was not until the development of the compound microscope in the 1830s that it was possible to begin to resolve cell structures. This was followed by a century of developments in optics to make possible polarization, phase and interference microscopy that could reveal structural information about the cell. These observations together with electron microscopy developed in the 1940s and newer methods of tissue preparation since the 1950s have made it possible to visualize cellular structures in molecular dimensions.

CELLULAR MEMBRANES

The cell, or plasma, membrane is described as a double layer of lipids and proteins (Figure 2a). Its molecular structure schematized by Danielli and

† Research aided in part by National Aeronautics and Space Administration NGR 39-002-011.

Physical principles of biological membranes

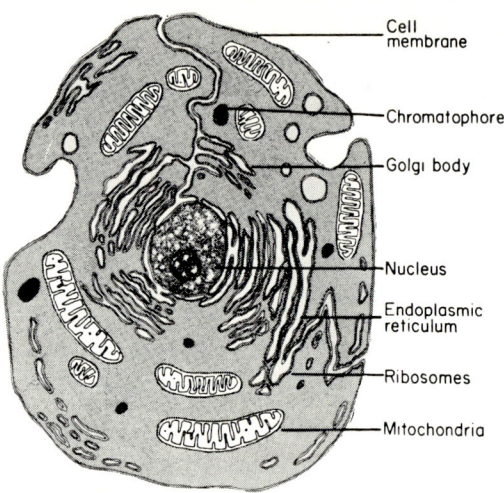

FIGURE 1 Schematic animal cell and its organelles.

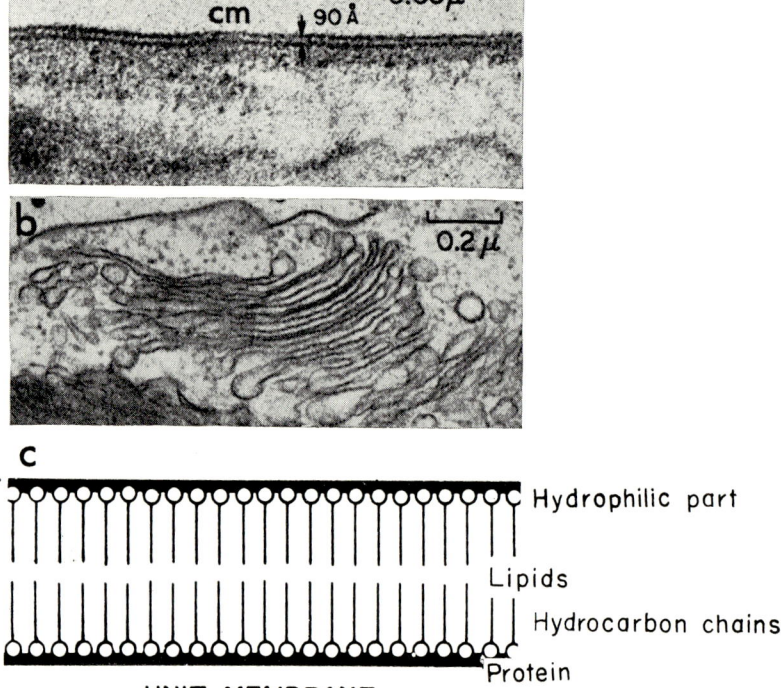

FIGURE 2 (a) cell membrane, *Euglena*. Electron micrograph. (b) golgi membrane system, *Euglena*. Electron micrograph. (c) hypothetical unit membrane.

Davson (1935) and later by Robertson (1966) has served as a model for all cell membranes (Figure 2c). An important question then is whether or not all organelle membranes are structurally and chemically similar to the cell membrane. However the molecular structure of the cell membrane and how it functions in allowing for the differential diffusion of ions and molecules is not completely understood.

The cell cytoplasm contains many organelle systems of complex membrane structures. One such system of membranes is the endoplasmic reticulum

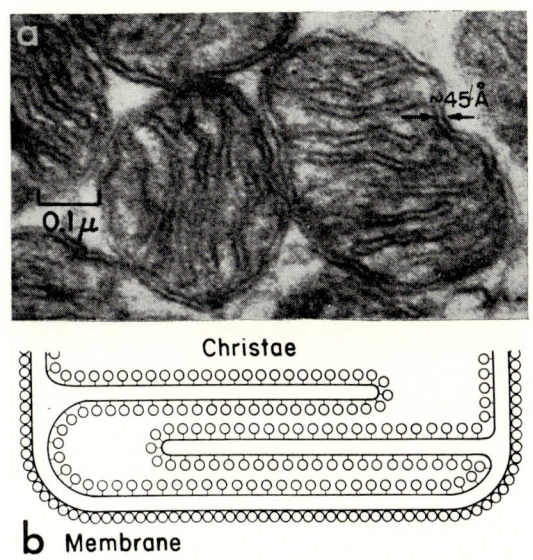

FIGURE 3 (a) mitochondria, inner segment of frog retinal rod. Electron micrograph, (b) mitochondrion, schematic.

(Figure 1) to which the ribosomes for protein synthesis are attached. Closely associated with the endoplasmic reticulum is another cluster of membranes, the golgi system (Figure 2b). The most numerous cytoplasmic organelles are the mitochondria (Figure 3a) which contain the respiratory cytochrome enzyme system. Lipids, mainly phospholipids, make up 30 per cent of the dry weight of mitochondria; of these the major phospholipids are phosphatidylethanolamine, phosphatidylcholine (lecithin) and cholesterol. Although mitochondria contain the enzymatic system for the biosynthesis of fatty acids, they do not possess all the enzymes required for assembly of the complex phosphatides (Lehninger, 1964). Structural information from electron microscopy shows that mitochondria possess two membranes, a limiting membrane and an inner membrane (Figure 3a). A system of ridges designated as *cristae* (Figure 3b) protrudes from the inside surface of the inner membrane. For more detailed information on the structure and function of the mitochondrian, refer to the discussions of Green and Chance in this conference.

Let us now explore the structural relationship between these cellular membranes and the structure of the photoreceptors.

PHOTORECEPTORS

The photoreceptors of plant and animal cells are specialized structures that contain a pigment or pigment system to capture light energy which initiates

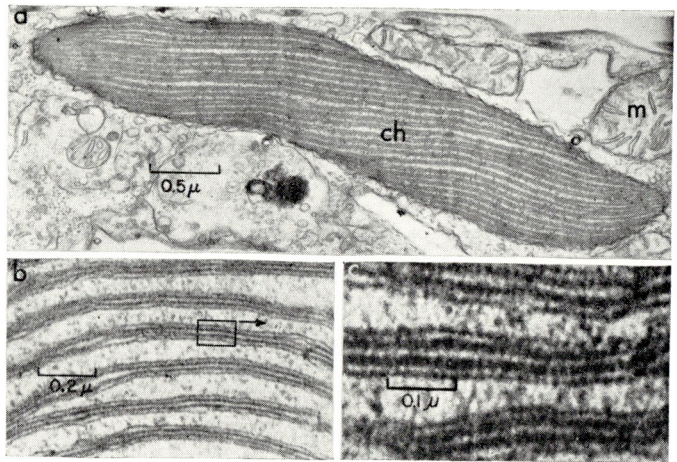

FIGURE 4 (a) *Euglena* chloroplast, ch; mitochondria, m. (b) enlargement of section of chloroplast. (c) further enlargement of area in (b). Electron micrographs.

phototropisms, photosynthesis and vision. In the process the photoreceptors function as transducers by converting or transferring this energy to chemical and/or to electrical energy. In plant cells the photoreceptors are chloroplasts for photosynthesis and in animal cells these are retinal rods and cones of the eye for vision.

Chloroplasts

The chloroplasts of *Euglena* for example, as seen by electron microscopy, consist of lamellae (Figure 4a). There are about 21 such lamellae in the *Euglena* chloroplast, each of which is a double layer of $2n$ surfaces. The total thickness of the lamellae is of the order of 250 Å and each membrane is of the order of 50 to 100 Å in thickness (see enlarged areas, Figure 4b, c). The electron-dense membranes are associated with the lipids and proteins, the less dense interspaces with aqueous proteins, dissolved salts and water-soluble enzymes. These designations are assumed from the staining and the chemical reactions of fixing agents (e.g., osmium tetroxide) within these structures.

The chloroplasts in a variety of plants contain from 35 per cent to 55 per cent protein, 18 per cent to 37 per cent lipids and 5 per cent to 8 per cent inorganic material on a dry weight basis. The chlorophyll pigments average

TABLE I The Composition of Spinach Chloroplast Lamellae (Quantasome)†

	Number of Molecules
Chlorophylls	230
chlorophyll *a*	160
chlorophyll *b*	70
Carotenoids	48
Lipids	
phospholipids	116
digalactosyl diglyceride	114 ⎫ 460
monogalactosyl diglyceride	346 ⎭
phospholipids	116
sulpholipids	48

† Taken in part from Park and Biggins (1964).

about 6 per cent and the carotenoid pigments about 2 per cent. In addition there is one cytochrome molecule and one ferredoxin molecule for about every 300 chlorophyll molecules (Wolken, 1967).

The major chloroplast lipids (Table I) are the mono- and di-galactosyl diglycerides (Benson, 1964; Park and Biggins, 1964). These mono- and di-galactosyl diglycerides, the carriers of α-linolenate, contain unusually high concentrations of polyunsaturated fatty acids (Wolken, 1968). The lipids are not only a structural part of the chloroplast membranes but may also be functional, for when the lipids are extracted from chloroplasts their photochemistry is inhibited.

To see how much chlorophyll was available to the chloroplast and what area of the total lamellar surfaces would be occupied by the chlorophyll molecules, the chlorophyll concentration per *Euglena* chloroplast was determined and the number and thickness of the lamellae were statistically evaluated (Table II). The cross-sectional area that would be available in the lamellae for the porphyrin (chlorophyllin) part of the chlorophyll molecule (see chlorophyll molecule, Figure 5) was found to be 222 Å (Wolken and Schwertz, 1953). The chlorophyll cross-section in the lamellae calculated for a variety of plant chloroplasts was also found to be of the order of 200 Å^2 (Elbers, Minnaert and Thomas, 1957). About 225 Å^2 is near the right cross-section for a porphyrin molecule spread on a water-air interface. As a result of the geometric and pigment analysis the schematic molecular network for the chloroplast (Figure 5) was conceived (Wolken and Schwertz, 1953). The

TABLE II Chloroplast, *Euglena gracilis*†

Diameter	1.23 μ (1.04 – 1.42)‡
length, D	6.50 μ (5.2 – 9.3)
number of dense layers, n	21 (18 – 24)
dense layer thickness, T	242 Å (180 – 303)
chlorophyll molecules, P	1.02 × 10^9 (0.88 – 1.36)
	1.34 × 10^9 (calculated from absorption obtained from a single chloroplast using the microspectrophotometer)

† Taken in part from Wolken and Schwertz (1953).
‡ Mean and extreme in the measurements.
  2.5 × 10^7, number of chlorophyll molecules per lamellar surface.
  4 × 10^{13}, number of chlorophyll molecules/cm^2.

chloroplast lamellar network shows that four chlorophyll molecules are united to form tetrads and oriented so that only one of the phytol tails is located at each intersection in the rectangular network (Figure 5). This arrangement has the advantage of leaving adequate space for at least one carotenoid molecule to every three chlorophyll molecules.

Since the molecular weights of the carotenoid molecules are one-half to two-thirds of the molecular weight of the chlorophyll molecules, a weight ratio, chlorophyll to carotenoid, of approximately 4 : 1 to 6 : 1 would be expected. On the other hand, the carotenoid molecules are slender linear molecules, ~5 Å in diameter, and therefore more than one molecule could conveniently fit into the 15 Å × 15 Å holes formed by the chlorophyll tetrads. However this would lead to very tight fitting which would be energetically improbable. One can therefore put a lower limit on the number of chlorophyll to carotenoid molecules of roughly 1 : 1 and a weight ratio of 2 : 1. Such close packing of the chlorophyll and carotenoid molecules in the lamellae would permit energetic interaction between them.

Cell and photoreceptor membranes

There are several possible ways in which the chlorophyll molecules could be oriented in the chloroplast lamellae. If the porphyrin part of the chlorophyll molecules lies at 0° as depicted in Figure 5, their greatest cross-section would be available for light capture. However if they are oriented at increasing angles to 45° the cross-sectional area available would be decreased by half to about 100 Å². Since the chlorophyll molecules in the chloroplast are probably

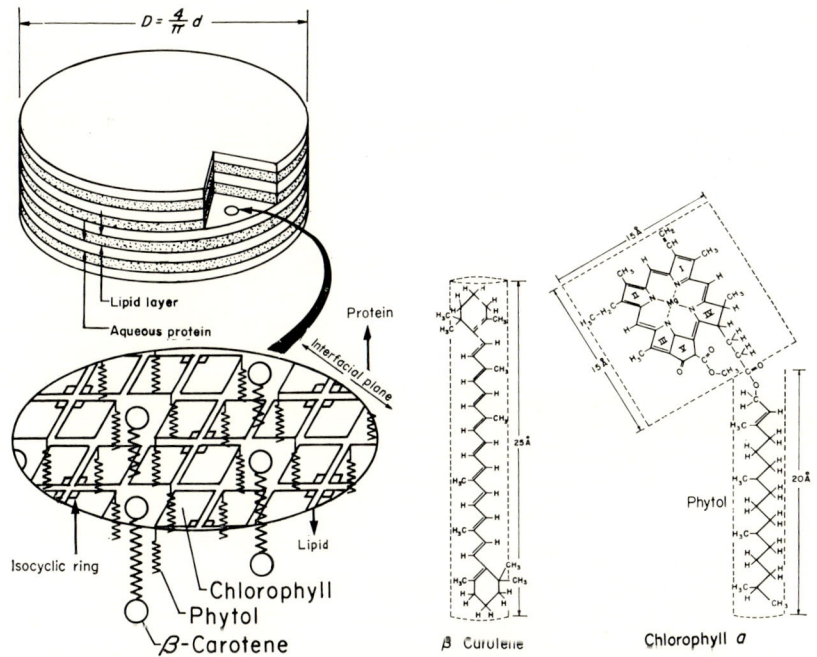

FIGURE 5 Schematized molecular network in or on the lamellae of the chloroplast.

in a dynamic state, they would arrange themselves for maximum light absorption. To do so they are probably held in a specific molecular configuration with its protein and lipid.

It will be noted in Table I that the mono- and di-galactosyl di-glycerides account for the major lipids in the chloroplast. It was suggested by Benson (1966) that these lipids can form a lipid or lipoprotein matrix for chlorophyll monolayers (Figure 4). Also from spatial considerations the ratio of two galactosyl diglyceride molecules to one chlorophyll molecule could stabilize all the chlorophyll molecules in the monolayer. That is there would be one phytol chain of chlorophyll for four cis-unsaturated acyl chains of -galactosyl

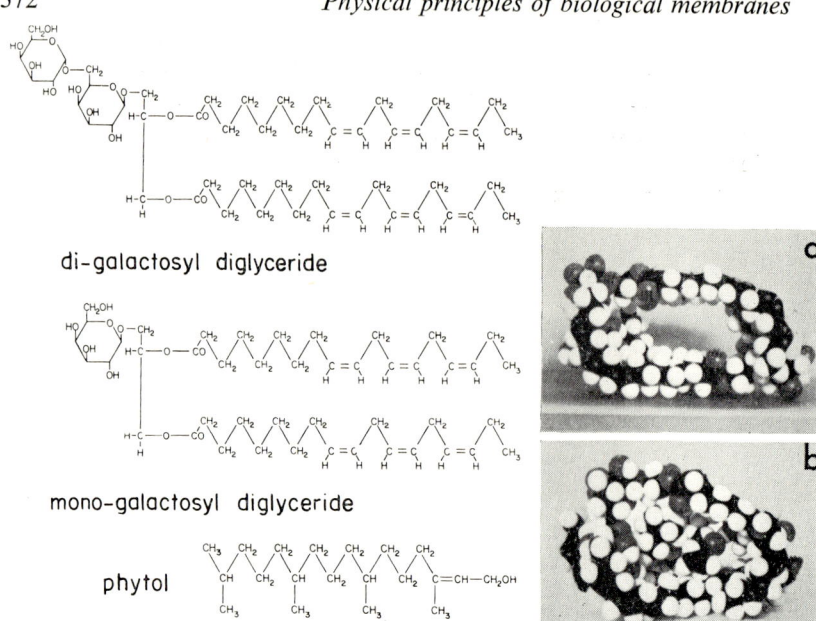

di-galactosyl diglyceride

mono-galactosyl diglyceride

phytol

FIGURE 6 Molecular structure of di-galactosyl diglyceride and monogalactosyl diglyceride, the major chloroplast lipid and phytol, the tail of the chlorophyll molecule. (a) molecular model of how the four chains of -galactosyl diglycerides would form a cis-unsaturated chain so that *phytol* could fit (b).

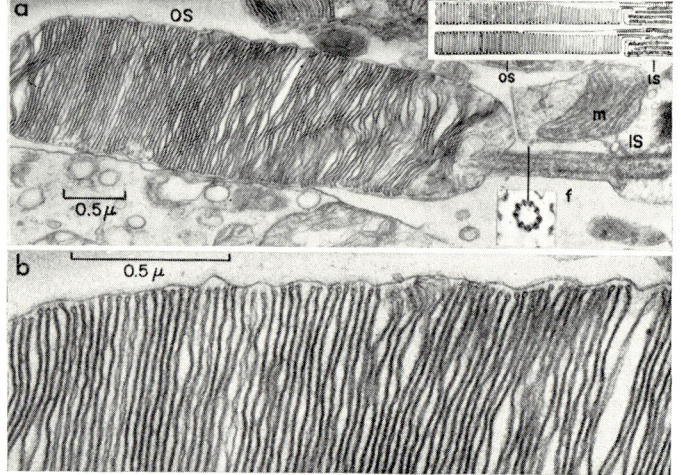

FIGURE 7 (a) cattle retinal rod, insert schematic retinal rod: outer segment, o.s.; inner segment, i.s.; flagellum, f; and insert cross-section. Electron micrographs. (b) enlarged area of (a) to show in greater detail the lamellae.

diglycerides (Rosenberg, 1967). Such a relationship is illustrated in Figure 6 and fits in with the molecular model proposed for the chloroplast lamellae (Figure 5).

Rod and Cone Structure

The nervous layers of the retina comprise the visual cells: the rods and cones, the bipolar cells and the ganglion cells. The rods and cones are differentiated structures of the retinal cells, each having an inner segment and what appears as a rod or cone shaped outer segment which contains all of the photosensitive visual pigments. The rod membrane appears as a continuum of the retinal cell membrane. Also the inner segments are packed mitochondria as seen in Figure 7. The outer segment appears to be connected to the inner segment by a fibril (cilium) that runs from the outer segment through the inner segment (Figure 7a). Embryologically the rods and cones are probably derived from flagella (De Robertis, 1956). The structure of the flagellum is similar in arrangement to that of the nine fibrils found in the cilia, flagella and spermatozoa tails of many plant and animal cells. The close association of the mitochondria of the inner segment with the central end of the fiber may be a significant factor in the functional chemistry of the rods and cones.

The rods and cones contain as their photosensitive visual pigments either retinal$_1$ or retinal$_2$ (the aldehydes of vitamin A_1 or A_2) complexed with a protein, opsin. Such extracted pigment-protein complexes are identified by their absorption spectra as rhodopsin (retinal$_1$ + rod opsin) or porphyropsin (retinal$_2$ + rod opsin) for the rods; and iodopsin (retinal$_1$ + cone opsin) or cyanopsin (retinal$_2$ + cone opsin) for the cones (Wald, 1961). The rods have their maximal sensitivity in the blue-green at about 500 nm; the cone sensitivity is transferred toward the red, lying in the yellow-green near 560 nm.

Chemical analyses of the rod outer segments in terms of wet weight indicate that the visual pigment, rhodopsin, accounts for 4–10 per cent, the lipids for 20–40 per cent, and the proteins for 40–50 per cent. For example, cattle rhodopsin constitutes 3.6 per cent of its wet weight and 13 per cent of its dry weight, whereas frog rhodopsin constitutes 10 per cent of its wet weight and more than 35 per cent of its dry weight. Using polarizing microscopy Schmidt (1935) demonstrated that the optical properties of the outer segment closely correspond with those of the myelin sheath of nerve. He suggested that since, from the optical properties, the long axis of the lipid molecules appears to run parallel to the axis of the outer segment, there must be planes of non-lipid material arranged at right angles to the long axis. This would account for the reversal in sign of birefringence when lipids are extracted. He then demonstrated that the molecules of the non-lipid discs lie transversely to the axis. Other observations on the dichroism of the outer segments already suggested

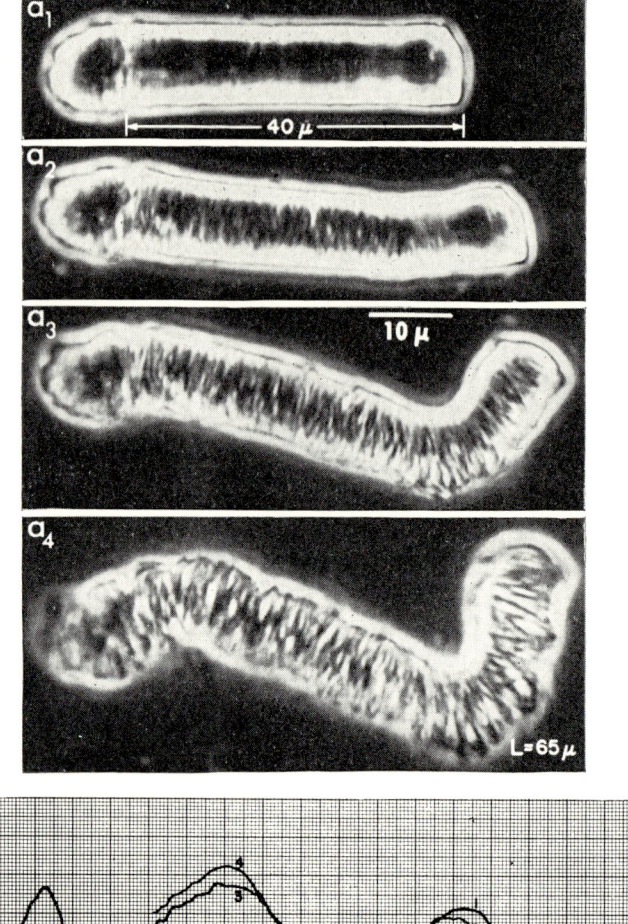

FIGURE 8 Freshly isolated frog retinal rod a_1, bleached with white light a_2, a_3, and a_4. Note change in length and structure. (b) spectra of a_{1-4} which accompanies these structural changes (i.e., bleaching of rhodopsin to retinal).

that the visual pigment is probably oriented in the protein regions of the retinal rod.

Electron micrographs of sonicated, fixed and shadowed guinea pig retinal rod outer segments (Sjöstrand, 1949, 1953) clearly showed that they were platelets of about 2μ in diameter with edges of 75 Å in thickness. Since these early electron microscopic studies all the vertebrate retinal rod outer segments that have been fixed and stained by various techniques and examined in the electron microscope show dense, double-membraned lamellae (platelets, discs, sacs) that are of the order of 200 Å in thickness. Each membrane (lamellae) of the discs is from 50–75 Å in thickness (Cohen, 1963; De Robertis and Lassansky, 1961; Fernández-Morán, 1961; Wolken, 1963, 1966b).

It was observed that during light excitation the rods expand and after dark adaptation contract (Detwiler, 1943). Isolated frog rod outer segments in frog Ringers solution or in vitreous also exhibit this effect upon light excitation. It can be seen in Figure 8a (1–4) that the rod almost doubles in length without any change in thickness. This change is accompanied by bleaching of rhodopsin (see Figure 8b). The photochemical and structural changes of fixed and sectioned rods before (Figure 8a(1)) and after (Figure 8a(4)) were examined with the electron microscope (Figure 9a, b). The electron micrographs showed that the lamellae expand like muscle fibers and suggested that in the process the light energy is transferred to mechanical energy and then to chemical energy (Wolken, 1966a).

To visualize a structural molecular model for the vertebrate retinal rods the geometry (length, diameter, number of lamellae and their thickness) was measured and tabulated from numerous electron micrographs. The rhodopsin concentration was determined from spectroscopic data of a single retinal rod or from extracts of a known number of rods. Both calculations show that there are of the order of 3×10^9 rhodopsin molecules per frog retinal rod and 4×10^6 rhodopsin molecules per cattle retinal rod. With these data on the geometry and pigment concentration, the surface area that the rhodopsin molecule would occupy in the lamellae was calculated (Table III).

Assuming that the lamellae are double layers of lipids and proteins, a double layer is then structurally represented as (lipo)protein with the lower molecular weight lipids occupying the interstitial spaces (Figure 10). It is assumed also that monomolecular layers of pigment molecules are at the (lipo)protein-lipid interfaces.

The cross-sectional area A which would be associated with each rhodopsin molecule and therefore with each pigment molecule is

$$A = \pi D^2/4P$$

where D is the diameter of the photoreceptor and P is the number of rhodopsin molecules in a single monolayer. The maximum cross-sectional area A for each rhodopsin molecule can be derived from the above equation where P is replaced by $N/2n$, in which N is the pigment concentration in molecules per retinal rod and n is the number of electron-dense lamellae per outer segment; then

$$A = \pi D^2 n / 2N$$

The cross-sectional areas calculated for cattle and frog rhodopsin are then 2500 and 2620 Å² respectively (Table III). The diameter of the rhodopsin molecule would be of the order of 50 Å. This would be about the right order of magnitude since a rhodopsin molecule (cattle, frog), if symmetrical, would

TABLE III Structural Data—Retinal Rod Outer Segments[†]

	D Diameter μ	n Number of dense layers per photo- receptor	N Number of rhodopsin molecules per rod	A Calculated cross- sectional area of rhodopsin Å²	d_m Diameter of rhodopsin molecule Å	M Calculated molecular weight rhodopsin[‡]
Frog	5.0	1000	3.8×10^9	2620	51	60,000
Cattle	1.0	180	4.2×10^6	2500	50	40,000

[†] Data taken from Wolken (1961).
  $n = 2n$ surfaces available.
  M is calculated on the basis that *opsin* is a protein, density 1.3; if a lipoprotein, the density would be closer to 1.1 and hence the molecular weight would be reduced.
[‡] HELLEN, J. (1969 *Biochem.*, **8**, 675), found that the molecular weight of cattle, frog, and rat rhodopsin is about 28, 000.

have a diameter of the order of 40 Å (Wald, 1954). The calculated area available for rhodopsin indicates that there would be sufficient space for all of the rhodopsin molecules on the surfaces of the lamellae.

DISCUSSION AND SUMMARY

All cellular membranes appear to be double layers (lamellae as seen in electron micrographs, Figures 2a, b, and 3a) of about 50 to 100 Å in thickness. Such structures for cellular membranes are associated with a property of the lipids, that is their capacity to form fairly uniform molecular layers as in myelin. This is attributable to the presence of hydrophilic, water-soluble groups at one end of the lipid molecule and to hydrophobic, fat-soluble groups at the other end to form lamellae. These lamellae would greatly maximize the surface area, minimize the organelle volume and serve to separate one part of a reaction from another.

The photoreceptors, the chloroplasts and the outer segments of the retinal rods and cones also consist of lamellae, a double layer of the order of 250 Å in thickness. Each lamella or membrane is of the order of 50 to 100 Å in thickness (Figures 4, 7, 9a and 11).

Arguments as to the number of lamellae or membranes and their molecular dimensions exist because of the chemical nature of fixation which is necessary for electron microscopy. Therefore the question of which fixative, osmium tetroxide (OsO_4) or potassium permanganate ($KMnO_4$), gives the most reliable structural information remains. The subtle chemical differences between these and other cell fixatives are yet to be resolved. Since all cellular

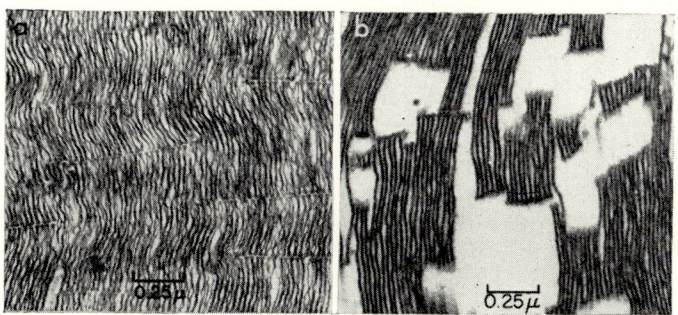

FIGURE 9 (a) retinal rod section of a_1 in Figure 8 and (b) section a_4, after bleaching. Electron micrographs.

organelles contain high concentrations of lipids, it is believed that osmium stains the polar surfaces of the oriented lipid molecules. However new microanalytical instrumentation with high resolving power is necessary to follow the chemistry and to determine the structure of these organelles in the living state. One such instrument which has been developed and applied to study the pigment-complexes in the photoreceptor organelles is the microspectrophotometer, see Figure 8b (Wolken, et al., 1968).

There is some experimental evidence to indicate that cell membranes in different organelles may differ from those of the *unit membrane* as depicted in Figure 2c (Branton and Park, 1968). It is of interest to note the ratio of the percent lipid to percent protein in various organelle lamellae which not only have functional differences (for example, myelin which is about 80 per cent lipid and 20 per cent protein, red blood cell membrane 40 per cent lipid and 60 per cent protein, mitochondrical inner membranes 25 per cent lipid and 75 per cent protein), but also structural differences. There is also the possibility

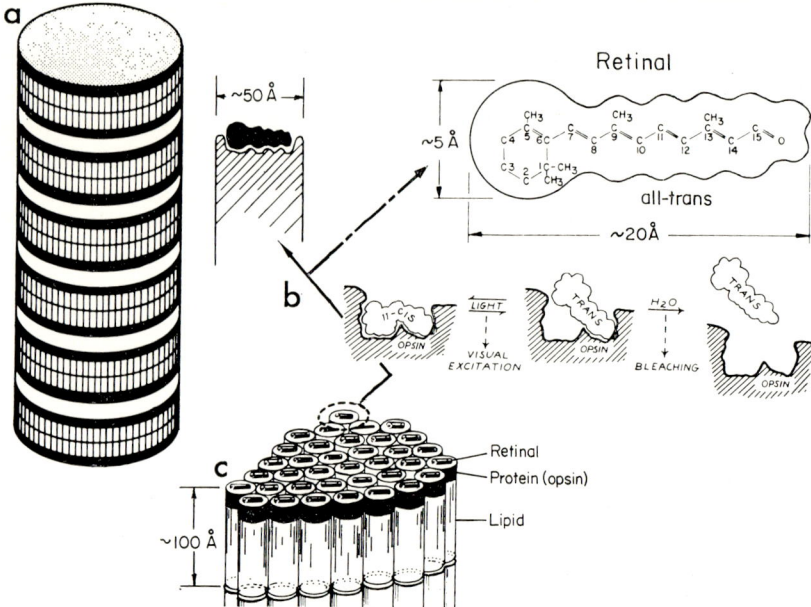

FIGURE 10 (a) schematic molecular model of retinal rod. (b) showing possible relationship of how 11-cis retinal is associated with its protein opsin and how light changes the configuration to the *all-trans* retinal in visual excitation (taken from Hubbard and Kropf, 1959). (c) showing how these rhodopsin molecules may be packed in the lamellar sacs.

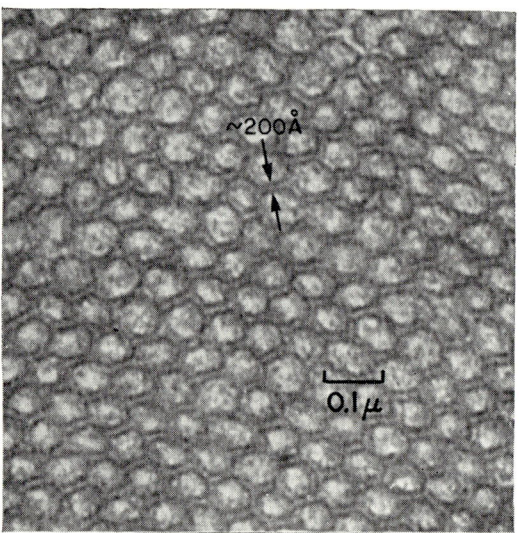

FIGURE 11 Structure of invertebrate arthropod rhabdomere (retinal rod) showing tubules which differ from the lamellar sacs, Figures 7 and 9a. Electron micrograph.

that the lamellae consist of globular subunits (Figure 4c; schematized in Figure 12) or a lipoprotein complex, micelles. Also, some note should be made here that the highly ordered photoreceptor structure may have as its analog that of a semiconductor.

Besides chlorophyll in the chloroplasts, all photoreceptors contain carotenoids: C_{40}—as in β-carotene, the precursor to vitamin A. Retinal, the

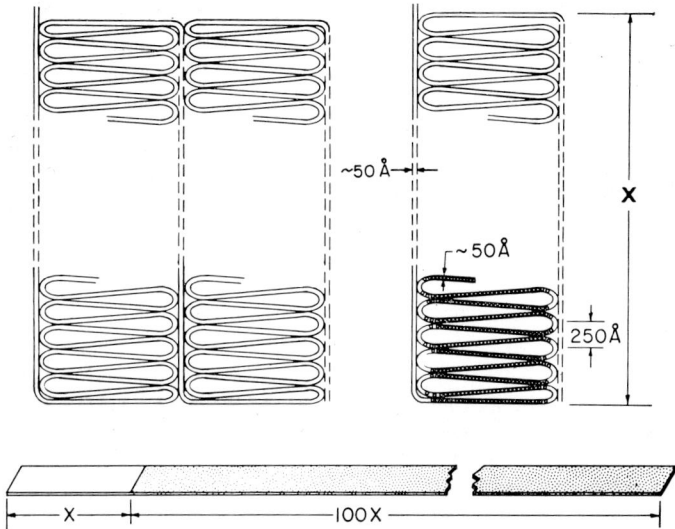

FIGURE 12 Schematic for possible structure of the retinal rod lamellae and the rhodopsin molecules (50 Å) that form the lamellae.

aldehyde of vitamin A is a C_{20}—carotene found in the retinal rods and cones of all eyes. Also note the resemblance in structure of the phytol part of the chlorophyll molecular to β-carotene and retinal (Figures 5 and 10). These polyene pigment molecules are part of the lipids of the lamellae and participate in the energy transfer process (Wald, 1968; Lundegärdh, 1967).

It has been suggested that temperature variation alone, within the physiological range, can alter the physical state of fatty acids and phospholipids. Luzzati and Husson (1962) and Stoeckenius (1962) have studied the phase changes which occur in brain phospholipids in vitro. They found that lipids may exist in both an expanded lamellar form and a condensed hexagonal form and that the transition from the lamellar to hexagonal forms occurs near 37°C. The lamellae exhibit properties of liquid crystals for they are temperature- and concentration dependent. It is of interest then that the chloroplast lamellae of *Euglena* are permanently destroyed at a temperature above

33°C within a generation and the result is a mutation, *Euglena* lacking chlorophyll and chloroplasts.

To measure an early receptor potential (ERP) in the retinal rods requires an intact rod and the rhodopsin molecules almost perfectly oriented in the lamellae (Wald, 1968). If the retinal rods are heated above 48°C which will disrupt these lamellae and hence the rhodopsin orientation, the ERP is gone. These studies strongly indicate that this highly ordered system of lamellae and the orientation of the pigment molecules are necessary for function. Therefore it becomes exceedingly important to learn the precise composition, expecially the kinds of lipids and proteins of these lamellae. The analysis of the chloroplast lipids (Table I) has suggested a possible relationship between phytol and the galactosyl diglycerides (Figure 6). However we know too little about the photoreceptor lamellar proteins, *plastins*, in the chloroplasts and *opsin* in the retinal rods and cones. This kind of information together with the chemistry and electrical properties is vital to understanding how the photoreceptor pigment-lipid-protein lamellar systems function in photoexcitation.

References

BENSON, A. A. 1964, *Ann. Rev. Plant Physiol.*, **15**, 1.
BENSON, A. A. 1966, *J. Amer. Oil Chem. Soc.*, **43**, 265.
BRANTON, D., and PARK, R. B. in *Papers on Biological Membranes*. Boston, Little, Brown & Co., 1968, pp. 1–20.
COHEN, A. I. 1963, *Biol. Rev. Cambridge Philos. Soc.*, **38**, 427.
DANIELLI, J. F., and DAVSON, H. 1935, *J. Cell. Comp. Physiol.*, **5**, 495.
DE ROBERTIS, E. 1956, *J. Biophys. Biochem. Cytol.*, **2**, 319.
DE ROBERTIS, E., and LASSANSKY, A. in *The Structure of the Eye*, G. K. Smelser, ed. New York, Academic Press, 1961, p. 29.
DETWILER, S. R. *Vertebrate Photoreceptors*. New York, Macmillan, 1943.
ELBERS, P. F., MINNAERT, K., and THOMAS, J. B. 1957, *Acta Botan. Neerl.*, **6**, 345.
FERNANDEZ-MORAN, H. in *The Structure of the Eye*, G. K. Smelser, ed. New York, Academic Press, 1961, p. 521.
HELLEN, J. 1969, *Biochem.*, **8**, 675.
HUBBARD, R., and KROPF, A. 1959, *Ann. N. Y. Acad. Sci.*, **81**, 388.
LEHNINGER, A. L. *The Mitochondrian*. New York, W. A. Benjamin, Inc., 1964.
LUZZATI, V., and HUSSON, F. 1962, *J. Cell Biol.*, **12**, 207.
LUNDEGARDH, H. 1967, *Nature*, **216**, 981.
PARK, R. B., and BIGGINS, J. 1964, *Science*, **144**, 1009.
ROBERTSON, J. D. 1966, *Ann. N. Y. Acad. Sci.*, **137**, 421.
ROSENBERG, A. 1967, *Science*, **157**, 1191.
SCHMIDT, W. J. 1935, *Z. Zellforsch. u. mikroskop. Anat.*, **22**, 485.
SJOSTRAND, F. S. 1949, *J. Cell. Comp. Physiol.*, **33**, 383.
SJOSTRAND, F. S. 1953, *J. Cell. Comp. Physiol.*, **42**, 15.
STOECKENIUS, W. 1962, *J. Cell Biol.*, **12**, 221.
WALD, GEORGE, 1968, *Science*, **162**, 230.

WALD, GEORGE, in *The Structure of the Eye*, G. K. Smelser, ed. New York, Academic Press, 1961, p. 101.
WALD, GEORGE, 1954, *Science*, **119**, 887.
WOLKEN, J. J. 1968, *J. Amer. Oil Chem. Soc.*, **45**, 241.
WOLKEN, J. J. *Euglena: An Experimental Organism for Biochemical and Biophysical Studies*. 2nd ed. New York, Appleton-Century-Crofts, 1967.
WOLKEN, J. J. *Vision: Biophysics and Biochemistry of Retinal Photoreceptors*. Springfield, Ill., Charles C. Thomas, 1966a.
WOLKEN, J. J. 1966b, *J. Amer. Oil Chem. Soc.*, **43**, 271.
WOLKEN, J. J. 1963, *J. Opt. Soc. Amer.*, **53**, 1.
WOLKEN, J. J., in *The Structure of the Eye*, G. K. Smelser, ed. New York, Academic Press, 1961, pp. 173–192.
WOLKEN, J. J., FORSBERG, R., GALLIK, G., and FLORIDA, R. 1968, *Rev. Sci. Instrum.*, **39**, 1734.
WOLKEN, J. J., and SCHWERTZ, F. A. 1953, *J. Gen. Physiol.*, **37**, 111.

DISCUSSION

GREEN   This very fascinating account of the photoreceptors in light-sensitive systems is now open for discussion. I was quite intrigued with the electron micrographs you showed. The studies of my laboratory would suggest that there is a tightening of the discoidal membranes of the rod outer segments in the course of bleaching. Have you ever observed that at all?

WOLKEN   No.

WALLACH   Is there any nuclear magnetic resonance spectroscopy during light activation?

WOLKEN   Yes, studies have been done on chloroplasts and retinal rods.

PAPER 17

The indication of a light induced electrical field by pigments incorporated in chloroplast membranes

W. JUNGE, H. M. EMRICH,† and H. T. WITT
*Max-Volmer-Institut, I. Instituet fur Physikalische Chemie,
Technische Universitaet Berlin*

SUMMARY

EVIDENCE IS provided that some of the light induced absorption changes in chloroplasts of green plants are due to an electrochromic response of pigments to a light induced electrical field across the thylakoid membrane. These absorption changes can be used as a molecular voltmeter between two aqueous phases of such a small size that they are not accessible to the common electrochemical measuring techniques.

The relevance of the light induced electrical field in photosynthesis is discussed and the extension of the measuring method to other biological systems proposed.

INTRODUCTION

The structure on which the primary processes of photosynthesis of green plants take place is a paucimolecular membrane inside the chloroplasts. According to widely accepted results of several authors (1, 2) this membrane is based on a unimolecular layer of lipids, a second layer of pigments and a third layer of structural protein (Figure 1). The pigment layer is composed of several types of chlorophyll and of carotenoids, which stick to the hydro-

† On leave from the Max-Planck-Institut fuer Biophysik, Frankfurt/M (Germany).

phobic side of the lipids. These membranes form disk shaped closed vesicles (3, 4, 5), so called thylakoids. The size of one thylakoid can be characterized by the fact that it contains in the order of several 10^5 chlorophyll molecules.

The primary processes of photosynthesis are initiated by the absorption of light energy by the bulk pigments in the second layer and by the final trapping of quanta by two special types of photochemically active chlorophyll molecules. The activation of the two light reactions is followed by three complex events:

(1) Electrons are transferred from water to a terminal electron acceptor $NADP^+$ (6). At least 9 intermediates of the electron transport system have been identified and the kinetic data of their interaction have been

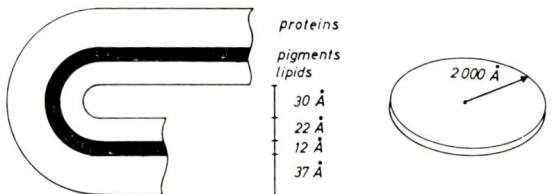

FIGURE 1 Gross structure and average dimensions of one thylakoid.

evaluated (for recent summaries see (7, 8)). The size of one electron transport chain can be characterized by the fact that the two photochemically active pigments are surrounded by a bulk of about 500 chlorophyll molecules (9). Thus one thylakoid includes several hundred electron transport chains.

(2) Protons are taken up by the chloroplasts (10). This proton uptake is accompanied by the release of other cations such as K^+ or Mg^{++} (11) and by swelling and shrinking of the thylakoids (12) which are at least partly due to an osmotic balance of ion fluxes.

(3) ATP is synthesized from ADP and inorganic phosphate (13).

The end products of the primary processes (NADPH and ATP) are consumed in the CO_2-fixation cycle (14).

It is still an open question, in which way part of the absorbed light energy may be transformed into the free energy necessary for the synthesis of ATP. This has to be specified since it has motivated the work reported herein.

In his hypothesis on this subject Mitchell (15) has postulated that on illumination a concentration gradient of protons and an electrical field may be set on across the photosynthetic membrane. The free energy thus stored in a difference of the electrochemical potential of the protons may be used by a membrane bound anisotropic ATPase, which translocates protons 'down-hill' this potential difference.

Several experimental results on the correlation between the light induced pH difference and the formation of ATP give support to Mitchell's hypothesis (16, 17, 18).

However, the question if there is any electrical field involved in the synthesis of ATP could not be answered by common electrochemical methods. Even microelectrodes are inadequate for measuring the electrical potential in the very small inner phase of one thylakoid.

So a direct response to any electrical field across the thylakoid membrane is reserved to indicators of a molecular size. A straightforward way to get an indicator for an electrical field is the study of electrochromic effects of the bulk pigments, that are incorporated in the thylakoid membrane.

Although related to the Stark effect, the splitting in an electrical field of the spectral lines of atoms in the gas phase, the term 'electrochroism' is used in the following in order to designate shifts in the unresolved spectra of larger molecules in the condensed state. The first two terms in a power series expansion of the frequency shift (Δv) of the absorption spectrum of a dye molecule are linear and quadratic in the field strength (**E**). They depend on the difference of the permanent dipole moments of the ground and the excited state ($\mu^o - \mu^2$) and on the difference in the polarizability ($\alpha^\dagger - \alpha^o$) respectively:

$$\Delta v = \frac{1}{h}[(\mu^\dagger - \mu^0) \cdot \mathbf{E} + \tfrac{1}{2}\mathbf{E} \cdot (\alpha^\dagger - \alpha^0) * \mathbf{E}]$$

Order of magnitude calculations (19) and experimental results on porphyrins (20) indicate that the shifts of the rotational and vibrational transitions are smaller than those contributed by the electronic transition. Therefore a nearly homogeneous shift of a whole absorption band can be expected. If the frequency shift is small as compared with the bandwidth the difference spectrum of a pigment between the two states with and without electrical field should be simply the derivative of the absorption spectrum.

Of course, one may think of other mechanisms that can give rise to spectral changes of a dye molecule exposed to a transient electrical field: a field induced change of chemical equilibrium as well as local conformational changes in the membrane, that may influence the interaction of pigments.

The question arises as how to detect absorption changes that may be due to a light induced electrical field across a very small membrane.

METHOD

The method of repetitive pulse photometry has been developed as an extremely sensitive tool for the detection of small absorption changes. This technique has been applied especially for those pigments which are involved in the primary processes of photosynthesis. Since details of this method are summarized in (21) only the principle may be reported.

A monochromatic beam of weak measuring light (wavelength) passes a cuvette with an aqueous suspension of spinach chloroplasts (Figure 2). The intensity of the transmitted light is detected and recorded. On repetitive excitation ($n \simeq 1000$) of the photochemically active chlorophyll molecules the n transient absorption changes are superimposed in a storage device and averaged. In this way the sensitivity and time resolution are increased by a

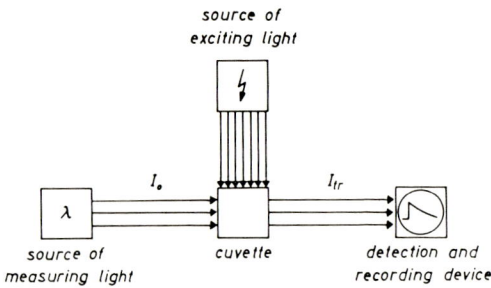

FIGURE 2   Scheme of a flash photometer.

factor $\sqrt{n}$ in comparison with single excitation. Mainly by this method the small absorption changes of 9 intermediates of the electron transport have been detected and the difference spectra clearly separated (7, 8).

Thus if there is any absorption change due to an electrical field across the thylakoid membrane, it has to be discriminated from the aforementioned absorption changes of the electron transport chain and from possible absorption changes due to light induced transients in the concentration of a special type of ions (e.g. H^+).

PROPERTIES OF THE FIELD INDICATING ABSORPTION CHANGES

Tests to discriminate between a field indicating and an electron transport indicating absorption change have been applied to a well known type of absorption change. This type with maxima at 478 nm (neg.), 515 nm (pos.) and 648 nm (neg.) was ascribed to a reaction of chlorophyll-b (22). It revealed

The indication of a light induced electrical field

some properties common with phosphorylation but different from electron transport (23, 24). It was proposed that it might be due to a protolytic reaction of chlorophyll-b.

A typical time course of the absorption change at 515 nm is depicted in Figure 3. On excitation with a short flash of light there is a rapid rise of absorption followed by a rather slow decay in the dark. If this absorption change indicates the rapid onset of an electrical field across the thylakoid membrane and the following decay of this field due to an intrinsic conductivity of the membrane (Figure 3), it has to react very sensitively to any change in the membrane's conductivity.

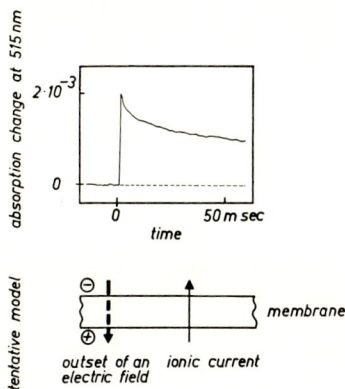

FIGURE 3 Typical time course of the absorption change at 515 nm (above) and tentative model for the interpretation of this absorption change (below).

There are two conclusive experiments that demonstrate that the absorption change at 515 nm does not reflect direct properties of the electron transport chains but fulfill the expectation for a field indicating absorption change perfectly. While the electron transport should work in principle even on membrane fractions that contain only a minimum set of about 500 chlorophyll molecules, the electrical field should disappear (respectively become not resolvable quickly) if the inner and the outer phase of thylakoids are short circuited by grave ruptures of the membrane.

Indeed, this behavior is reflected by the 515-absorption change (Figure 4) on treatment of chloroplast membranes with the permeability increasing antibiotic gramicidin D. In contrast to this there is no influence on the rate of (uncoupled) electron flow.

On the other hand only one pore per each thylakoid ($\simeq 10^5$ chlorophyll molecules) that is permeable for ions should yield an acceleration of the field decay. This has been tested (25) with the antibiotic gramicidin D which in rather low concentrations increases the permeability of several biological membranes for alkaline ions (26, 27). Again the expected effect, an accelleration of the field decay on gramicidin treated chloroplasts, is reflected by the 515-absorption change (Figure 5). Moreover, a minimum concentration of only one molecule of gramicidin on $2 \cdot 10^5$ chlorophyll molecules induced an acceleration of only one half of the amplitude of the absorption change while

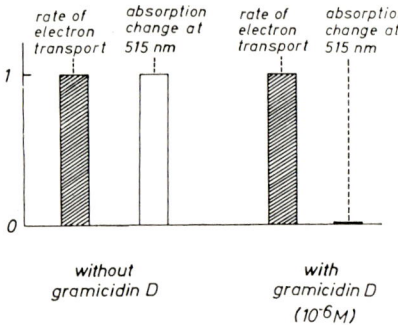

FIGURE 4 Response of the rate of (uncoupled) electron flow and of the absorption change at 515 nm to relatively high concentration of gramicidin D. (Chlorophyll concentration: 10^{-5} M; rate of electron flow in continuous illumination; absorption change on flash excitation). (8).

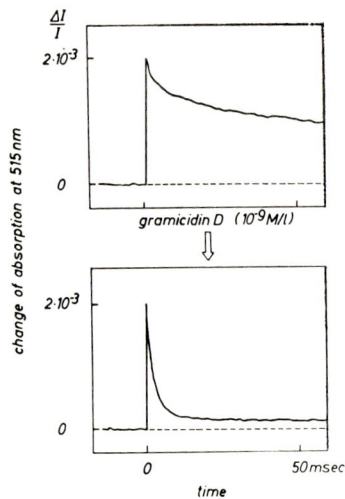

FIGURE 5 Acceleration of the absorption change at 515 nm by small concentrations of gramicidin D. (Chlorophyll concentration: 10^{-5} M).

the other half decayed slowly as before (25). This demonstrates that the absorption change at 515 nm reflects a property that is defined on one whole thylakoid ($\simeq 10^5$ chlorophyll molecules). This is conclusive to discriminate between the 515-absorption change and absorption changes of the electron transport chain, and it is consistent with the assumption, that the 515-absorption change indicates an electrical field across the thylakoid membrane.

It has still to be demonstrated that the absorption change at 515 nm does not reflect transient changes of the local concentrations of only one special type of ion. In this context it would be conclusive if the absorption change is reactive only to the electrical charge of permeating ionic species but not to their chemical nature.

In fact the absorption change at 515 nm can be accelerated if the concentration of chemically different species of ions is increased (e.g. K^+, Mg^{++}, Cl^-). This has been found for thylakoids where the membrane permeability was increased by an osmotic shock (25, 28). Whereas on gramicidin treated chloroplasts the acceleration was dependent on the concentration of alkaline ions (25), which reflects the well known property of gramicidin (26, 27).

If the decay rate of the absorption change at 515 nm reflects the discharge of the thylakoids capacity by ion fluxes across the thylakoid membrane, the observed first order decay rate ($1/\tau_{\frac{1}{2}}$) should be proportional to the sum of several conductivity terms taken over all permeating ionic species.

$$\frac{1}{\tau_{\frac{1}{2}}} \sim \sum_i \sigma_i$$

Ciani has shown theoretically that even on paucimolecular membranes the conductivity (σ_{K^+}) contributed by a given ion (e.g. K^+) can be approximated by the value contained in the Nernst–Planck equation (29).

$$\sigma_{K^+} = u_{K^+} \cdot \bar{c}_{K^+}$$

In this equation u_{K^+} is the specific conductivity coefficient for the given ion (K^+) and $\bar{c}_{K^+}$ an appropriate average of the concentration of this ion in the membrane phase.

The relationship between the decay rate of the field and the conductivity becomes very simple if the contribution of one ionic species is dominating. This can be realized on gramicidin treated chloroplasts provided that the concentration of e.g. potassium is higher than the concentration of any other alkaline ion. Under these conditions the following relation has to be expected:

$$\frac{1}{\tau_{\frac{1}{2}}} \sim u_{K^+} \cdot \bar{c}_{K^+}$$

Thus the specific conductivity u_{K^+} and therefore the decay rate $1/\tau_{\frac{1}{2}}$ should increase with increasing gramicidin concentration (the number of pores for

K^+) if the potassium concentration is held constant. Indeed this behavior is reflected by the decay rate of the 515-absorption change (Figure 6, right side). On the other hand, the gramicidin concentration being constant, it has to be expected that the average concentration ($\bar{C}_{K^+}$) of potassium in the membrane and therefore the decay rate increases with the potassium concentration in the suspension. That this holds in a certain concentration range is shown in Figure 6, left side (28).

Thus the properties of the absorption change at 515 nm are in agreement with the assumption, that this absorption change indicates a light induced electrical field across the thylakoid membrane. The question arises as to the mechanism of this indication.

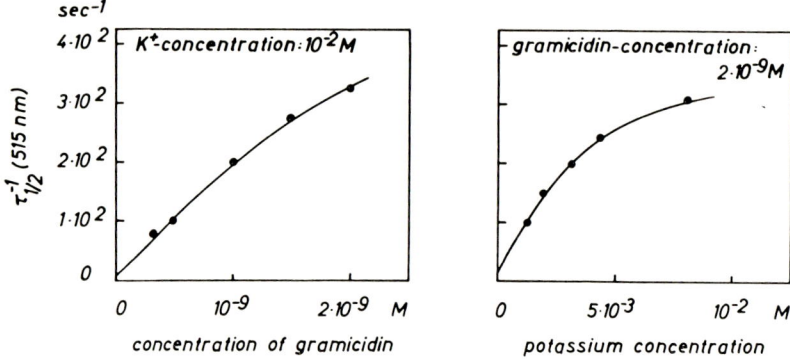

FIGURE 6 Dependence of the decay rate of the absorption change at 515 nm ($\tau_{1/2}^{-1}$) on the concentration of gramicidin (left side) and on the concentration of potassium (right side). (28).

If the response to an electrical field at the wavelength 515 nm is due to an electrochromic effect, similar absorption changes can be expected in all those spectral regions where the bulk pigments absorb. Since these pigments stick directly to the non-aqueous lipid layer in the thylakoid membrane across which the field only can exist (see Figure 1), they are exposed to the stray-field at the membrane. Indeed the overall spectrum of the field indicating absorption changes (Figure 7) reveals maxima in the blue where the carotenoids and chlorophylls absorb and in the region of the red bands of chlorophylls *in vivo* (chl-b_{653}, chl-a_{673}, chl-$_{683}$, chl-a_{695}).

Only those components of the complex transient absorption changes at each wavelength have been included in this difference spectrum, which reveal the same characteristic properties as the absorption change at 515 nm. This spectrum can be obtained by a kinetic analysis of two difference spectra with and without gramicidin for instance.

The superposition of spectral shifts due to several pigments makes the analysis of the spectrum very difficult. However, at least the maxima at 648 nm (neg.) and 665 nm (pos.) that are nearly antisymmetrical to the center frequency of chlorophyll-b *in vivo* (653 nm) correspond to the expectation for a difference spectrum due to a small homogeneous electrochromic shift of the chlorophyll-b band to the red. (The minor height of the 660-change may be due to the superposition with the negative lobe of a similar differential spectrum centered around chlorophyll-a_{673}.)

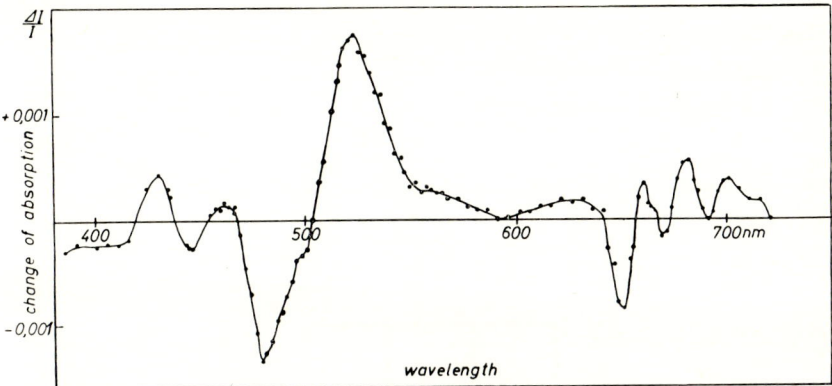

FIGURE 7 Separated difference spectrum of the field indicating absorption changes on spinach chloroplasts. (37).

Under the assumptions of a homogeneous shift of the whole chlorophyll-b *in vivo* band and using the spectral profile of the *in vitro* band (30) the wavelength of the shift can be calculated from the height of the maximum at 648 nm of the difference spectrum. It corresponds to a shift of about 0, 1 Å.

However, this is the order of magnitude that has been measured for the shift of the Soret band of porphyrins (*in vitro*) due to an electric field of about 10^6 V/cm (20).

THE ROLE OF THE LIGHT INDUCED ELECTRICAL FIELD IN PHOTOSYNTHESIS

The field indicator properties of the absorption changes at 515 nm have been used for detailed studies of the role of the light induced electrical field in photosynthesis.

The results have been published elsewhere (25, 31, 32, 33, 36, 37, for a summary see (8)) therefore only some of them shall be briefly summarized in the following scheme (Figure 8). In a single short flash of light (time of duration 10^{-4} sec) the two light reactions promote the transport of one electron from water to $NADP^+$. Additionally each of the two light reactions sets on one half of the electrical field strength across the thylakoid membrane and translocates one proton into the inner phase of the thylakoid (32). The time for the onset of the electrical field measured as the rise time at 515 nm on excitation with an extreme short saturating giant laser pulse is extremely fast ($\leq 2 \times 10^{-8}$ sec (7, 33)).

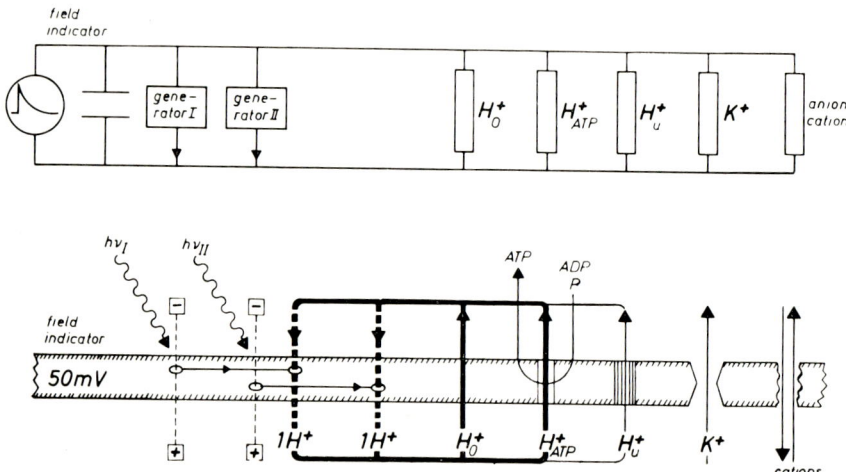

FIGURE 8 Model for the role of the light induced electrical field in photosynthesis.

Although the stoichiometric coupling between field onset and the proton translocation does not imply a kinetic coupling it can be concluded that the field strength corresponds to a charge separation of two elementary charges per electron transport chain. Using X-ray diffraction data (34) on the area of one electron transport chain ($\simeq 10^5$ Å) and on the thickness of the insulating lipid layer (see Figure 1), the order of magnitude of the voltage across the membrane can be calculated. The voltage amounts to 50 mV on excitation with the single short flash (32). (This corresponds to a field strength of about 2×10^5 V/cm.) In continuous saturating illumination this value can increase about fourfold to about 200 mV and in the steady state it decreases to 100 mV (32). (These conclusions are based on the assumption that the absorption change at 515 nm is a linear indicator of the field strength. This can be concluded from (32) and it has been demonstrated in a wider range in (35).

In the scheme (Figure 8) three artifically induced ways to accelerate the field decay have been indicated by thin lines (osmotic shock increases the membranes conductivity for different types of ions (e.g. Cl^-, Mg^{++}), gramicidin treatment for alkaline ions only, so called uncouplers for protons).

Relevant for *in vivo* photosynthesis are only the events on intact membranes (indicated by thick lines). Studies on the correlation between the absorption change at 515 nm and the synthesis of ATP have shown, that the field decay after several milliseconds of excitation is accelerated if ATP is synthezised (36). This gives support to the hypothesis of Mitchell that the energy used for ATP synthesis is gained by the translocation of protons 'down hill' the gradient of their electrical potential across a membrane. That means the translocation of protons back from the positively charged inner to the outer phase of the thylakoid.

CONCLUSIONS

Evidence has been provided that some of the light induced absorption changes in chloroplasts of green plants are due to an electrochromic response of pigments to a light induced electrical field across the photosynthetic membrane. These absorption changes have been used as a molecular voltmeter between two aqueous phases of such small dimensions that they are not accessible to the common electrochemical measuring techniques.

The photometric detection of an electrical field across a microscopical membrane (in combination with the repetitive technique) should be extendable to other biological systems, since it has two valuable properties: its extremely fast response to transients in field strength and its ability to detect fields that are localized in very small regions of space.

Literature

1. KREUTZ, W., and MENKE, W. 1962, *Z. Naturf.*, **17b**, 675.
2. MUEHLETHALER, K. Itern. Congress of Photosynth. Res. June 4–8, 1968 Freudenstadt.
3. MENKE, W. 1960, *Experientia*, **16**, 537.
4. MUEHLETHALER, K. 1960, *Z. Wiss. Mikrosk.*, **64**, 444.
5. GIBBS. S. P. 1960, *J. Ultrastruct. Res.*, **4**, 127.
6. VISHNIAC, W., and OCHOA, S. 1951, *Nature*, **167**, 768.
   TOLMACH, L. J. 1951, *Nature*, **167**, 946.
   ARNON, D. I. 1951, *Nature*, **167**, 1008.
   SAN PIETRO, A., and LANG, H. M. 1956, *Science*, **124**, 118.
7. WITT, H. T. In: Fast Reactions and Primary Processes in Chemical Kinetics (Nobel Symposium V). Ed. Claesson, p. 261, 1967, Almqvist & Wiksell, Stockholm; Interscience Publ. New York, London, Sydney.

8. WITT, H. T., RUMBERG, B., and JUNGE, W. 1968, 19. Mosbach Colloquium, S. 262, Springer-Verlag, Berlin, Heidelberg, New York.
9. EMERSON, R., and ARNOLD, W. 1932, *J. Gen. Physiol.*, **16**, 191.
10. JAGENDORF, A. T., and HIND, G. 1963, Photosynth. Mech. of Green Plants NAS-Nat. Res. Councils Publ. **1145**, 699.
11. DILLEY, L. A., and VERNON, L. P. 1965, *Arch. Biochem. Biophys.*, **111**, 365.
12. PACKER, L. 1962, *B.B. Res. Com.*, **9**, 355.
13. ARNON, D. I., ALLEN, M. B., and WHATLEY, F. R. 1954, *Nature*, **174**, 394.
14. CALVIN, M. 1962, *Angew. Chem.*, **74**, 165.
15. MITCHELL, P. 1961, *Nature*, **191**, 144.
    MITCHELL, P. 1966, *Biol. Rev.*, **41**, 445
16. JAGENDORF, A. T., and URIBE, E. G. 1966, *Proc. NAS-USA*, **55**, 170.
17. RUMBERG, B., REINWALD, E., SCHRÖDER, H., and SIGGEL, U. 1968, *Naturwiss.*, **55**, 77.
18. REINWALD, E., SIGGEL, U., and RUMBERG, B. 1968, *Naturwiss.*, **55**, 221.
19. LABHARD, H. 1967, *Adv. in Chem. Phys.* XIII, p. 179, Interscience Publ., London.
    LIPTAY, W., 1969, *Angew. Chem.*, **87**, 195.
20. MALLEY, M., FEHER, G., and MAUZERALL, D. 1968, *J. Mol. Spectr.*, **25**, 544.
21. WITT, H. T. In: Fast Reactions and Primary Processes in Chemical Kinetics (Nobel-Symposium V). Ed. s. Claesson, p. 81, 1967, Almqvist & Wiksell, Stockholm; Interscience Publ. New York, London, Sydney. RUEPPEL, H., and WITT, H. T. 1969, 'Methods in Enzymology', in: *Fast Reactions* (Ed. S. P. Colowick and N. O. Kaplan.) Acad. Press Inc., New York, p. 316.
22. RUMBERG, B. 1964, *Nature*, **204**, 860.
23. WITT, H. T., DOERING, G., RUMBERG, B., SCHMIDT-MENDE, P., SIGGEL, U., and STEIHL, H.'H. 1966, Brookhaven Symposia in Biology, **19**, 161.
24. RUMBERG, B., SCHMIDT-MENDE, P., SIGGEL, U., and WITT, H. T. 1966, *Angew. Chem.*, **2**, 522.
25. JUNGE, W., and WITT, H. T. 1967, *Ber. Bunsenges.*, **71**, 923; 1968, *Z. Naturf.*, **23b**, 244.
26. BANGHAM, A. D., STANDISH, M. M., and WATKINS, J. C. 1965, *J. Molec. Biol.*, **13**, 238.
27. CHAPPELL, J. B., and CROFTS, A. R. 1967, *BB-Library*, **7**, 293.
28. JUNGE, W. 1968, Dissertation TU-Berlin.
29. CIANI, S. 1965, *Biophysik*, **2**, 368.
30. BELLAMY, W. D., and LYNCH, M. E. 1963, G.E. Res. Lab. Rep. 63-RL-3469 G.
31. JUNGE, W., REINWALD, E., RUMBERG, B., SIGGEL, U., and WITT, H. T. 1968, *Naturwiss.*, **55**, 36.
32. SCHLIEPHAKE, W., JUNGE, W., and WITT, H. T. 1968, *Z. Naturf.*, **23b**, 1571.
33. WOLFF, CH., BUCHWALD, H.-E., RÜPPEL, H., and WITT, H. T. 1967, *Naturwiss.*, **54**, 489.
    WOLFF, CH., BUCHWALD, H.-E., RÜPPEL, H., WITT, K., and WITT, H. T. 1969, *Z. Natur.*, **24b**, 1033.
34. MENKE, W., 1962, *Z. Naturf.*, **17b**, 675.
    KREUTZ, W. 1968, *Z. Naturf.*, **23b**, 520.
35. REINWALD, E., STIEHL, H.-H., and RUMBERG, B. 1968, *Z. Naturf.*, **23b**, 1616.
36. RUMBERG, B., and SIGGEL, U. 1968, *Z. Naturf.*, **23b**, 239.
37. EMRICH, H. M., JUNGE, W., and WITT, H. T. 1969, *Z. Naturf.*, **24b**, 1144.

DISCUSSION

GREEN   I would like to ask whether one could interpret your experiments very simply in terms of a conformational change in the system which is intrinsic to an energy cycle of the membrane. When gramicidin is present, the energized system becomes discharged and ions are moved. When ADP is present, the energized system becomes discharged and ATP is synthesized. This is a rather simple interpretation. Is there any reason why the particular interpretation I am suggesting should not be considered? It accounts for all the facts very simply.

JUNGE   As yet it is very hard to discriminate whether the ion fluxes are caused by conformational changes or vice versa. This is partly due to the fact that ion fluxes are coupled to osmotic phenomena, which induce conformational changes in a very trivial sense.

However, the Nernst–Planck-like behavior of the ion fluxes that we studied by means of absorption changes gives evidence that the ions are driven by an electrical field across the thylakoid membrane. Whether this field is set on by a conformational change or not has to be studied in the future.

ADAM   You have a very fast rise of the absorption change with 2×10^{-8} seconds. Now, do you think that this is an indication of an electric field built up by reaction of the reactive center to the light? I guess you do. Then, this time is very fast indeed, since, for capacitive changes of the axon membrane potential you have microseconds as the duration in which the field is established. So, I see a discrepancy of two orders of magnitude.

JUNGE   This discrepancy between the axon and the chloroplast may be due to a difference between both systems. The field in the chloroplast is set on in the course of photochemical reactions, which are very rapid.

ADAM   How do you imagine that this field comes about? Is it a movement of ions or is it something else?

JUNGE   As yet there is no evidence favoring a special mechanism of the field onset. One may imagine that the translocation of either a proton or an electron charges the membrane's capacitance.

As Professor Onsager has shown, the translocation of a proton across the membrane can be sufficiently rapid.

On the other hand one may assume that a slower hydrogen translocation follows the field onset. The field may be set on by a rapid electron transport from the inner to the outer side of the membrane. This may be followed by the uptake and neutralization of a proton from the outer phase, a transport of one hydrogen to the inner side and the final release of one proton to the inner phase. By this antitransport: e^-, H the field can be set on very rapidly, but the net translocation of a proton may be much slower. This has been discussed in Refs. 8 and 25.

ADAM If I may add one comment, I would find it rather improbable that such a fast rising electric field can come about by ionic movements. I would think that these are rather electronic shifts, maybe in charge transfer complexes.

JUNGE Well, that is one of the possibilities I tried to point out.

GREEN I can't resist making a comment here. This is a general comment for so many systems of this kind. On the one hand one has physical methods of great precision where very precise calculations are made, and on the other hand complete ignorance about the system one works with, so that you have this contrast of incisive physical methods in total darkness with the most elementary features of the system itself.

PAPER 18

Concept of the reactive site in biological transport†

HALVOR N. CHRISTENSEN

The Department of Biological Chemistry, The University of Michigan, Ann Arbor, Michigan.

WHEN ENZYMOLOGISTS conceived that only a circumscribed region of the enzyme molecule takes hold of the substrate to destabilize it, they introduced the term, *active center*. The term has various forms; the form *reactive site* has evolved, probably because it is now realized that this region lies more at the periphery than at the center of the molecule, and because the action is a mutual one between enzyme and the substrate.

Concurrently, the tertiary structure of non-enzymic, globular proteins has come to be recognized also to present the high likelihood of binding sites with considerable degrees of specificity for classes of small molecules. Such sites may bind any of a variety of substances: in some cases an amino acid, in others a fatty acid, a bile salt, or even a coenzyme such as pyridoxal phosphate, generally without any of these substances undergoing any obvious destabilization or increase in reactivity. Such binding sites are, however, characteristically lost on denaturation.

Transport has been shown to describe binding sites of the same general character just described, again with a three-dimensional specificity which almost demands that *macromolecules*, and indeed *proteins*, present these reactive sites at the surfaces of biological membranes.‡ Some of my colleagues

† Manuscript prepared while Nobel Guest Professor, The University of Uppsala. The research reported here was supported in part by a grant (HD 01233) from the Institute for Child Health and Human Development, The National Institutes of Health, United States Public Health Service. Insulin assays in rat plasma were provided by the collaboration of Drs. Stefan Fajans and Sumer Pek.

‡ Two previous discussion of the title subject (1, 2), and a consideration of its kinetic aspects (3), have been published elsewhere. These will provide additional bibliographic references.

have argued that these transport proteins will in fact prove to be ordinary enzymes in extraordinarily strategic positions—and so they may perhaps be in some cases—whereas I have argued (1, 2) that more characteristically they are probably not enzymes, but another type of protein, one which we have termed, without particular originality, *transport-mediating proteins*, or more briefly, *transport mediators*. That is to say that its destabilization is not the only catalytic event that may occur to a small molecule that becomes protein-bound in the context of the living cell. Instead it may be delivered unchanged to a phase other than the one from which it was received.

Whereas the study of the transport of inorganic ions has kept investigators aware that no persistent structural change need arise from transport, it is the ease with which analogs to the organic substrate of transport can be discovered or constructed that has pushed forward the concept of the reactive site as the characteristic receptor and releasing component in biological transport. As long as such studies of the effect of varying the structure of the substrate remained qualitative in nature, they gave oversimplified pictures as to the number of transport mediators present in the plasma membrane. When quantitative inhibition analysis came to be applied, however, it became clear for the case of the amino acid (which I shall regard as illustrative of what may be expected for transport in general) that a second (4), then a third (5) mediating system had to be taken into account—and perhaps one or two others (6–9)—just for amino acids carrying no net charge. These are in addition to some systems specific to a single amino acid observed in bacteria. Unfortunately for any easy analysis, no one-to-one specificity relation could be shown between amino acid and mediator; instead three of the systems each serve for a considerable group of amino acids, in such an odd, overlapping pattern that alanine divides its uptake by the Ehrlich cell among three systems, with, I regret to say a small unassigned component remaining (10). For the interpretation of these relations, a code of sorts had to be broken.

The substantial degree of complexity inherent in the exposition of these systems distressed some of our colleagues enough to motivate us to redouble the evidence for the correctness of the picture, and to show that most of these several transport systems really operate separately (8–13). As a result we can present them as discrete entities, occurring regularly in many tissues and species. We suppose their importance is complementary to that of the amino acids whose migrations they effectuate.

Let me cite briefly one of the strategies followed in confirming the nature of a complex transport interaction between two analogs (two amino acids, of course). Amino acid *B* inhibits the cellular uptake of amino acid *A*; but the Michaelis–Menten curve describing the inhibition appears to be the sum of at least two rectangular hyperbolas. Amino acid *A* inhibits the uptake of *B*, and shows just the same kind of curve. Are they competing for transport by two distinct mediating agencies?

Let us first modify the structure of *B* to *B'*, say by N-methylating it (11). We now find that its inhibition of the uptake of *A* represents accurately a single rectangular hyperbola, but it leaves a substantial portion of the migration of *A* completely uninhibited. We conclude that by modifying *B* we have made it completely unreactive with one of the two transport systems serving for *A*, especially since we find that *A* is now able to inhibit all the uptake of *B'* in a homogeneous (hyperbolic) manner.

In a second stage of simplification we will modify the structure of *A* in the same manner. Now we find that both aspects of the interaction (that of *A'* on the uptake of *B'*; that of *B'* on the uptake of *A'*) trace simple rectangular hyperbolas, encompassing all the mediated uptake of each. The competitive interaction has thus been brought to the simplest case.

This example exaggerates our ability to detect heterogeneity in transport by tracing the Michaelis–Menten curve, or by deriving a linear transformation of that curve (14). Before we can regard a transport activity as due to a single agency, we must manipulate many conditions beside the substrate concentration—and then the label must be regarded as provisional.

These methods have necessitated the design of amino acid analogs that serve to explore to its limits the scope of the receptivity of the transport-mediating site under description. I want to spend most of my remaining space considering the behavior of some of the model substrates obtained in that effort. Three types of behavior have had special interest for this program, namely a structural unsuitability for,

1. one of the transport-mediating systems present;
2. all of the transport-mediating systems present;
3. all but one of the transport-mediating systems present.

I have listed these in the order of the increasing value I give to the substrates described.

Although the behavior of the following model amino acid has been amusing to observe,

this substance, often referred to as *cycloleucine*, escapes only one of the transport-mediating agencies. It cycles about the animal organism for months, filtered at the glomerulus, resorbed by the renal tubule, not appreciably catabolized or anabolized (15). But the closely related α,α-diethylglycine,

$$\begin{array}{c} CH_3CH_2 \\ CH_3CH_2 \end{array}\!\!\!\!\!\times\!\!\!\!\!\begin{array}{c} COO^- \\ NH_3^+ \end{array}$$

is more useful, because it is a *poor* substrate for all the ordinary transport systems.

An even more complete transport negativity was obtained with α,α-dicyclopropyl glycine, apparently by a crowding of the area about the α-carbon:

$$\begin{array}{c} \triangle \\ \triangle \end{array}\!\!\!\!\!\times\!\!\!\!\!\begin{array}{c} COO^- \\ NH_3^+ \end{array}$$

It moves very slowly into and out of cells, the two rates, relative to concentration, being very nearly equal (16). Its use has permitted us to challenge the adequacy of the list of known types of absorption from the small intestine (17).

The most useful type of model substrate is the third one on our list, namely that specific to a single, transport system. The most specific model we know for the ubiquitous transport System *A* is α-(methylamino)-isobutyric acid (MeAIB)†:

$$\begin{array}{c} CH_3 \\ CH_3 \end{array}\!\!\!\!\!\times\!\!\!\!\!\begin{array}{c} COO^- \\ N^+H_2CH_3 \end{array}$$

For it we have so far detected no effect on any other amino acid transport system.

In a preliminary search for model structures that might instead prefer System *L* and avoid System *A*, we encountered the substance shown at the left in Figure 1, 1-amino-3-methylcyclohexanecarboxylic acid, although it also fell short of the objective (4). Because its methyl group had contributed to the selectivity of that amino acid, it seemed to us desirable to stabilize the position taken by the methyl carbon atom by making it part of a bridge. A product of the intention is indeed specific to the *L* system (10). This substance, 2-aminobicyclo[2, 2-1]heptane-2-carboxylic acid, abbreviated BCH$^+$, is of sufficient interest to receive much of the remainder of my attention.

† Three abbreviations, BCH, AIB and MeAIB have been used, to designate 2-aminobicyclo [2, 2, 1] heptane-2-carboxylic acid, α-aminoisobutyric acid, and its N-methyl derivative.

BCH appears to be devoid of reactivity with Systems *A* and *ASC*, and also with transport systems for glycine, β-amino acids and diamino acids, respectively. But it appears to react characteristically with systems for branched-chained amino acids in all species tested.

A further value for BCH is that its structure may be disposed in four different ways. Of the four isomers, we expect to find only one universally reactive. The study of this compound, and that of some related structures suggested by it, should be highly informative as to the positions which the atoms of the amino acid sidechain may and must take in reacting with System *L*, and as to the positions not suitable for System *A*. We can trace in this structure carbon atoms corresponding to those of leucine (encircled) or to those of isoleucine (asterisks) or of valine. Two or three extra carbon atoms in BCH serve to hold those carbon atoms in rigidly prescribed positions.

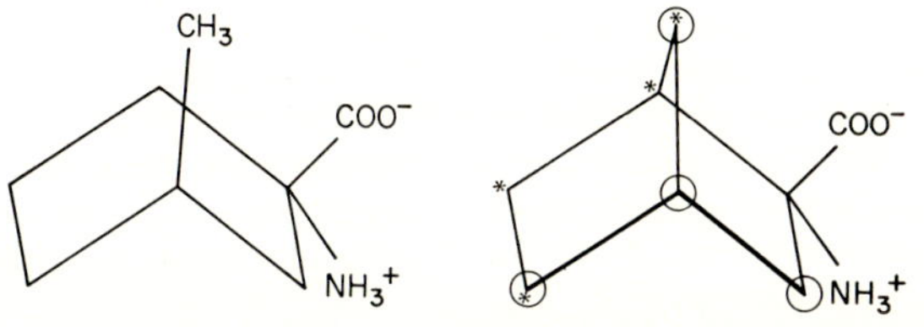

FIGURE 1 Structure of two model substrates, as discussed in the text. The encircled atoms show how one can trace the skeleton of leucine, the asterisks that for isoleucine, both rigidified in space by 2 additional carbon atoms. Only one of 4 isomers is represented by each formula. Reproduced from the *Journal of Biological Chemistry* (10) with permission.

By the use of atomic models one can show that the seven carbon atoms in the bicyclic ring structure are in positions not greatly different from those which the corresponding carbon atoms of dicyclopropylglycine can occupy. Yet the difference is decisive for entry into the *L* transport system.

Figure 2 illustrates that the two related synthetic pathways to BCH gave quite different proportions of the two isomeric forms, which we suppose are the *endo* and the *exo* isomers. These are designated *a*, for the one eluted more rapidly from the sulfonic acid resin of the amino acid analyzer, and *b*, for the one more greatly retarded in elution. The Bucherer synthesis from norbornanone gave 8 per cent of *a* and 92 per cent of *b*; the Strecker synthesis gave the proportions 69 per cent of *a* and 31 per cent of *b*. This result was not unexpected, since major differences have been noted previously between the products of these two reactions with various substituted cyclohexanones, although the literature disagrees as to which is the *endo* and which is the *exo*

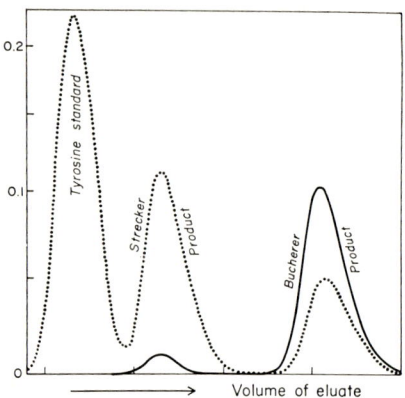

FIGURE 2 Difference in proportions of the 2 isomers in BCH prepared by 2 different methods. The tracing from an amino acid analyzer is shown under elution by 0.2 M citrate, pH 4.25 at 50°. The vertical scale is a logarithmic scale of absorbancy units. About 160 ml of eluate were collected before this record began, and 56 ml during the 43-min interval shown. Reproduced with permission from the *Journal of Biological Chemistry* (10).

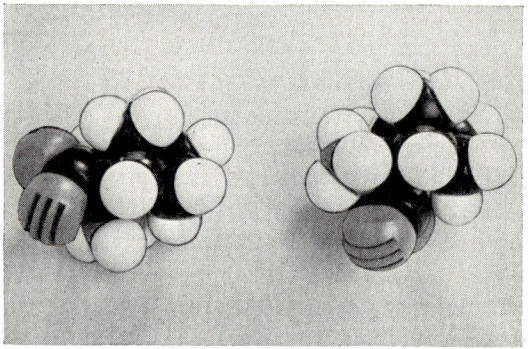

FIGURE 3 Atomic models to contrast endo and exo isomers of BCH. The isomer at the left has the carboxylate group axial, the amino group equatorial, somewhat as shown in Figure 1. Photograph by H. S. Tager.

isomer (18, 19). The difference lies in whichever group, the amino or the carboxyl, is *axial* and which is *equatorial* (Figure 3). No doubt the transport differences to be reported here arise from these differences in orientation.

When we have in hand a specific model substrate for each known transport system, the analysis of the migration of an amino acid, or the determination of which transport systems are active in a given tissue, should be simplified. The second goal implies a congruity among various species as to the nature

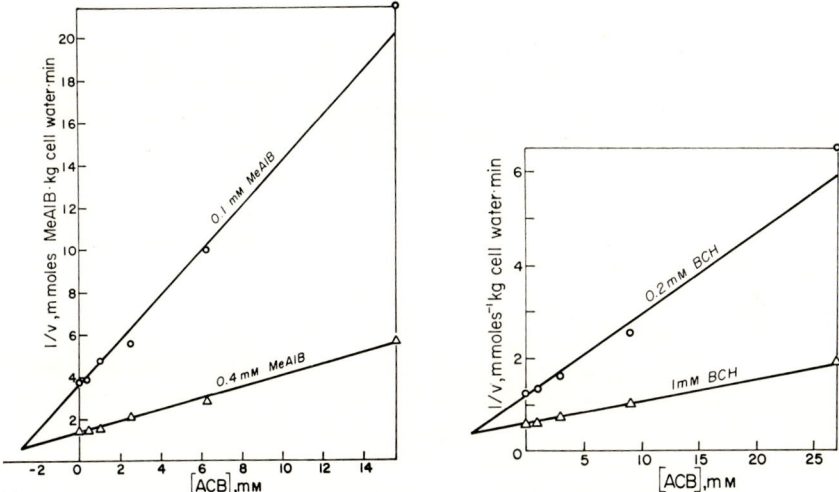

FIGURE 4 Determination of the K_i values of 1-aminocyclobutane-carboxylic acid as an inhibitor of the uptake by the Ehrlich cell of α-(methylamino)-isobutyric acid (left) and of BCH (right). Uptake was observed during 1 min for the first case and during 30 sec for the second, both from Krebs-Ringer bicarbonate medium at 37°, pH 7.4. Reproduced with permission from the *Journal of Biological Chemistry* (10).

of the transport systems, a hypothesis which I will support here by results with BCH.

First let me use Figure 4 to illustrate a way in which a set of model amino acids can be used to check how a given amino acid divides its uptake among the mediating agencies. In this case we were testing the unknown amino acid, 1-aminocyclobutane carboxylic acid, which was available to us only in unlabeled form. In Figure 4 it has been tested as an inhibitor of the uptake of MeAIB and BCH by the Ehrlich cell. The Dixon plots show K_i values of 3 mM for the *A* system, and 4 mM for the *L* system (10). Given such similar reactivity with the two systems, we may be sure that a direct kinetic plot of the uptake of the cyclobutyl amino acid would have given no clue for heterogeneity in its uptake.

Such model substrates as MeAIB and BCH have also proved invaluable for investigating the mediated exodus of ordinary amino acids from cells. In that context their use avoids uncertainties that otherwise arise from the difficulty of producing full saturation of any system for exodus. By their use, exodus from the Ehrlich cell could be shown to occur largely by the same routes serving for entry of the same amino acids (16). The results emphasize again that it is not correct to suppose that cells in general have large aqueous pores penetrating their normal plasma membrane. Exodus by reversal of the processes of mediated entry appears often to approach thermodynamic reversability; the role of the thermodynamics of irreversible processes may accordingly be smaller than is supposed under the pump-and-leak hypothesis.

In general BCH has proved repeatedly decisive as a probe for site L. It is for example the only known L system substrate that does not inhibit the uptake of cationic amino acids in various cells; hence that interaction can be dissociated from System L.†

The next several figures show that the difference between the structure of the a and b isomer is highly significant for transport. Figure 5 shows that the uptake of b-BCH by the Ehrlich cell is five times as fast as that of a-BCH. Figure 6 shows that both the human and the avian erythrocyte discriminate between the two isomers. A characteristic feature is the much greater 'overshoot' shown early in the course of uptake by isomer b than isomer a. Presumably an exchange for endogenous amino acids contributes to the uptake; as the levels of such amino acids decline through exodus from the cell, the contribution of exchange to 'uphill' transport decreases (10).

Figure 7 shows that in *E. coli* the discrimination between the two isomers is complete. The squares along the base line show that no uptake at all of a-BCH could be recorded. The same result was seen even when 30 min was allowed. Our preparation of a-BCH, incidentally, showed no inhibition of the uptake of b-BCH. The latter shows an initial rate of uptake lower than that of leucine, although the V_{max} is substantially higher. The inhibition shown between leucine and BCH for uptake by *E. coli* was consistent with their entry by the same system (10).

The more complete discrimination between the two isomers shown by the system of the L type in *E. coli*, compared to that in animal cells, is no doubt related to the circumstance that in *E. coli* this system is largely specific to the branched-chain aliphatic amino acids, and nearly devoid of reactivity with phenylalanine and methionine. The latter serve, in contrast, as substrates

† H. N. Christensen and M. E. Handlogten.

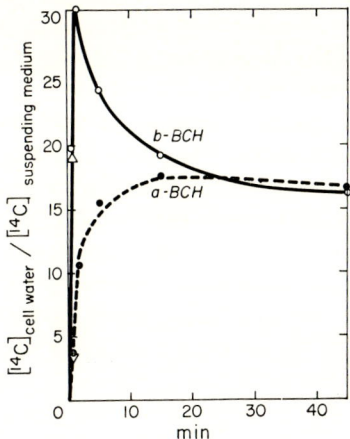

FIGURE 5 Comparative courses of the uptake of the isomeric forms of BCH by the Ehrlich cell. Uptake from 6.2 μM solutions from Krebs-Ringer bicarbonate medium at pH 7.4 and 37°. The triangles, (▽) for a-BCH and (△) for b-BCH show the negligible effect of 50 mM α-(methylamino)-butyric acid on the uptake at 30 sec. Reproduced with permission from the *Journal of Biological Chemistry* (10).

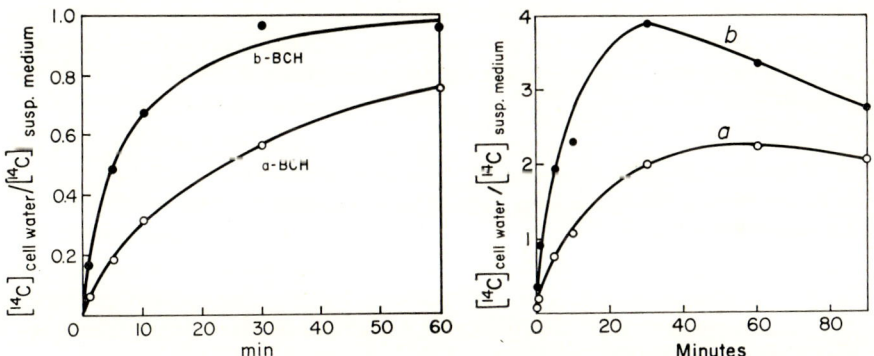

FIGURE 6 Comparisons of the uptake of a-BCH and b-BCH by the human (left) and pigeon (right) red blood cell. Uptake was measured from 7.4 μM solutions of b-BCH-^{14}C in Krebs-Ringer phosphate medium at pH 7.4, at 27° for the human red cell and at 37° for the pigeon red cell. The initial rates were 2.5 times (left) and 3.6 times (right) as fast for the *b* isomer as for the *a* isomer. Reproduced from the *Journal of Biological Chemistry* (10) with permission.

for the L system in various animal cells. Hence the system that discriminates most closely in favor of the branched aliphatic sidechains is complete in its preference for b-BCH (10).

If BCH provides a good challenge of the scope of the reactivity of the L system, it should also provide a challenge for the presumptive transport role of the leucine-binding protein isolated from *E. coli* by Piperno and Oxender (20, 21). This protein, and a sulfate-binding protein obtained from *Salmonella typhimurium* by Pardee and his associates (22), provide the first directly

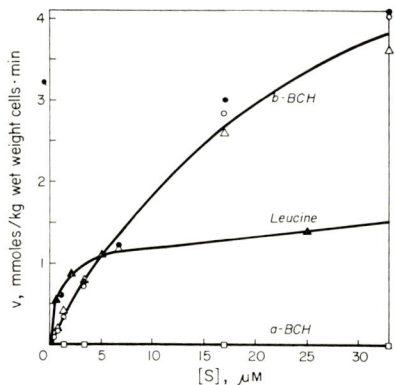

FIGURE 7 Concentration dependence of the uptake of b-BCH and leucine by *E. coli*. Failure of a-BCH to be taken up or to inhibit uptake of b-BCH. Uptake from 0.01 M potassium phosphate, pH 6.9, at 37° during 30 sec. The solid circles represent the uptake of b-BCH in the presence of 16.7 μM a-BCH; the other circles and triangles represent observations with b-BCH or leucine present alone; and the squares along the baseline are for a-BCH. The results correspond to values of about 10 μM and 1 μM for the K_m of b-BCH and leucine, respectively, and of 3.5 and 1.4 mmoles per kg cell water $\cdot$ min for V_{max}, respectively. Reproduced from the *Journal of Biological Chemistry* (10) with permission.

studied examples for reactive sites that may well serve in transport. The leucine-binding protein is released from the transport-positive bacterial cell by a controlled osmotic shock in the cold. The cells suffer a loss in transport activity that can be directly correlated with the appearance of the binding protein in the suspending fluid. Other aspects tend also to identify the protein with the transport system. For example, growth of the cells in the presence of leucine causes a loss in the capacity for transport of branched-chain amino acids that can be correlated with the amount of the binding protein recoverable from the cells on osmotic shock.

We have now observed that transport of BCH is lost under the same conditions as transport of leucine. BCH does react with the binding protein (Figure 8), the *b* isomer more strongly than the *a* isomer, but not strongly enough to correspond to its strength of inhibition of leucine transport in the intact bacterial cell. This situation does not disprove the participation of the binding protein in leucine transport, but it raises important questions as to

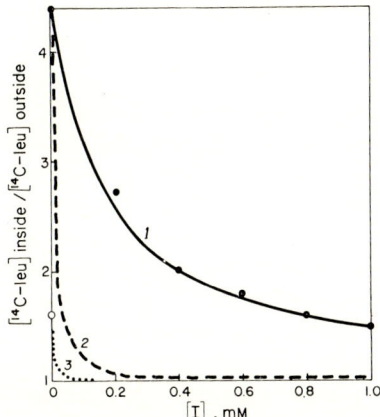

FIGURE 8   Failure of BCH to inhibit to the predicted extent the binding of leucine by the binding protein of *E. coli*. The 5 solid circles show the inhibition exerted by BCH at 5 concentrations on the binding of 0.05 μM leucine in a solution calculated from its binding action to contain the protein at 3.4 μM, in 0.01 M potassium phosphate pH 6.9. Binding is measured by the attainment of distribution ratios (ordinate scale) higher than *one* for the amino acid concentration inside the dialysis bag relative to that outside. Curve 1 predicts the inhibitory effect to be expected if the dissociation constant of the BCH-protein complex is 175 μM, curve 2, if it were 10 μM, the K_i value found for it as an inhibitor of leucine uptake. Curve 3 corresponds to the K_i of 1 μM found for leucine. The open circle shows that 5 μM leucine was highly inhibitory to leucine binding. Reproduced with permission from the *Journal of Biological Chemistry* (10).

mechanistic details. Pardee and his associates noted that various anions likewise do not take the same order in their inhibition of sulfate uptake by *Salmonella typhimurium* as they do in their reactivity with the sulfate-binding protein isolated from that organism (22).

The pattern of transport preference for isomer *b* over isomer *a* was extended by studies of their distribution in the rat. Both entered various tissues rapidly and concentratively, isomer *b* being preferred slightly over isomer *a*. These compounds were catabolized to an immeasurably small extent. As for certain other model amino acids, their renal tubular resorption was almost complete, so that the dose was excreted only very slowly (23).

I want to turn to the case of a curious and significant reaction of neutral amino acids with the receptor site for cationic amino acids, observed both in the Ehrlich cell and the rabbit reticulocyte. This action could be dissociated from an occupation of the sites for neutral amino acids by the use of specific model substrates such as MeAIB and BCH. The action of externally present neutral amino acids includes not only an inhibition of the uptake of cationic amino acids (Figure 9) but also a stimulation of their exodus from the cell (Figure 10). In the Ehrlich cell both processes are saturable at the same concentrations of neutral amino acids; and *both require* Na^+ *as a cosubstrate*, (Figures 9, 10) although the corresponding effects of cationic amino acids (uptake, inhibition of uptake, stimulation of exodus) do not. Given sufficient Na^+, essentially all mediated uptake of cationic amino acids can be blocked; in the presence of such neutral amino acids, Na^+ becomes an *inhibitor* of cationic amino acid uptake (Figure 9). The neutral amino acid thus generates a binding spot for Na^+ at the Ly^+ site, whereby Na^+ is caused to exchange for internal diamino acids. Undoubtedly, for the *surrogate* substrate, Na^+ *plus diamino acid*, the sodium ion takes the place of the distal amino group on the diamino acid transport site. For this effect the omega hydroxy amino acid, homoserine (Figures 9, 10) is the most active neutral amino acid encountered; but phenylalanine, leucine, methionine and many others are effective. We see in this way how ubiquitous metabolites can complete a binding site for Na^+ transport, where previously there was no suitable binding site.†

Investigators have occasionally supposed that the reactive sites for transport might arise in a unique way, by generation *between* or *among* protein molecules where these come into contact in the structure of the plasma membrane (24). One can conceive that a succession of similar sites reaching through the thickness of the membrane might be generated in that way, as visualized by Stein and Danielli in 1956 (25). Such a plan may seem to have advantages in the ease with which a distortion might be propagated through the succession of sites, so that the transport substrate could be successively received and discharged by each site, with successive increases in its chemical potential. But that phenomenon presumably can also be produced within a single protein molecule, perhaps by helical sections in which serine hydroxyl groups and the like are exposed at suitable intervals.

A situation in which transport sites are created cooperatively by 2 protein molecules could explain the persistent failure of investigators to find in most

† Unpublished results, H. N. Christensen, M. E. Handlogten, and E. L. Thomas.

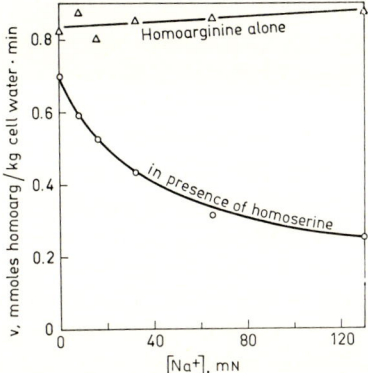

FIGURE 9 Inhibition of homoarginine uptake by the Ehrlich cell when a neutral amino acid is present, but not when it is absent. Uptake during one min from a 0.2 mM solution of ^{14}C-homoarginine in Krebs-Ringer medium in which various amounts pf the Na$^+$ have been replaced by choline, in the presence and in the absence of 25 mM homoserine. Similar results were obtained with phenylalanine and various other neutral amino acids.

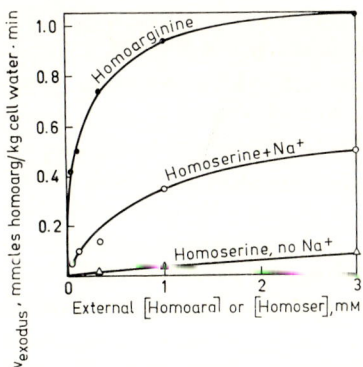

FIGURE 10 Acceleration of exodus of previously accumulated ^{14}C-homoarginine from the Ehrlich celli in the presence of homoserine, with or without Na$^+$ present. Comparative action of external homoarginine. Homoarginine was first accumulated to a calculated 3 mM internal concentration by a 15-min incubation. After washing the cells, exodus was observed during 1 min at 37° into Krebs-Ringer bicarbonate medium, or the same medium in which choline replaced Na$^+$. Unlabeled homoarginine or homoserine was present in the medium at the indicated levels. The absence of external Na$^+$ had instead an enhancing effect on exodus into homoarginine-containing solutions. Results similar to those with homoserine were obtained for phenylalanine and various other neutral amino acids.

cases any distinctly measurable quantities of the expected binding site on breaking up the membrane of animal cells—the binding site may disappear when the membrane proteins are separated from each other. This supposition is, however, not one we are yet forced to entertain; the maximal concentrations of transport receptor sites observed on the surface of the intact cell are also immeasurably small when one looks for stoichiometric binding of the substrate there. Hence there is no need to suspect that the sites are lost when the plasma membrane is broken.

One may speculate that in the biological context the difficulty of maintaining the essential bonding intervals for the reception of the substrate might be excessive if two or more separate protein molecules were to contribute chemical groups to the site. We now know that a site within a single protein subunit can be subject to distortion by events occurring at another point on the same protein molecule, or even by an event on an adjoining protein subunit. Indeed binding sites restricted to a single protein subunit could likewise disappear when that protein is solubilized from the membrane, either through a denaturation by the reagents used, or by an inability of the protein to retain its effective conformation once removed from the membrane matrix. Racker has used the term *allotrophy* for changes in the behavior of macromolecules on their removal from a natural context (26).

In any event, the similarity of the definition of the sites for transport to those of enzymes and other globular proteins also dissuades me from accepting this proposal as to their origin and the corresponding pessimistic explanation for the difficulties encountered in observing their activity once the plasma membrane is broken.

As a possibly related matter, we were surprised to observe that BCH mimics the action of leucine in stimulating insulin release and in producing hypoglycemia in the rat, particularly in animals that had received tolbutamide (Table I). This action was produced by the *b* isomer but not by the *a* isomer of BCH, in agreement with the transport results. From this action we may conclude that effector molecules do not need to enter metabolism to stimulate the release of this hormone. All they apparently need to do is to enter and occupy a recognition site. We can go further and note that the effective substrate can be a rigid molecule, not likely to respond in its shape on reaction with the effector site. In the meantime Renold and Lambert have noted that both AIB^2 and BCH stimulate insulin release from an organ culture of the fetal pancreas of the rat.† Considering that several of the substances that stimulate insulin release also have their transport accelerated by insulin, we can speculate that the binding sites at which insulin release is triggered may well be transport sites (23).

† Personal communication from Dr. Albert Renold.

TABLE I Hypoglycemic and insulin-releasing action of BCH in the rat. An iso-osmolar solution of 8 mmoles of BCH, AIB or NaCl/kg body wt was injected intraperitoneally into fasted rats weighting 60 to 80 g. After 40 min the plasma insulin was measured by a double-antibody radioimmunoassay, through the kindness of Drs. Stefan S. Fajans and Sumer Pek. After 60 min blood glucose was estimated using glucose oxidase. Tolbutamide-primed animals had received 100 mg of this agent either 2 hr previously, or in 5 equal does of 20 mg each at 12-hr intervals during the preceding 3 days. See ref. (23) for further details.

Preparation of Rats	Change in Blood Glucose $\Delta_{control} - \Delta_{BCH}$ mg/100 ml mean ± S.D.	Serum Insulin Increase μ units/ml mean ± S.D.	
		Control	BCH
Untreated‡	20 ± 7.7 (5)†	0.75 ± 2.81 (4)	5.6 ± 4.15 (7)
Tolbutamide, 2 hr.	42 ± 6.7 (3)		
Tolbutamide, 3 days‡	44 ± 8.5 (4)	1.2 ± 2.44 (5)	8.3 ± 4.47 (10)

† Numbers in brackets are number of rats studied.
‡ AIB did not affect blood glucose under these conditions.

TABLE II Failure of insulin to stimulate the uptake of BCH by the isolated, uncut diaphragm of the rat.

Thiocyanate was present simultaneously for estimation of the extracellular portion of the tissue water. The distribution ratio is the calculated concentration reached in 1 hr in the cellular water, divided by that in the suspending medium. Crystalline bovine zinc insulin was added to 0.55 units per ml. Four observations with BCH in each case. Results with AIB (2 tests) show the responsiveness of the system. Experiments by A. M. Cullen in the author's laboratory.

	Distribution Ratio mean ± S.D.	
	BCH	AIB
Control	0.76 ± 0.125	0.49, 0.49
Insulin added	0.68 ± 0.026	1.95, 1.54

If that should not prove to be the case, we have perhaps taken an excursion from the area of transport models into a different subject, namely structural models for the reception of messages by the cell. We note that BCH fails to respond to insulin addition by accelerated uptake by the rat diaphragm (Table II). Hence, to suppose that the substances that stimulate insulin release are the same ones that have their transport stimulated by insulin is too simple an interpretation. Nevertheless, we want to extend our attention to further analogous structures not subject to metabolic alteration, for the comprehension they may give us as to how cellular receptor sites work, both for material and informational transport.

Perhaps I may finally be permitted to speculate, if those who are most dedicated to the description of the transport of the alkali-metal cations would briefly turn their attention to the relation of their subject to the migration of organic metabolites and the role therein of the reactive sites of protein structure, that some unexpected progress toward their own goals might well follow.

References

1. Christensen, H. N. 1960, *Adv. in Protein Chem.*, **15**, 239.
2. Christensen, H. N. 1967, *Perspectives in Biol. and Med.*, **10**, 471.
3. Christensen, H. N. 1969, *Adv. in Enzymology*, **32**, 1.
4. Oxender, D. L., and Christensen, H. N. 1963, *J. Biol. Chem.*, **238**, 3686.
5. Christensen, H. N., Laing, M., and Archer, E. G. 1967, *J. Biol. Chem.*, **242**, 5237.
6. Vidaver, G. A., Romain, L. F., and Haurowitz, F. 1964, *Arch. Biochem. Biophysics.*, **107**, 82.
7. Winter, C. G., and Christensen, H. N. 1965, *J. Biol. Chem.*, **240**, 3594.
8. Wheeler, K. P., and Christensen, H. N. 1967, *J. Biol. Chem.*, **242**, 1450.
9. Eavenson, E., and Christensen, H. N. 1967, *J. Biol. Chem.*, **242**, 5386.
10. Christensen, H. N., Handlogten, M. E., Lam, I., Tager, H. S., and Zand, R. 1969, *J. Biol. Chem.*, **244**, 1510.
11. Christensen, H. N., Oxender, D. L., Laing, M., and Vatz, K. A. 1965, *J. Biol. Chem.*, **240**, 3609.
12. Inui, Y., and Christensen, H. N. 1966, *J. Gen. Physiol.*, **50**, 203.
13. Antonioli, J. A., and Christensen, H. N. 1969, *J. Biol. Chem.*, **244**, 1505.
14. Christensen, H. N. 1966, *Fed. Proceedings*, **25**, 850.
15. Christensen, H. N., and Jones, J. C. 1962, *J. Biol. Chem.*, **237**, 1203.
16. Christensen, H. N., and Handlogten, M. E. 1968, *J. Biol. Chem.*, **243**, 5428.
17. Antonioli, J. A., and Christensen, H. N. 1968, *Am. J. Physiol.*, **215**, 951.
18. Munday, L. 1961, *J. Chem. Soc.* (*London*), p. 4372.
19. Cremlyn, R. J. W., and Chisholm, M. 1967, *J. Chem. Soc.* (*London*), p. 2269.
20. Piperno, J. W., and Oxender, D. L. 1966, *J. Biol. Chem.*, **241**, 5732.
21. Penrose, W. R., Nichoalds, G. E., Piperno, J. R., and Oxender, D. L. 1968, *J. Biol. Chem.*, **243**, 5921.

22. PARDEE, A. B., PRESTIDGE, L. S., WHIPPLE, M. B., and DREYFUSS, J. 1966, *J. Biol. Chem.*, **241**, 3962.
23. CHRISTENSEN, H. N., and CULLEN, A. M. 1969, *J. Biol. Chem.*, **244**, 1521.
24. WHITTAM, R., Transport and diffusion in red blood cells. William and Wilkins Co., 1964, pp. 174–5.
25. STEIN, E. D., and DANIELLI, J. F. 1956, *Discussions Faraday Soc.* **21**, 238.
26. RACKER, E. 1968, *Scientific American*, **218** (2), 34.

DISCUSSION

CHRISTENSEN   There are several points that I made which I intended to be slightly provocative—perhaps I can call your attention to those so that you may wish to respond to them. I don't think there is much evidence for holes or pores in the plasma membranes. I believe that the indications are substantial that transport is produced by somewhat the same kind of binding site that is typical of enzyme action; yet the binding protein does not need to destabilize the molecule and therefore is probably not an enzyme. It is more like other globular proteins in this binding property. Binding of specific substances has often been shown for globular proteins without their having any obvious biological significance. The binding of tryptophan by serum albumen is an example. There are many others. It appears to me highly likely that this binding activity is used biologically not only for catalysis, i.e., the destabilization of molecules; but also for their selection for introduction into another phase, i.e., for specific transport. This process may very well occur by, say, conformational changes in such binding molecules which are a portion of the membrane. Let us take the case of hemoglobin as an example. When O_2 is bound, we recognize movements of parts of the molecule, with respect to other parts, of a magnitude of seven Angstroms or more. These may be the kinds of repositionings that can produce a crossing in a single, decisive movement; or perhaps instead, a series of smaller movements may produce transport by multiple stages. They would represent a catalytic action of a special kind.

FOX   I wonder if Dr. Christensen would elaborate on the pattern of action of these overlapping transport mediators. They presumably have some differences in their pattern that one could recognize.

CHRISTENSEN   Yes, I'll be glad to try a quick summary.† There is in all animal cells studied so far, a sodium dependent system (A) of wide reactivity. A second type of reactivity (ASC) is limited largely to amino acids such as serine, cysteine, alanine and other three or four carbon aliphatic, hydroxyl-aliphatic and mercaptoaliphatic amino acids. There is also a separate system for glycine, at least one, a separate transport system for the beta amino acids B-alanine and taurine—that is the only case in which the sulphonic acid group will serve in place of the carboxylate group. In addition, there is at least one system for the anionic and one for the catonic amino acids. You can't quite count them on one hand, but the total number is modest. We have to add to this list the numerous much more specific systems that the microbiologists find in microorganisms where the specificity may be to a single

† A fuller summary appears in reference no. 3 of my bibliography.

amino acid. But one also finds broad-scope systems in the bacterial cell and in yeast.

Question Are there coupled amino acid systems, so that you transfer energy from one system to the other?

CHRISTENSEN Yes, there are. Such an amino acid as methionine, for example, is a substrate for the A system, which appears to be energized at least partially by the sodium ion gradient, and also for another system (L) which is not so energized. This permits methionine and various other amino acids to transfer energy from one system to another. Hence, if we say that the A system is a satellite system to the alkali-metal ion pump, then the L system is a satellite to that satellite system, or a third order or tertiary transport system with respect to the receipt of energy. Finally, I have discussed here another case where through the mediation of sodium ion, the gradients established for the cationic amino acids apparently contribute to gradients in the distribution of neutral amino acids, and vice versa. I think we have a complex association of fluxes here. It's been shown, of course, for sugars, too. What all this means, I believe, is that the movement of, let us say, the sodium ion in the direction of the concentration, gradient in the normal context is very often conserving energy that one would not suppose it were conserving, because at the same time organic metabolites indigenously present, are also being relocated.

Question The reason I asked the question was mainly because you mentioned a binding protein which has been isolated, and one wonders whether the transport characteristics and the binding characteristics of the binding protein itself are in agreement in this kind of context. It's very difficult to tell if there is a coupling between two substances, because you can't tell whether stimulation of transport by a substance and its binding protein are the same thing or separate processes.

CHRISTENSEN This work by Oxender and Piperno[†] has now been published in detail. I have referred to a number of associations indicating that this protein participates directly in transport. One of them that I did not mention is that the concentrations at which the transport of leucine, isoleucine and valine are half-saturated are essentially equal to the dissociation constants of the binding protein for these three amino acids at the same ionic strength. I have reported, however, another amino acid for which the dissociation constant appears not to be in agreement with its K_m for transport. The ability to recover the binding protein is associated with the presence of the transport activity. When the binding protein has been released, the cell is still viable

† See references 20, 21 in the bibliography of my paper.

but one fails to find the transport activity, in accordance with the appearance of the binding protein in the suspending solution. There are present, simultaneously, other transport systems, for example, one for which alanine is a typical substrate which is not lost on 'cold shock' and which can still be detected in the shocked cell. In that case, one finds no alanine binding activity in the supernatant solution. The transport system for the branched chain amino acids is repressed when *E. coli* is grown in the presence of leucine; in that case one cannot recover leucine-binding protein.

Membrane-like properties in microsystems assembled from synthetic protein-like polymer

SIDNEY W. FOX, ROBERT J. McCAULEY,
PHILIP O'B. MONTGOMERY,† TAKESHI FUKUSHIMA,
KAORU HARADA, and CHARLES R. WINDSOR

*Institute of Molecular Evolution, University of Miami,
Coral Gables, Florida and † University of Texas Southwestern Medical School,
Dallas, Texas*

THE RESEARCH in our laboratory has been concerned with models of primordial cells (1). Models of the protocell must, however, in our view, be discernibly related to contemporary cells. According to this thesis any model of a primitive cell must therefore be also a model of the contemporary cell, although the inverse relationship is not a necessary corollary.

The kinds of microsystems and precursor macromolecules which are used in this research are first described. Then we present some of the aspects of the behavior of the boundaries of these microsystems, such as can be investigated as models of the contemporary unit.

PROTEINOIDS

The protein-like polymers are obtained by simply heating mixtures of amino acids containing sufficient proportions of aspartic acid, glutamic acid, lysine, or mixtures of these. When this reaction is carried out at, for example 170° for six hours, the polymers which result are found to have molecular weights of many thousands, some proportion of each of the amino acids common to contemporary protein, and most of the properties which are ordinarily associated with contemporary proteins (2). Inasmuch as this similarity could be recognized at the outset, these polymers have been referred to as proteinoids (3).

Some have believed that in the event that amino acids could be condensed simultaneously to yield protein-like polymers, the resultant products would be highly disordered (4). What we find is quite different, namely, a sharply limited heterogeneity (5). No one has had a random assortment of macromolecules to compare with these products, but theoretically one would expect the elution pattern of such an assortment to be a horizontal straight

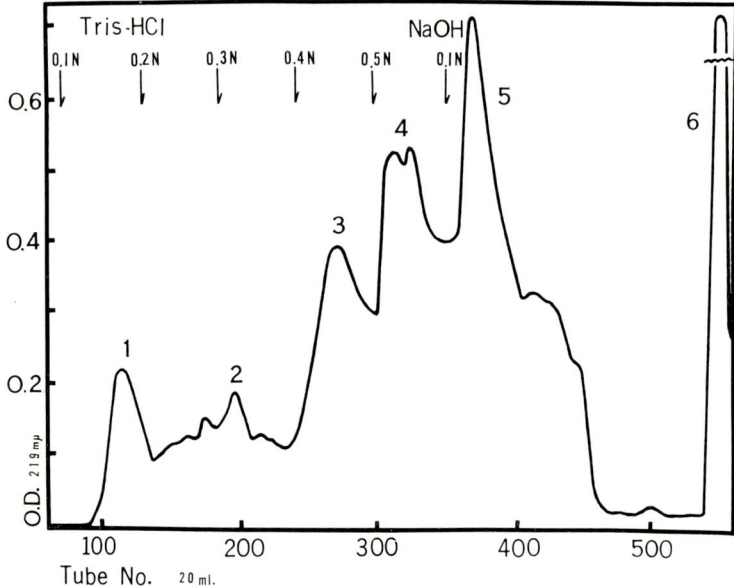

FIGURE 1 Elution pattern of (amidated) 1:1:1-proteinoid fractionated on DEAE-cellulose (5).

line (Figure 1). Figure 1 shows, however, that the polymer has separated into six relatively discrete fractions.

The total hydrolyzates and the partial hydrolyzates of fractions 3, 4, and 5 from DEAE-cellulose have been examined. The heterogeneity is thus analyzed further. The upper three chromatograms (Figure 2) are of the total hydrolyzates. They show that the amino acid profiles, while quantitatively somewhat different in the leucine area, are very much alike. The bottom three chromatograms are 'fingerprints', or partial hydrolyzates, examined on the amino acid analyzer. One may note for these three also a very considerable similarity. Fifteen of the forty peaks are amino acids, twenty-five are peptides.

We must conclude from these results that both the composition and the sequence is highly uniform throughout the polymer. Such an interpretation is consistent with at least six other kinds of evidence which have been published (5).

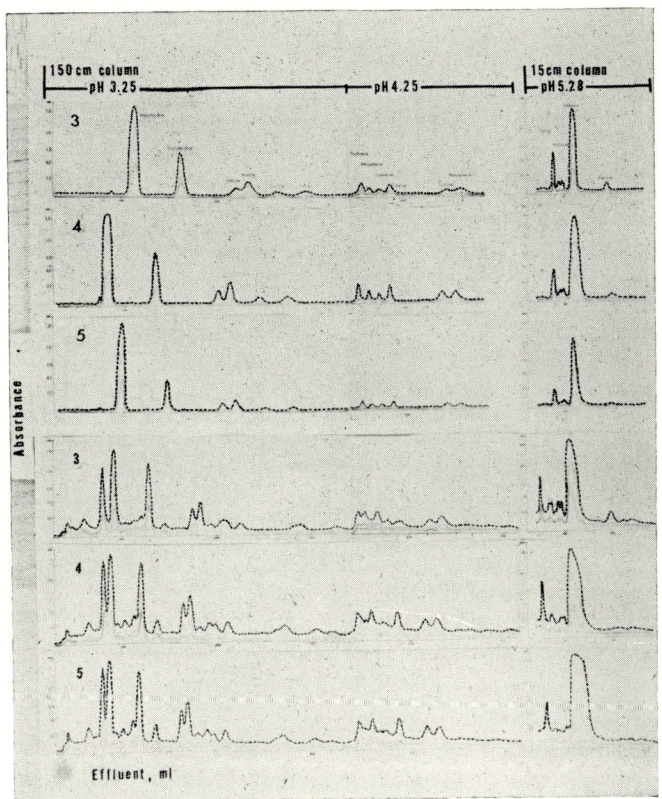

FIGURE 2   Chromatograms of total hydrolyzates (top three) and partial mineral acid hydrolyzates (bottom three) of materials designated by peaks 3, 4, and 5 from DEAE-cellulose, prepared as in Figure 1 (5).

The fact that these polymers possess catalytic activities is brought out in Table I. Such catalytic activities have been reported from six laboratories in over fifteen journal publications (1, 6). These papers describe four types of reaction. The reactions are hydrolysis, decarboxylation, amination, and deamination and have been shown to apply to a total of more than ten

TABLE I Catalytic activities found in thermal proteinoids.

Reaction and substrate	Authors and date
Hydrolysis	
p-Nitrophenyl acetate	Fox, Harada, and Rohlfing, 1962; Rohlfing and Fox, 1967; Usdin, Mitz, and Killos, 1967
Activity destroyed by heating in buffer solution	
p-Nitrophenyl phosphate	Oshima, 1968
Adenosine triphosphate	(Through Zn salt) Fox and Joseph, 1965
Decarboxylation	
Pyruvic acid	Krampitz and Hardebeck, 1966
α-Ketoglutaric acid	Hardebeck, Weber, Wood, and Fox, 1967
Glucuronic acid	Fox and Krampitz, 1964
Amination	
α-Ketoglutaric acid	Krampitz, Diehl, and Nakashima, 1967, 1968
Deamination	
Glutamic acid	Krampitz, Diehl, and Nakashima, 1968

TABLE II Properties of thermal proteinoids in common with those of contemporary proteins

Limited heterogeneity
Qualitative composition
Quantitative composition (except serine, threonine)
Range of molecular weights (4,000–10,000)
Color tests
Inclusion of nonamino acid groups
Solubilities
Salting-in and salting-out properties
Precipitability by protein reagents
Hypochromicity
Infrared absorption maxima
Recoverability of amino acids on mineral acid hydrolysis
Susceptibility to proteolytic enzymes
Some catalytic activity
Nutritive quality
Tendency to assemble into microparticulate systems
Many catalytic activities
Inactivatability by heating in aqueous solution
Hormonal activity (melanocyte stimulation)
Binding of polynucleotides (by basic proteinoids)

substrates. For amination or deamination both proteinoid and cupric or cuprous ion, respectively, are required. This is one kind of evidence which indicates the presence of reactive centers. Such evidence can be compared with the results and discussion of Dr. Christensen at this conference.

In Table II are listed all of the properties which the proteinoids as a group have been found to have in common with the proteins as a group (1). The last one listed in the table, the tendency to form microsystems, is our immediate concern here.

PROTEINOID MICROPARTICULATE SYSTEMS

The proteinoid microparticles, or microsystems, are made by a process simpler than that by which the polymer was produced. This process consists of adding water to the polymer. The best result is obtained by heating the

TABLE III   Thermal polyamino acids yielding regular microparticles.

Polymer	Nature of unit
Proteinoid	Spherule
Aspartic acid-glutamic acid	Spherule
Aspartic acid-lysine	Spherule
Aspartic acid-leucine	Spherule
Aspartic acid-methionine	Spherule
Aspartic acid-glutamic acid-leucine	Spherule
Glutamic acid-glycine	Oblate spherule
Alanine-aspartic acid-glutamic acid-glycine-diaminopimelic acid-glucosamine	Nonuniform spherule
Polyglycine	No spherule
Polyaspartic acid	No spherule

proteinoid with water or aqueous solution and allowing the hot solution to cool (7). When this is done, vast numbers of microspheres of the type seen in Figure 3 are formed. The proteinoid microspheres are uniform in size, quite stable, and are made from synthetic polymer which permits a wide range of controlled variation.

In Table III are presented the results with a series of different polymers which formed units in water. Many thermal polyamino acids have been tested. Most of them easily yield spherules, others give erythrocyte-shaped units, and some do not easily form structures. Some polyamino acids other than those made thermally have been found to possess this property.

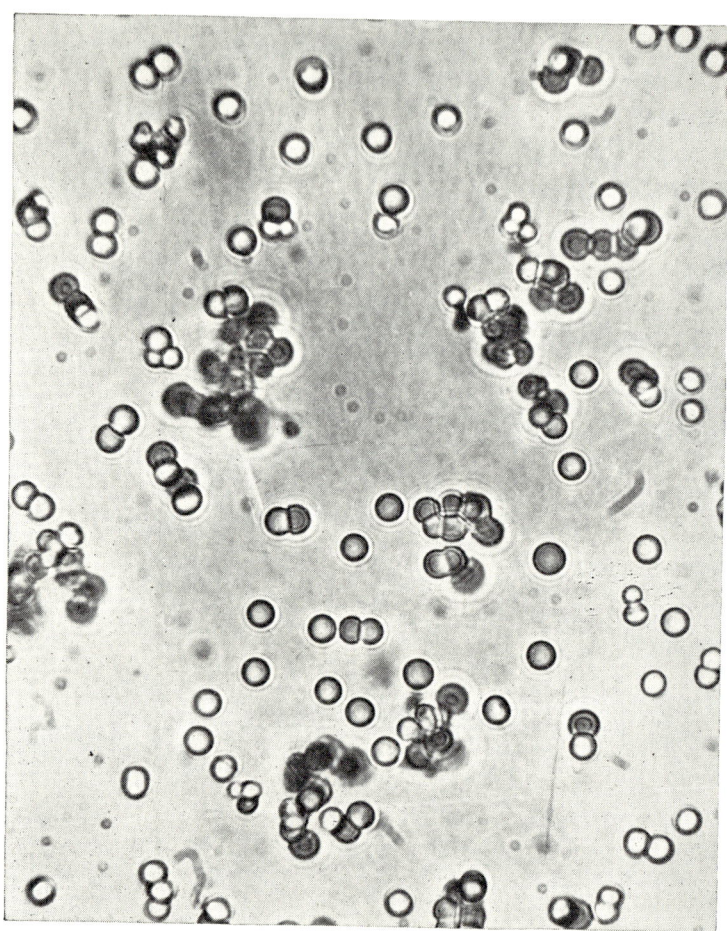

FIGURE 3   Proteinoid microspheres. Approx. 2 μ in diameter.

HISTOCHEMICAL DIFFERENTIATION

The microspheres produced as shown in Figure 3 stain gram-negative. If one include a sufficient proportion of lysine-rich proteinoid, they stain gram-positive (8). This kind of experiment (Figure 4) demonstrates a special sort of controllability of this system and exemplifies a virtue of these microsystems as models. The difference between gram-negative and gram-positive is controversial (8). Gram-positiveness has for example been attributed to the presence in cells of magnesium ribonucleate (9), a claim which has been

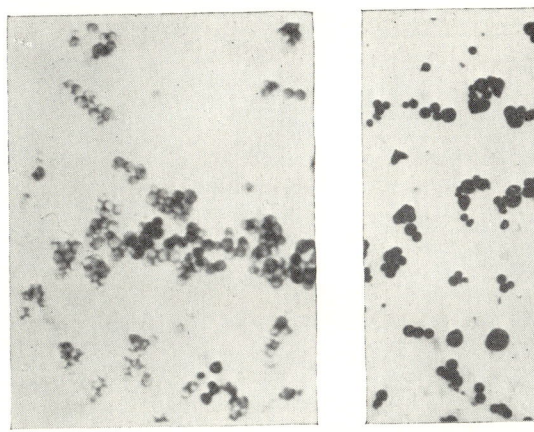

FIGURE 4 Proteinoid microspheres staining gram-negative (on left) containing lysine-rich proteinoid and staining gram-positive (on right; 8).

disputed (10). The virtue of the model from synthetic polymer is that it provides a relatively sure understanding of the nature of what has gone into it, and we can be certain that no ribonucleate is present. The dispute about the contribution of magnesium ribonucleate in contemporary gram-positive cells (10) is clouded by a lack of assurance.

SIZE

Figure 5 illustrates that one of the factors controlling the size of the microsystem is the concentration of salt in the solution from which they are made (7). Some other factors are nature of proteinoid, ratio of quantity of proteinoid to aqueous solution, and presence of other materials, e.g. lipids.

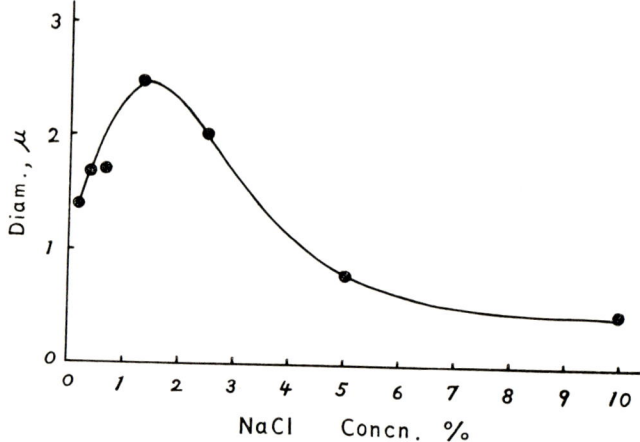

FIGURE 5 Variation in size of proteinoid microspheres prepared in solutions of sodium chloride of varying concentration. Adapted from tabulated data (7).

VARIATION IN SIZE WITH TRANSFER TO SOLUTIONS OF DIFFERENT CONCENTRATIONS

Table IV records experiments in which microspheres have been transferred to solutions hypertonic to, or hypotonic to, those in which they were made. Shrinking or swelling, respectively, is observed. This type of response suggests an osmotic behavior.

ULTRASTRUCTURE

Figure 6 shows the electron micrograph of a section of proteinoid microsphere (11) and also that of a PPLO (12). Figure 6c reveals another phenomenon in another electron micrograph. The experiment in this case involved raising the pH in the suspension of the microspheres. Under these circumstances the polymer in the interior diffuses out through the boundary, revealing a double layer.

Figure 7 shows the same kind of process through a quartz optics microscope using light of wavelength 2665 Å. This process is selective, inasmuch as the analyses of the boundaries are similar to the analyses of the polymer in the interior (Table V). Other experiments have shown that the polymer accumulates outside of the microsphere (Figure 8).

TABLE IV Effect of 'hypertonic' and 'hypotonic' sodium chloride solutions.

Solution in which spherules prepared	Range of diameter when made	Solution to which spherules transferred	Range of diameters after transfer
2.0 N NaCl	2.8–3.2 μ	0.2 N NaCl	3.4–3.9 μ
0.2 N NaCl	3.4–3.6 μ	2.0 N NaCl	2.0–2.1 μ

Solution to which particles transferred in each experiment was saturated with proteinoid.

TABLE V Analyses of hydrolyzates of interior and boundary polymers from proteinoid microspheres.

Amino acid	Total	Boundary
Lysine	1.57	1.42
Histidine	0.55	0.69
(Ammonia)	5.43	8.20
Arginine	0.84	1.33
Aspartic acid	69.0	67.3
Threonine	trace	0.69
Serine	trace	0.34
Glutamic acid	12.2	9.44
Proline	0.38	0.35
Glycine	1.33	1.60
Alanine	2.73	2.57
Valine	0.81	0.98
Methionine	0.94	0.76
Isoleucine	0.38	0.55
Leucine	0.77	1.06
Tyrosine	0.95	1.01
Phenylalanine	1.02	1.06
Alloisoleucine	0.41	0.21
Unknown (leucine equivalent)	0.43	0.65

All figures in mole per cent.

426   *Physical principles of biological membranes*

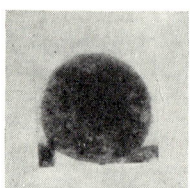

FIGURE 6(a)

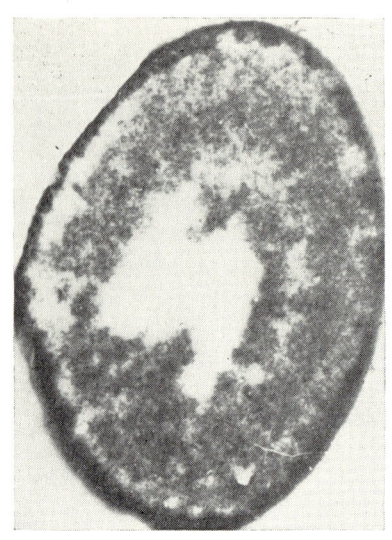

FIGURE 6(b)

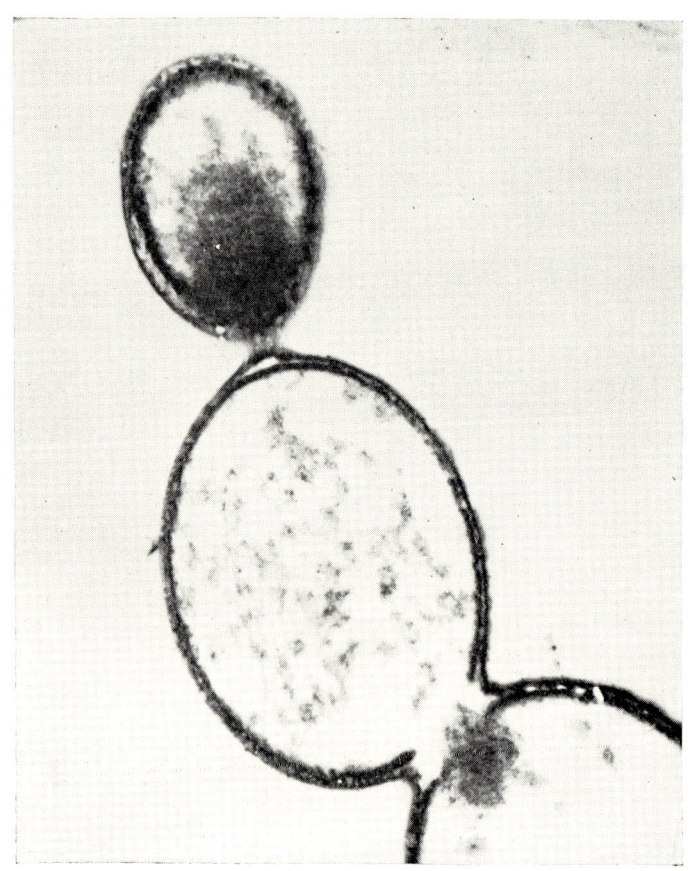

FIGURE 6(c)

Membrane-like properties in microsystems

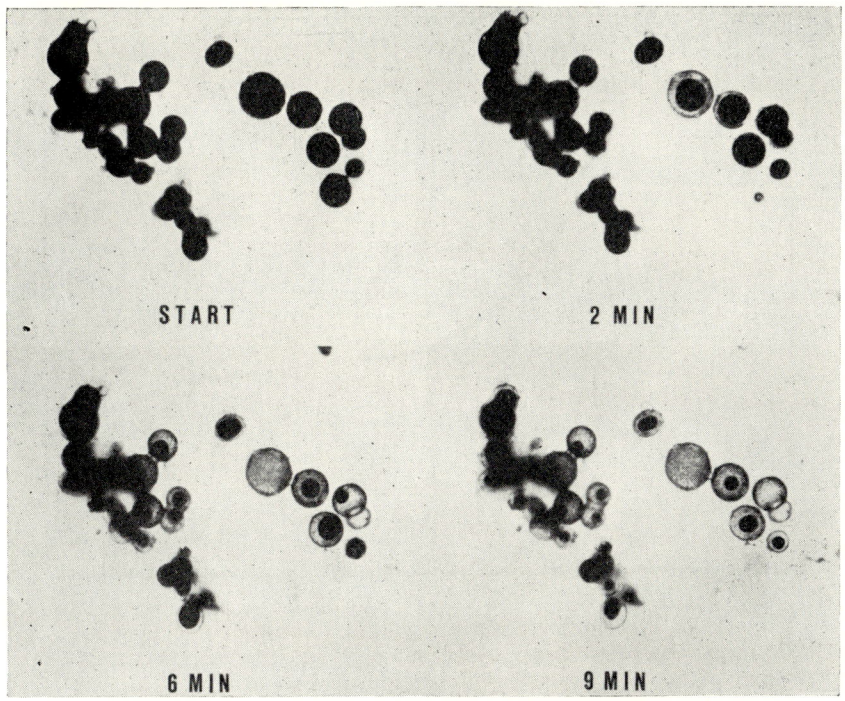

FIGURE 7 Time-lapse photomicrographs of 2:2:1-proteinoid microspheres in suspension treated with 2 parts of McIlwain's buffer at pH 6.5. Photographed in 4 second exposures with light of 2665 Å through quartz optics.

FIGURE 6 (a) Electron micrograph of section of *Mycoplasma hominis* $\frac{1}{3}\mu$ (12); (b) electron micrograph of section of proteinoid microsphere, $2\frac{1}{2}\mu$ (11). Electron micrograph on left courtesy of Dr. D. M. Anderson.

FIGURE 6c Double layer in electron micrograph of proteinoid microsphere in suspension in which pH has been slightly raised (11).

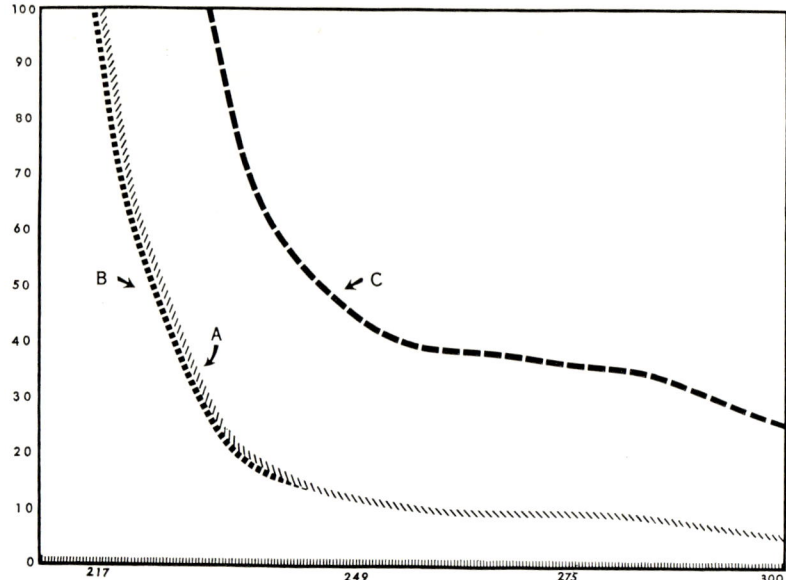

FIGURE 8 Spectrophotometric tracings of absorption of light by (a) initial supernatant from which microspheres separated, (b) supernatant plus McIlwain buffer in concentrations equal to those at end of experiment, (c) supernatant after 1 hr. of diffusion to exterior, leaving boundaries as residues. No microspheres were observed in this experiment to dissolve entirely.

SELECTIVE RETENTION

Without employing an increase in pH, selectivity may also be observed. When proteinoid microparticles are produced in solutions containing 2.0 per cent of each of the saccharides mentioned in Table VI, and then washed in a standard way four times with water, one obtains the analyses recorded. The monosaccharides washed out completely. The polysaccharides are retained. The microsystem is thus selective. How much of this activity is attributable to the boundary alone is not known.

Table VII confirms the result of Table VI by use of radioactively labelled carbohydrates.

TABLE VI Content of carbohydrate in proteinoid microspheres after four standard washings with water.

Carbohydrate	Content Carbohydrate/proteinoid
Fructose	0.0%
Glucose	0.0
Glycogen	0.7
Starch	0.5

Determined by a method using anthrone. Microspheres produced in 2.0 per cent solution of carbohydrate in each case.

TABLE VII Content of radiocarbohydrates in proteinoid microspheres after four standard washings with saturated proteinoid solution.

Carbohydrate	Amount retained d. p. m.	
	In fourth wash	In microspheres
Glucose-^{14}C	29	68
Glycogen-^{14}C	190	1860
Starch-^{14}C	27	636

Microspheres produced in 2.0 per cent solutions of carbohydrate.

CONCLUSION

The relationship to lipids is of particular interest for the modelled evolution of microspheres. The fact that structures composed of polyamino acid alone can be electron micrographed can be interpreted in the context of contemporary cells. One inference that results is the possibility that protein alone is sufficient to explain osmophilicity in contemporary cells. Perhaps more fundamental is the fact that the hydrocarbon side chains of amino acid residues in protein or proteinoid are themselves lipid in character and confer upon the system lipid quality.

Advantages of this system include the fact that the macromolecules may be controllably varied in precise ways and that the resultant systems and their properties may be correlated to these variations. Another advantage is that diffusion outward can be studied as well as diffusion inward.

The membrane-like properties of these systems can, also, be investigated within a total cell-like microsystem which has many other properties of contemporary cells (1, 2).

Aside from the utility of these microsystems as models of contemporary cells, they are of unique interest as models of primordial cells, or protocells. The desirability of studying them in this context has been stated eloquently by Stanier, Doudoroff, and Adelberg (13): 'Cellular organization, the universal basis of biological structure in the present living world, represented the greatest of all evolutionary breakthroughs. . . . It is for this reason that "life" now appears to us to be such a distinctive form of matter.'

Acknowledgement

The authors thank the National Aeronautics and Space Administration for research grants nos. NGR-10-007-008 and NAS 9-8101. Contribution no. 123 of the Institute of Molecular Evolution.

References

1. Fox, S. W. 1969, *Naturwissenschaften*, **56**, 1–9.
2. Fox, S. W. 1968, in Encyclopedia of Polymer Science and Technology, H. Mark, *et al.*, eds., Vol. 9, 284–315, Interscience.
3. Hayakawa, T., Windsor, C. R., and Fox, S. W. 1967, *Arch. Biochem. Biophys.*, **118**, 265–272.
4. Oparin, A. I. 1957, The Origin of Life on the Earth, 290, Academic Press, Inc.
5. Fox, S. W., and Nakashima, T. 1967, *Biochim. Biophys. Acta*, **140**, 155–167.
6. Rohlfing, D. L., and Fox, S. W. 196? *Advan. in Catal.*, **20**, 373–418.
7. Fox, S. W., Harada, K., and Kendrick, J. 1959, *Science*, **129**, 1221–1223.
8. Fox, S. W., and Yuyama, S. 1963, *J. Bacteriol.*, **85**, 279–283.
9. Henry, H., Stacey, M., and Teece, E. G. 1945, *Nature*, **156**, 720–721.
10. Lamanna, C., and Mallette, M. F. 1950, *J. Bacteriol.*, **60**, 499–505.
11. Fox, S. W. 1965, *Nature*, **205**, 328–340.
12. Anderson, D. M., and Barile, M. F. 1965, *J. Bacteriol.*, **90**, 180–192.
13. Stanier, R. Y., Doudoroff, M., and Adelberg, E. A. 1963, The Microbial World, 2nd Edition, 59, Prentice-Hall, Inc.

DISCUSSION

EISENMAN   I have a question: and then, depending on your answer, I might like to make a comment. The question is this: 'Have you seen any particular pairs of amino acids that tend to be neighbors?'

Fox   There hasn't been enough of that kind of sequence determination to answer your question. I could take a minute to tell you what has been done in that direction, but it would only be a peripheral answer to your question.

EISENMAN   I reported some observations in 1963 on the response of cation selective glass electrodes to certain amino acids (glycine, alanine, phenylalanine and histidine). These can be found on pp. 189–191 of *Conferences on Cellular Dynamics* (L. D. Peachey, editor, Published by N.Y. Acad. Sciences Interdisciplinary Communications Program, New York, 1968). The intriguing finding was that systematic selectivity differences could be seen among these amino acids for a series of glass electrodes, the acidity of whose fixed negative charges was varied. In other words, systematic changes in the selection among these four amino acids occurred as the acidity of the ion exchange sites of the glass was varied.

Fox   Is this in the substitution series?

EISENMAN   It varies from glass to glass. Thus, for a particular glass, the sequence of decreasing effectiveness was histidine, phenylalanine, alanine, glycine; while for another glass, the sequence could be switched to alanine, histidine, glycine and phenylalanine; and so on. Of course, I tried to reconcile these data within a theoretical framework, which makes sense intuitively. I found that one could correlate, indeed 'predict', the sequence from simple considerations of the pK_2 values constants of the amino groups. Starting from the notion that different carboxylic groups will have different negative 'field strengths' (indicated by their different pK_2 values, one would expect certain carboxylic groups to prefer particular amino groups, whose positive 'field strength' differences are indicated by their differing pK_2 values). If I remember it correctly, the correlation was that the carboxylic groups which were the weakest acids would be expected to interact preferentially with those bases which shed protons most easily. Although these differences may represent relatively small energies, one would expect some differences in the frequency of nearest neighbor amino acids on this basis.

CHRISTENSEN   Can you select a size or type of solute molecule which shows the phenomenon of saturability in entering or leaving these *serials*? It appears that it takes the glucose some time to get out and the glycogen a great deal longer. What about the nature of the kinetics, are they saturation kinetics?

Fox   We just don't have the data on that. Sorry.

List of participants

PROFESSOR GEROLD ADAM
University of Munich
Institute for Physical Chemistry
Goethestrasse 33
8 Munich 15, Germany

DR. WILLIAM BARRY
National Institute for Mental Health
9000 Rockeville Pike
Bethesda, Maryland 20014

PROFESSOR MARTIN BLANK
Department of Physiology
College of Physicians and Surgeons
Columbia University
630 West 168th Street
New York, New York 10032

PROFESSOR BRITTON CHANCE
The School of Medicine
Biophysics and Physical Biochemistry Department
University of Pennsylvania
Philadelphia, Pennsylvania

PROFESSOR HALVOR N. CHRISTENSEN
Medical School
Michigan
Ann Arbor, Michigan

DR. KENNETH S. COLE
Department of Health, Education and Welfare
National Institutes of Health
Bethesda, Maryland 20014

DR. H. E. DERKSEN
Physiology Laboratorium der Rijks Universiteit te Leiden
Leiden, The Netherlands

DR. FREDERICK DODGE, JR.
Department of Biophysics
Rockefeller University
New York, New York 10021

PROFESSOR WALTER DROST-HANSEN
Institute of Marine Science
University of Miami
Virginia Key, Florida

PROFESSOR GERALD M. EDELMAN
The Rockefeller University
New York, New York 10021

PROFESSOR GEORGE EISENMAN
Department of Physiology
The University of California
Los Angeles
California 90024

DR. HARRY FOX
Office of Naval Research
U.S. Department of the Navy
Washington, D.C.

PROFESSOR SIDNEY FOX
Institute of Molecular Evolution
University of Miami
521 Anastasia Avenue
Coral Gables, Florida

PROFESSOR DAVID E. GREEN
Codirector
Institute for Enzyme Research
The University of Wisconsin
Madison, Wisconsin 53706

PROFESSOR TERRELL HILL
Vice Chancellor—Sciences
University of California
Santa Cruz, California 95060

DR. G. J. IVERSON
Center for Theoretical Studies
University of Miami
Coral Gables, Florida 33124

DR. REED M. IZATT
Department of Chemistry
Brigham Young University
Provo, Utah 84601

DR. WOLFGANG JUNGE
Technische Universitat Berlin
Max-Volmer Institut fur physikalische Chemie
1000 Berlin 12, Germany

DR. DONALD R. KALKWARF
Research Associate
Radiological Physics Section
Battelle Memorial Institute
Richland, Washington 99352

PROFESSOR ORA KEDEM
Atomic Energy Commission
Nuclear Research Center-Negev
Beer Sheva, Israel

PROFESSOR BEHRAM KURSUNOGLU
Center for Theoretical Studies
University of Miami
Coral Gables, Florida 33124

DR. J. LAM†
Center for Theoretical Studies
University of Miami
Coral Gables, Florida 33124

PROFESSOR PETER LÄUGER
Department of Biology
University of Konstanz
Konstanz, Germany

DR. GERHARD MACK
Center for Theoretical Studies
University of Miami
Coral Gables, Florida

PROFESSOR LORIN J. MULLINS
Head, Department of Biophysics
University of Maryland
Baltimore, Maryland 21201

MR. KAROL MYSELS
Associate Director of Research
R. J. Reynolds Tobacco Company
Winston-Salem, North Carolina

DR. SHINPEI OHKI
Health Sciences Center
State University of New York at Buffalo
Amherst, New York 14226

PROFESSOR LARS ONSAGER
Chemistry Department
Yale University
New Haven, Connecticut 06520

PROFESSOR ARNOLD PERLMUTTER
Department of Physics and Center for Theoretical Studies
University of Miami
Coral Gables, Florida 33124

DR. HAROLD E. PODALL
Office of Saline Water
U.S. Department of the Interior
Washington, D.C.

PROFESSOR MORRIS ROCKSTEIN
Department of Physiology
University of Miami
Coral Gables, Florida

DR. H. ROSENWASSER
Naval Air Systems Command
U.S. Department of the Navy
Washington, D.C.

PROFESSOR MILTON R. J. SALTON
School of Medicine
New York University Medical Center
New York, New York

† Now at National Research Council, Ottowa, Canada.

List of participants

Dr. R. Schlögl
Max-Planck Institut fur Biophysik
Kennedy Allee 70
Frankfurt, Germany

Professor Francis O. Schmitt
Neurosciences Research Program
Massachusetts Institute of Technology
Brookline, Massachusetts

Dr. Fred M. Snell
Dean
Graduate School
State University of New York at Buffalo
Buffalo, New York

Dr. Ichiji Tasaki
National Institute of Mental Health
9000 Rockville Pike
Bethesda, Maryland 20014

Dr. Donald F. H. Wallach
Biochemical Research Laboratory
Massachusetts General Hospital
Boston, Massachusetts 02114

Professor Jerome J. Wolken
Director
Biophysical Research Laboratory
Carnegie-Mellon University
Pittsburgh, Pennsylvania 15213